Science in Culture:
The Early Victorian Period

Science in Culture:
The Early Victorian Period

SUSAN FAYE CANNON

Dawson
and
Science History Publications
New York • 1978

First published in the United States by
Science History Publications
a division of
Neale Watson Academic Publications, Inc.
156 Fifth Avenue, New York 10010

(CIP data on final page)

© Neale Watson Academic Publications, Inc.   1978

Published in Great Britain
by Wm. Dawson & Sons, Ltd.
Cannon House Folkestone,
Kent, England
ISBN 0–7129–0895–1

Printed in U.S.A.

To Don Hegstrom
for a better way of understanding

Table of Contents

Preface

"...to be labouring only for the torment of little boys and girls, always struck me as a hard fate; and though I know it is all very right and necessary, I have often wondered at the person's courage that could set down on purpose to do it."

"That little boys and girls should be tormented," said Henry, "is what no one at all acquainted with human nature in a civilized state can deny; but on behalf of our most distinguished historians, I must observe, that they might well be offended at being supposed to have no higher aim; and that by their method and style, they are perfectly well qualified to torment readers of the most advanced reason and mature time of life."

—Northanger Abbey, ch. 14

Austen is right, as usual. Reading analytic history is surely an acquired taste; more suitable for the philosopher than for the young. And we do have a higher aim when we do it. What actually happened is what history is about. Any narrative, however, depends upon a host of previously assembled facts; and what facts have been assembled, how they have been assembled, indeed what are facts ("things done") at all, depends upon the previously existing collection of ideas, prejudices, interests, and generalizations which other historians have accepted as proper initial conditions for spinning out their patterns of the period.

My interest in England in the first two-thirds of the 19th century is for its own sake. I like it there. I like the people. As Oscar Wilde has one of his characters say, "In the whole world there is no king who has peacocks like unto my peacocks."

This book is the result of my attempts to produce ideas and generalizations with which we can assemble and arrange facts to show off the beauty of my collection of peacocks. It is not a complete scheme. There are many good approaches in the field already; the ones described herein are the ones I wish to advocate as not fully exploited.

My initial poles are two: a basic intellectual orientation of the period; and an actually existing group of nameable individuals. Then I develop four analytic tools in five chapters. Two new and growing intellectual orientations are described: the very general one I call "Humboldtian science," and the specific one of a new entity conceived not as a specialization but as an attempt at framing *the* basic science—it came to be called "physics." Then I use the founding of the British Association for the Advancement of Science as my major demonstration subject for the analytic tool called "professionalization," and also for how the role and functions of societies were determined or influenced by other forces in society as well. This last is only indicated, not set forward in a scheme of its own.

In Chapter Eight, then, I remark on the orientation needed by *us* to approach the history of science as part of history itself. Much has changed since 1964, when I published an article on the same subject; much remains the same. Finally comes an overall construct of my own, which I call "the Truth-Complex and its demise." It is not final, not too detailed; perhaps it is a good way-station, one that is on the right route.

The surgeon has an array of cutting tools, and uses each according to the feel of the precise situation. Proposing some particular tools suitable for this culture, and indicating their use, is what this book is about. Some of my tools are of the 1960's, have been re-ground and re-sharpened and seem to work as precisely as ever; a number are new, not quite so well tested, but promising.

In case there is any confusion about bibliography, I here note that before 1977 I published under the names Walter F. Cannon and W. Faye Cannon.

I apologize to all of the fine historians whose works are not referred to in my notes. Such omissions are inevitable in today's scholarly world. If each of you will send me a copy of your monograph and also make sure that a copy of this book is purchased, then the sales will be high enough to warrant a second edition and I promise to mention each of you in its notes.

I make less apology to all of the sociologists of science, from Robert Merton on, whose works are not cited. This is because I hope to enter into discussions with them in future publications. For now, I will simply note that the historian who feels that the approach of, say, Philip Elliott, *The Sociology of the Professions* (London, 1972) is too limiting, now can take his start from the feast of approaches indicated by the various essays in Ina Spiegel-Rösing and Derek de Solla Price, *Science, Technology and Society: A Cross-Disciplinary Perspective* (London, 1977). Or he can see that he has more to gain than lose in flexibility by re-reading Arnold Thackray, "Natural Knowledge in Cultural Context: The Manchester Model," *American Historical Review, 79* (1974), 672–709.

Now comes the part of this Preface I enjoy: thanking the people who have helped me in these pursuits.

I am grateful to the Editors of the following journals for permission to use published articles as the basis for some of my chapters:

Journal of the History of Ideas, for "The Normative Role of Science in Early Victorian Thought," *25* (1964), 487–502, the basis of Chapter One.

The Journal of British Studies, for "Scientists and Broad Churchmen: an Early Victorian Intellectual Network," *4* (1964), 65–88, the basis of Chapter Two.

History of Science, for "History in Depth—the Early Victorian Period," *3* (1964), 20–38, the basis of Chapter Eight.

The Smithsonian Journal of History, for "P. S. If I find out what Truth is, I'll drop you a line," *2* (1967), 1–24, the basis of Chapter Nine.

I am grateful to the National Science Foundation, which supported me from March 1962 to August 1963 (Grant G-22497, "The Scientific Community in England, 1820–1860").

Then, to all librarians, those sitting at the back desks as well as those up front, thank you even if I forgot to write down your name at the time.

Among institutions, pride of place goes to the Royal Society of London, which gave me a desk and a welcome from March 1962 to August 1963, as well as on numerous shorter visits. Thanks go especially to its late librarian, I. Kaye, and to a member, Sir Gavin de Beer.

Among fellow scholars: Dr. Nathan Reingold. We have argued at lunches, seminars, coffee breaks, even at parties. Dr. Reingold wants to establish the right answer, not his answer. What more is there to say?

Thanks (in no particular order) to: Jack Goodwin and Charles Berger of the Smithsonian Library; Roger Prouty, Joseph Pequigney, and William Stahlman; Elaine Jancourtz, Frank Haber; Richard Shryock, Sandra Herbert, Peter Vorzimmer, Conway Zirkle; Marie White, Margaret Faw Cannon, the Duke Divinity School Library; Michael Wolff; Adrian Lyttleton and the members of his class at All Souls; Miss Lauter, Mr. Halcrow, Mr. Rawlings, Mr. Dodwell, Mr. Lee, N. Russell, Mr. Vernon; Dr. H. R. Calvert of the Science Museum, the Astronomer Royal and Dr. P. S. Laurie of the Royal Greenwich Observatory; the Councils and Libraries of the Royal Astronomical Society, the Geological Society of London, the Geological Survey, the Cambridge Philosophical Society, St. John's and Trinity Colleges (Cambridge), Somerville College, the Royal Institution.

Also, Arthur Molella, Jacob Gruber, Robert Butts, Michael Hoskin, Maurice Crosland, Jerome Ravetz, Jack Morrell, Douglas Bassett; Mr. Barr and Miss Langley of the York Minster Library, Alexander Ospovich, the Cambridge University Library, Mr. Gautrey, Jon Hodge, the American Philosophical Society Library, the British Museum (Natural History) Library, the Science Museum Library; my predecessors as Curators of Surveying at the Smithsonian; Herman Fries, Angus Armitage, Mrs. E. D. Shorland, Judy Victor, Stephen Goldfarb, Whit Bell, Gordon Davies, Margaret Sone, Mary Kofron; the Royal Photographic Society, Mrs. F. Johnston, the Kodak Museum, P. S. Schultz, the National Maritime Museum (Greenwich), A. W. H. Pearsall, Alan Jeffreys, Lionel Madden,

Donna Elliott, Anastasia. And of course William Smeaton, and Martin Rudwick, and Robert Robson and Michael Hoskin, and Frank Greenaway, and George Basalla, and Malcolm Ogden, and Barney Finn, and Jane Finn, and all the Finns, and Léonie Scott-Matthews, and the Troubadour Coffee House.

I would like to hear from the rest of you. A simple postcard will do:

"I told you about/showed you/made available to you _____
(subject or material)
on or about ____ in _____ while talking to you/eating with you/entertaining you/carrying out my job."
(date) (location)

I will enjoy being reminded of all the good times I have had and people I have met in the course of composing the book.

Chapter One

Science as Norm of Truth

So protean a thing as natural science played many roles in the second quarter of the 19th century. For example, it served to baptize fresh-air fun. One could roam the mountains or the moors, protected from the pressure of Evangelical duty, if one brought back beetles or bits of rock, for the study of nature was the study of one of God's two great books. We listen to the Reverend Augustus Hare in 1830, wishing he shared the virtues of a fellow clergyman: "... your care of the parish, your love of natural science, your activity, your unremitting endeavours to improve the condition of the poor."[1] The English love of roaming through the countryside was present long before Wordsworth, but in the absence of pilgrimages, it was useful to have the justification of geological field trips.

In metaphysics, science served to stabilize the world. Evangelical preachers such as Thomas Chalmers (1780–1847) emphasized the radical instability of the world, depending as it does for its very existence on the mere wish of God. Moreover, what collisions with comets, what subterranean upheavals, what noxious exhalations from the bowels of the earth, might wipe out all life even in the normal course of nature, did not God prevent such events from happening! On the other hand, empiricist philosophers liked to abolish all semblances of sensible continuity among events: there are no causes, laws are only observed regularities, mere contingency is all the world has to offer. This position commended itself to at least one Anglican theologian, J. B. Mozley at Oxford.[2]

As opposed to both of these schools, natural science portrayed a world ruled by "Newtonian" law, one which would not be deranged by its own actions, one in which the earthquake and volcano were subservient to general laws and beneficent purposes, and one in which the concept *force* revealed the chains which, in spite of Hume, do bind the world together—force, which was so real, so tangible, that around mid-century scientists transmuted it into *energy* and awarded it a reality equal to—for some, superior to—matter itself.[3]

Other roles for science proposed by various Victorian writers have been treated by George Foote, and I shall not repeat his survey of Victorian arguments concerning science as power, science as the model of pro-

1

gressive knowledge, science as an element of culture.[4] Nor shall I attempt to expand on R. K. Webb's observation, in *The British Working Class Reader,* that the technological revolution in printing particularly helped attempts to educate the working classes in science, both theoretical and applied.[5] Instead I shall illustrate what is perhaps the fundamental difference in the role of science between early Victorian times and our own day. That is, for cultured early Victorians, natural science provided a norm of truth. There cannot, as Victorians were fond of saying, be two truths, and the norm by which proposed truths were judged was, explicitly or often implicitly, the norm of natural science.

Let me note that I shall use the term "science" in its restricted modern sense, meaning the physical sciences (including the biological sciences). This usage was only beginning to take hold in England in the 1830's—against a fair amount of opposition—and a contemporary author must be read carefully to determine what he is actually talking about when he says "science." For such as John Henry Newman, we expect theology to be included. A more common problem is "moral science," which does not mean merely "ethics," but all of the sciences relating to human conduct, very definitely including "political economy" for someone like Thomas Macaulay.[6] The new "moral sciences" tripos at Cambridge included modern history and international law.

Within the natural sciences there was a hierarchy of prestige, so that we can specify which one of the sciences was the norm for all other sciences and thereafter for all other truths. As we might expect, it was Newton's science. Let us be quite clear what that science was. Addresses to the British Association for the Advancement of Science made the point explicitly: "Several mathematical and mechanical Sciences have since [Newton's] time made their appearance in the world, claiming to be the younger sisters of Physical Astronomy." Yet these have not "filled the civilized world with the noise of their fame." Astronomy is still the "only perfect science." As for the others, "the summit of their ambition, and the ultimate aim of the efforts of their votaries, is to obtain their recognition as the worthy sisters of the noblest of these sciences—Physical Astronomy."[7] Not dynamics merely, then; certainly not "physics," a word which had an indefinite varying meaning: but *physical astronomy.*

The fact that two of the quotations given are taken from speeches by an Anglican clergyman reminds us how different the English Enlightenment was from the French. Sheltered under Newton's great name, science and religion had developed a firm alliance in England, symbolized by that very British person, the scientific parson of the Anglican Church. In the 1820's the great Laplace was visited by a young English geologist, Adam Sedgwick. Laplace, as bitterly anti-clerical as ever, praised the English for keeping the clergy in their place and warned Sedgwick, as a fellow-scientist, to keep alert lest the Church try to rise again, as it had done in France. Sedgwick was somewhat taken aback. He was, after all, the *Reverend* Adam Sedgwick, who could not bear to hear even geology praised unless the speaker ended by showing how science led one's thoughts up to the Great First Cause.[8] The seniority of astronomy fitted in well with this Christian desire, and Thomas Chalmers was only the most impressive of the writers who made much of the immensity of God's heavens and the intrinsic littleness of Man, whom God has nevertheless graciously taken notice of. Christian writers have often liked evidence of man's intrinsic lowness; we remember that Dante put man near the center of the universe because the center is the bad place, to which dull matter and dross descends. It is the *bottom* of the universe; the Devil is at the exact center, and any direction up is good. The new stellar and nebular universe had no equivalently wicked center; but it certainly did add emphasis to the question, "What is man, that Thou art mindful of him?"

It was because of this alliance between science and religion that science could be allowed its exalted role. The alliance was shattered in 1859 by Darwin's *Origin of Species,* and that is why I have limited this discussion to the early Victorian period. Darwinian science was denied normative status by moralists and theologians. This refusal can be seen in *Lux Mundi,* the book of theological essays from Oxford edited by Charles Gore in 1889 as part of an attempt to re-establish a viable Anglo-Catholic position. The essayists were proud of their liberality in accepting, often welcoming "evolution" in general; and quietly dismissed Darwinian evolution in particular. Other than that, their position with respect to science was somewhat backward compared to that of Anglican natural theologians in the 1830's and 1840's; and in one matter was reactionary. Gore himself claimed that "nature was to find its crown and justification" in religious man. He thus tended to reverse the assertion of natural theology that the world is equally made for all of God's creatures.[9]

The result of the refusal of such opinion-leaders to accept Darwinian science was a multi-normative world. Indeed the scientific hierarchy itself was in danger, as biology was pitted against physics, geology against astronomy, intuition against statistics, in the Darwinian debates. There ceased to be any longer a universal norm. Art, literature, philosophy, politics, theology, each went its own way in search of truth. The early Victorian denied the possibility of even two truths, but the late Victorians had to live with many, as we do.

It seems possible that this disintegration of the "Truth-Complex," as I call it, hastened acceptance of the image of the scientist as explorer or pioneer, rather than as possessor of truth. For example, Karl Pearson in his

Grammar of Science of 1892 said, "The pioneers of science are pushing forward into unoccupied and difficult country."[10] (Note the difference between this and the romantic image of the scientist as Romantic quester. Pioneer country is not infinite in scope. It will eventually be occupied and tamed.)

In order to illustrate the prestige of science before 1859, I shall use the tactic of uncovering examples where they might not be expected, indeed in some cases where the author was not interested in science but was merely using it casually by way of illustration—as one might say, "It's as true as that two and two make four." Thomas Chalmers' *Astronomical Discourses* were so widely known that it was only fitting, in a memoir of 1847, that he and his colleague Welsh at the University of Edinburgh were presented as complementing each other in such terms as, "Each had his own tint and magnitude; but the two close-united made a double star, which looked like one; and now that they have set together, who will venture to predict the rising of such another?"[11]

This is perhaps no more significant than Thomas de Quincey's stray scientific image in *Confessions of an English Opium-Eater* (1856 version), when he says that for years he was "receding or approaching, round my great central sun of opium. Sometimes I ran perilously close into my perihelion; sometimes I became frightened, and wheeled off into a vast cometary aphelion, where for six months 'opium' was a word unknown."[12]

On the other hand, Thomas Carlyle's psalm of praise to Force in *Sartor Resartus,* beginning, "Knowest thou any corner of the world where at least Force is not?" is almost too serious. Carlyle is actually preaching science itself (or at least what Carlyle believes to be science), so that he who does not believe in God will at least believe in the wholeness of the universe.[13] A better example for my purpose, is in Carlyle's *Past and Present* of 1843, where he uses the rotation of the earth as an example of how we must be shaped to perfection, preferably by work:

> Destiny, on a whole, has no other way of cultivating us. A formless Chaos, once set it *revolving,* grows round and ever rounder; ranges itself, by mere force of gravity, into strata, spherical courses; is no longer a Chaos, but a round compacted World. What would become of the Earth, did she cease to revolve? In the poor old Earth, so long as she revolves, all inequalities, irregularities disperse themselves; all irregularities are incessantly becoming regular.[14]

One of my favorite examples, not for its importance but for the book to which it refers, comes from young John Stuart Mill's essay on Bentham in the *Westminster Review* in 1838. He compares the development of English law to geological history:

> The deposits of each successive period [are] not substituted but superimposed on those of the preceding. And in the world of law no less than in

the physical world, every commotion and conflict of the elements has left its mark behind in some break or irregularity of the strata: every struggle which ever rent the bosom of society is apparent in the disjointed condition of the part of the field of law which covers the spot: nay, the very traps and pitfalls which one contending party set for another are still standing, and the teeth not of hyenas only, but of foxes and all cunning animals, are imprinted on the curious remains found in the antediluvian caves.[15]

The hyenas and their teeth-marks make possible a positive identification. Mill's image is taken, whether directly or indirectly, from William Buckland's *Reliquiae Diluvianae* of 1824. Even fourteen years later, Mill thought his audience would recognize and respond to the "hyena story," as Buckland called it.[16]

Another animal which Buckland helped to popularize was the extinct giant sloth of South America, the megatherium. It survived in popularity even after the introduction of ichthyosaurs, megalosaurs (Buckland's own discovery), and the whole host which Richard Owen dubbed "dinosaurs." Thus in Disraeli's *Lothair*, Theodora complains, "But I cannot read in a railroad, and the human voice is distressing to me amid the whirl and the whistling, and the wild panting of the loosened megatheria who drag us."[17]

These are random usages I have enjoyed noting. Now let me turn to a serious case developed at some length, one which suggests that the normative role of science was not merely inherited from the past but was actually growing in the early 19th century. Let us consider the conversion of Wordsworth.

In considering a long-lived poet, it is certainly a mistake to ask, "What was his view of science?" Poets change their intellectual positions at least as much as do other men, perhaps more so. Young Wordsworth expressed the 18th-century pieties he learned at school in such words as these: "Science with joy saw Superstition fly/Before the lustre of Religion's eye."[18] Yet his suspicion of scientists in his early creative years is obvious. Such lines as, "Our meddling intellect/Misshapes the beauteous forms of things:—/We murder to dissect;" or, "Philosopher! a fingering slave,/One that would peep and botanize/Upon his mother's grave;" have a sting and venom that no amount of explaining away will quite neutralize.[19]

Nor was Wordsworth's suspicion absent from his more thoughtful poems. We recall his "Star-Gazers" of 1807, in which he considered a crowd waiting to look through a telescope: "Whatever be the cause, 'tis sure that they who pry and pore/Seem to meet with little gain, seem less happy than before."[20] And there is a very intense passage in the 1805–06 version of the *Prelude*, where, when all the poet's hopes were shattered by the betrayal of the ideals of the French Revolution, he turned his thoughts, "knife in hand," to dissect the living body of society. This misguided attempt left him

5

in even worse mental shape; sick, "wearied out with contrarieties," he resigned himself to mathematics and the possibility of clear proof. It was from *this* despondent depth that Coleridge and Dora Wordsworth rescued him and restored him to his true self. Mathematics, however respected, was thus represented as the exact opposite of the poet's own humanity. By the end of his life Wordsworth had changed his opinion, as we shall see, and in the 1850 *Prelude* this particular reference to mathematics was covered over by a vaguer reference to "abstract science."[21]

Another passage in the *Prelude* complimenting mathematics contrasts it explicitly to the poet's mind "beset with images." It is therefore only a recreation for the poet, although it is of great power to the scientist who reasons, "From star to star, from kindred sphere to sphere,/From system on to system without end."[22] Stellar—that is to say, Herschelian—astronomy is noble, but it is not for the poet.

Wordsworth's characterization of the processes of anatomy as unpleasant, even murderous, is interesting if we contrast it to early 17th-century usage. For poets in that century, "to anatomize" was a mild and instructive process. Oliver in *As You Like It* says, "I speak but brotherly of him, but should I anatomize him to thee as he is, I must blush and weep...." Later, Jacques says, "The wise man's folly is anatomized/Even by the squandering glances of the fool." Ben Jonson's Asper, in *Every Man Out of His Humour,* speaks of a mirror, "Where they shall see the times deformitie/Anatomiz'd in every nerve, and sinnew." And in a quite religious poem, "Elegie on my Muse," Jonson describes the Saviour, our supreme Judge, "Who knows the hearts of all, and can dissect/The smallest Fibre of our flesh; he can/Find all our Atomes from a point t'a span."[23]

Jonson and Shakespeare, it would seem, are using as a reference-point the neat and instructive results displayed in a book of anatomical drawings, as though they were obtained by X-ray vision. Wordsworth, on the other hand, has in mind the actual gory process of dissection itself, including the killing of an animal for the purpose of studying its anatomy. At any rate he escaped the dilemma caused by the Romantic objection that dissection shows not what a living animal is like, but what a dead one is like: it reveals the parts but not the functioning whole. One answer to the Romantic objection in the latter part of the 19th century was, not to abandon anatomy, but to increase vivisection—presumably not what the Romantics wanted.

On science in general, it is useful to recall Wordsworth's preface to the second edition of *Lyrical Ballads,* with its opposition of poetry to "matter of fact, or science," and its patronizing note that *if* men of science should do anything that really matters, the poet will be willing to transfigure it in verse. Until then, he will not "break in upon the sanctity and truth of his picture by transitory and accidental ornaments...."[24] As Roger Sharrock has pointed out, this preface was Wordsworth's (and Coleridge's) reply to Humphry Davy's introductory lecture at the Royal Institution in 1802. It was more favorable to the scientist than their earlier writings had been, but it by no means conceded Davy's claim that the pursuit of science was a stimulus to the true creative imagination. If science is only "matter of fact," the superiority of the poet is obvious.

The "curious mixture of patronage and distrust" of science, in Sharrock's phrase,[25] is understandable as a defensive position. Chemistry was popular; *Lyrical Ballads* was not. Davy was a far more brilliant, influential, and (probably) widely-read figure than were the two poets in the next twenty-odd years. It was not the case that chemistry needed the support of poetry; it was that the new poets were afraid of being swamped by an increasingly powerful intellectual movement which they could not understand or handle in their new poetic approach.

Wordsworth may not have taken the preface very seriously, but it describes his practice fairly well. As Sharrock points out, the preface says that science has not yet transformed the poet's view of life; and this was certainly true of Wordsworth's imagery. He almost always used classical terminology in his astronomical references, for example: the signs of the Zodiac, personified planets, etc.; and these give some poems just that archaic prettiness that he in general wished to combat. By way of contrast, we note that Wordsworth used so recent a scientific theory as the diurnal motion of the earth (Copernicus, 1543) largely for its shock effect.[26] Yet by 1837 this poet was able, in the lines on the statue of Newton added to the *Prelude,* to express the noblest of all tributes to the man of science. How did this change come about?

Partly, of course, through Wordsworth's general conversion to conservatism and, with it, to (or at least toward) the orthodoxy of the Established church. That orthodoxy included a strong respect for the use of natural science in natural theology, which Wordsworth accepted but only as the end-product of a dialectical struggle. Thus in determining his position with respect to science in the 1830's, one should not analyze only, say, poem XI of "Poems Composed or Suggested During a Tour in the Summer of 1833," but poems XI, XII, XIII, XIV as a sequence in the course of which Wordsworth modifies his initial position. The same is true, for historical science, of the sequence IV, V, VI of "Memorials of a Tour of Italy, 1837."[27]

At the level of intellectual analysis, the excellent monograph of Ben R. Schneider, Jr., on *Wordsworth's Cambridge Education* (which is much wider in scope than its title suggests) shows that in rejecting a "universe of death" Wordsworth was rejecting the extreme French interpretations of

the 18th century in favor of something rather closer to what British followers of Newton—such as James Thomson—had upheld.[28] To be sure,
Wordsworth's something that was "far more deeply interfused" in the
world was not, in his Tintern Abbey phase, exactly God; but his position
was intellectually unstable (or fuzzy, if you prefer), so that it was not surprising that he could find more or less of a home in the Truth-Complex,
once he realized that science in that context did not insist on abolishing
moral feelings, conscience, and the sense of the sublime.

But also Wordsworth was influenced by personal contacts. In 1820 his
brother Christopher was named Master of Trinity College, Cambridge,
and William once more came into relation with the university scene, only
this time not as an unpolished odd-spoken undergraduate[29] but as a master
in his own right. He found himself admired by a brilliant and energetic
group of young men, the Fellows of Trinity, who were as Christian, as enthusiastic about nature, as fascinated by the poetry of Wordsworth and
Coleridge, as the sage could wish. And many of them were scientists. William soon became so friendly with them that he did not, for example, hesitate to scold them when they favored Catholic Emancipation.[30] When they
wished to promote an astronomer as parliamentary candidate for Cambridge, he wrote an epigram for use in the contest.[31]

One of the Fellows was, indeed, a family friend. Young George Airy
was a *protégé* of the Wordsworths' friend, Thomas Clarkson. When George
went to Trinity as an undergraduate, Dora Wordsworth, worrying that
Christopher did not keep his family informed about his health, set up a
spy system whereby George's reports on the Master's physical condition
eventually wound up at the Wordsworth home at Rydal Mount.[32] By
1830 "Orson the hairy, alias Professor Airey," as William called him, was
professor of astronomy at Cambridge and was soon to become Astronomer Royal.[33]

More important was Wordsworth's contact with the professor of geology, Adam Sedgwick, in whom Wordsworth found an irresistible companion. Sedgwick, a hearty Yorkshireman, *was* one of those Dalesmen whom
Wordsworth praised so highly. He loved to roam in the hills as much as did
the poet himself; and he loved to have Wordsworth with him. Sedgwick too
saw God through nature, rejected materialism and utility, believed in final
causes ruling the world. So we have the romantic but true picture of the
aged sage and the young geologist tramping over the Lake District together, the geologist quietly geologizing while the poet tried out romantic
thoughts.[34] Eventually Sedgwick had to object to the passage in the *Excursion* which seemed to ridicule his own profession. (Why young Englishmen
like Sedgwick, Darwin, and Ruskin were impressed by the *Excursion* is still a
mystery to me; but they were.) Wordsworth answered:

8

You perhaps don't remember that the Pocket Hammerers were complained of not by me in my own person, but in the character of a splenetic Recluse; I will, however, frankly own to that a certain extent I *sympathized* with my imaginary personage, but I am sure I need not define for you how far, but no farther, I went along with him. Geology and *Mineralogy* are very different things.[35]

The contrast of geology (good) to mineralogy (foolish) is interesting, and I shall return to it later.

With this exposure to scientists in the early and middle 1820's as a new part of his experience, Wordsworth was ready when in 1827 he met the brilliant young Irish mathematical scientist William Rowan Hamilton, who impressed him as had no other mortal save Coleridge. The story of their friendship, and of how Hamilton encouraged Wordsworth to write his eulogy of Newton, has been told before.[36] I shall simply give the two versions, from the *Prelude* of 1805–06 and from that of 1850, to show how much Wordsworth changed. Note that he has not merely added lines to the later version. By changing details he has altered the nature of the experience. The early version is:

> And, from my Bedroom, I in moonlight nights
> Could see, right opposite, a few yards off,
> The Antechapel, where the Statue stood
> Of Newton, with his Prism and silent Face.

The poet is only recording the surroundings of his room. If anything is suggested about Newton, it is a slight sense of lack of communication. The later version, in contrast, suggests a tryst by moonlight, a fellowship between two persons who have known what it was to be awake, alone, late at night:

> And from my pillow, looking forth by light
> Of moon or favoring stars, I could behold
> The antechapel where the statue stood
> Of Newton with his prism and silent face,
> The Marble index of a mind for ever
> Voyaging through strange seas of Thought, alone.[37]

The Romantic poet has at last fully acknowledged the romance of the scientific quest. He has equated Newton with the author of the *Rime of the Ancient Mariner*.

It is not too surprising to find an alliance between science and literature at Trinity College, Cambridge. Let us now move to a less propitious setting. Wordsworth was much in favor of the Oxford Movement, for which he believed he had helped pave the way by his *Ecclesiastical Sonnets*

9

of 1822.[38] And Oxford's professor of poetry, John Keble, was an ardent Wordsworthian. We may next consider the three leaders of the Oxford Movement (whose members were called in contemporary writings Puseyites, Tractarians, or New High Churchmen—but not Ritualists, which were a different movement which Pusey disliked). In increasing order of importance the leaders were: Keble, John Newman, and Edward Pusey.

Wordsworth's own writings were far from popular at first. Less than half of the original printing (500 copies) of *Lyrical Ballads* was sold before the publisher disposed of the rest to another publisher and returned the copyright to Wordsworth, it being considered valueless.[39] In 1825, Longmans was still balking at an edition of 1,000 copies for Wordsworth. In 1838, writing to Sir Robert Peel on the need for extended copyright, Wordsworth said that he had made more from his long-published writings in the last five or six years than in the thirty preceding ones. Writing to William Gladstone he gave the figure of nearly £1,500 in the past three years.[40]

In contrast, Keble's *Christian Year* was an immediate best seller. It was published in 1827 and had twenty-five editions by 1834; I have seen a *seventy-fifth* edition of 1863.[41] Wordsworthian nature imagery, therefore, reached a mass audience not in the tortuous setting of Wordsworth's thought, but as part of the more direct and conventional Christianity of the *Christian Year.* We do not expect to find much use of science in Keble's poems. What is of interest is how science and natural theology fit into this Christian best seller whenever they are mentioned.

The answer is that Keble is quite conventional:

> When round thy wondrous works below
> My searching rapturous glance I throw
> Tracing out Wisdom, Power and Love
> In earth or sky, in stream or grove,—[42]

Here is no suspicion of natural theology. Rather, the works of God are compared to the Bible:

> There is a book, who runs may read,
>   Which heavenly truth imparts,
> And all the lore its scholars need,
>   Pure eyes and Christian hearts.
>
> The works of God above, below,
>   Within us and around,
> Are pages in that book, to show
>   How God himself is found.

This is the beginning of the poem for Septuagesima Sunday. Its accompanying text is: "The invisible things of Him from the creation of the world are clearly seen, being understood by the things which are made. *Romans* i. 20."[43]

Elsewhere we are given a list of what pilgrims are not seeking: "Not for light Fancy's reed,/Nor Honour's purple meed,/Nor gifted Prophet's lore, nor Science' wondrous wand."[44] In this ascending list of merely human rewards, science is at the top. Keble is rather less restrained in his use of astronomical images than is Wordsworth; once, for example, he compares Christ to a shooting star.[45] On one theoretical problem, however, he lags behind his master: twice he refers to the sun's heat as inexhaustible.[46] Wordsworth in 1817 in his "Vernal Ode" had recognized that stars and planets are free only from the "semblance" of decline:

> What if those bright fires
> Shine subject to decay,
> Sons happy of extinguished sires,
> Themselves to lose their light, or pass away
> Like clouds before the wind....[47]

Wordsworth's point is that all material things decay, but that the system of Nature survives. He seems to say, however, that *species*, such as the bee, are at least as old as the original golden age of the ancients (his text, at the beginning of the Ode, is from Pliny).

Keble follows Wordsworth rather than his scientific contemporaries. For him, the flowers of the field are "relics of Eden's bowers" and are unchanged in form since the beginning of the world.[48] He has chosen to ignore paleontology, with its evidence of extinct forms. Note that I say *chosen* to ignore. This might not be true for Wordsworth in 1817, or for Keats; perhaps paleontology had not been brought to their attention in any striking way. But the Tractarians were Oxford colleagues of the energetic William Buckland, and noted in particular his *Reliquiae Diluvianae* of 1824. Among other things, Buckland tried to suggest the possibility of the Mosaic Flood by geological evidence. The book was a rather important one, scientifically speaking, but some of Buckland's contemporaries were more sensitive to the remarks concerning the Flood than they were to the scientific findings announced, Newman wrote to Pusey long afterwards: "I feel quite what you say about Buckland's 'Reliquiae.' It has made me distrust every theory of Geology since."[49]

It is interesting to note how none of this suspicion was carried over into the poems of the *Christian Year*. Keble simply ignored paleontology, but this did not indicate to the reader a distrust of natural science and of natural theology along with it.

For Newman it was quite otherwise. When that admirer but not imitator of Butler said, "Theology and Physics cannot touch each other, have no intercommunion, have no ground of difference or agreement, of jealousy or of sympathy," he did not mean what a modern theologian might mean. He meant rather that the fact of Moses' flood is known, is a given truth which religion presents to science as basic information to be incorporated into its body of data. Buckland's search for physical proof of this already established fact was therefore not only unnecessary but also impertinent. For natural theology in a common contemporary form, that is, what Newman scornfully called "physical theology," he had no use at all: "Really it is no science at all, for it is ordinarily nothing more than a series of pious or polemical remarks upon the physical world viewed religiously."[50]

Newman felt that natural science was relatively unimportant because, "With matter it began, with matter it will end; it will never trespass into the province of mind." And science does not help us in dealing with human nature. To maintain "that the mind is changed by a discovery, or saved by a diversion, can thus be amused into immortality,—that grief, anger, cowardice, self-conceit, pride, or passion, can be subdued by an examination of shells or grasses, or inhaling of gases, or chipping of rocks, or calculating the longitude, is the veriest of pretences which sophist or mountebank ever professed to a gaping auditory." In one excellent sentence Newman thus managed to suggest the unimportance of zoology, botany, chemistry, geology, and astronomy. Nor was he more generous to men of science than he was to their subjects. Galileo, he believed, deliberately went out of his way to insult the received interpretation of Scripture in a wanton and arrogant attack. Thus he deserved his punishment, although the resulting publicity was unfortunate. And in general, Newman observed, there has been a tendency for experimental philosophers to move toward infidelity, either in its atheistic or its Protestant version.[51] Newman's unsympathetic view of Galileo was not unprecedented; William Whewell, in his influential *History of the Inductive Sciences* of 1837 was also unsympathetic toward Galileo personally. But Whewell believed that experimental science led philosophers *toward* God, especially when seen in the light of physical theology.[52]

Newman's general departure from the older Anglican tradition, while he was still in the Anglican church, has been ably described by H. L. Weatherby.[53] To his analysis I will add that in nothing did Newman depart more definitely than in his opinion of natural science and natural theology. He is surely a reluctant witness for my case. If even Newman must resort to considering science as norm of truth, then this is indeed good evidence of the status of science. And he does so: not, we assume, because he believes it, but because it is the *only language available* with which to praise theology sufficiently highly. Thus he ends his first discourse in *The Idea of a University* with the following summary: "Religious doctrine is knowledge, in as full a sense as Newton's doctrine is knowledge. University Teaching without Theology is simply unphilosophical. Theology has at least as good a right to claim a place there as Astronomy." On the other hand, he observes, if a man does not believe that religious facts are true "in the sense in which the general facts and the law of the fall of a stone to the earth is true, I understand his excluding Religion from his University."[54] In other words, common-sense truth is *defined* by the statement that $s = \frac{1}{2} g t^2$ is true!

Equally interesting is Newman's use of scientific comparisons in the *Essay on the Development of Doctrine* of 1845—a very different use from that in Butler's *Analogy*, as Owen Chadwick has pointed out.[55] Newman is not so sure about analogies between development in nature and development in religion; he rather uses comparisons to the development of men's thought about nature—not the processes of nature, but the processes of the history of science. In the introduction he admits that his explanation is a hypothesis, but so too were the explanations of astronomers from Ptolemy to Newton. Nor should one object to an explanation simply because it is new, any more than one should have objected on that ground to the theory of gravitation when it was first announced. In Chapter III, "Method of Proof," he points out that we normally use principles on faith. "Thus most men take Newton's theory of gravitation for granted, because it is generally received ... and if phenomena are found which it does not satisfactorily solve, this does not trouble us, for a way there must be of explaining them, consistently with that theory, though it does not occur to ourselves." The conclusion is that "the *onus probandi* is with those who assail a teaching which is, and has long been in possession."[56]

In Chapter IV, "Instances in Illustration," Newman asserts that "this line of argument is not unlike that which is considered to constitute a sufficient proof of truths in physical science. An instance of this is furnished in a work on Mechanics of the past generation." He then uses a quotation from Wood's *Mechanics* to show that neither derivation from self-evident principles nor exact evidential proof is possible for the laws of motion, which we nevertheless accept.[57]

Another interesting use of science occurs in Chapter V, "Genuine Developments Contrasted With Corruptions." There are seven tests of a genuine development, and in every case but one (the fifth), Newman makes his idea clear by referring to what might be considered the materials of "physical theology."

Finally, Newman defines development as the full elaboration into systems of what was really present in the idea before. Development is the development of men's grasping of the Idea. This presentation is interestingly similar to Whewell's explanation of how the central Idea of a

13

scientific discipline is gradually mastered by scientists.[58]

The Tractarian leaders, even if they could not completely do without it, were not subjectively well disposed toward science in the 1830's and early 1840's, and science did not then flourish at Oxford.[59] Attendance at (voluntary) scientific lectures, always small, went almost to nothing as stricter classical examinations left the students with less leisure time for innocent amusements. William Buckland all but deserted Oxford after being made Dean of Westminster by Peel, and even refused to support the science activities described below, feeling that the cause was hopeless. He wrote: "Some years ago I was sanguine...as to the possibility of Natural History making some progress at Oxford, but I have long since come to the conclusion that it is utterly hopeless."[60]

But a striking change took place after young Henry Acland appeared on the scene in 1845—dare we call him a symbolic replacement for Newman? As an undergraduate he had been one of those who walked with Newman. Ill health forced him to give up his plans for a medical career. He therefore accepted the position of Lee's Reader in Anatomy at Christ Church, Oxford, even though this amounted to abandoning all of his prospects for worldly success.

Christ Church was the college where Buckland and Edward Pusey, not only Tractarian leader but also professor of Hebrew, had been fellow canons and good neighbors for many years in spite of their intellectual differences. Acland's first duty, as he saw it, was to call on Pusey and ask him two questions:

> 1. Am I right in believing that you, Mr. Newman, Mr. Keble, and your friends disapprove of Physical Science as a branch of education at Oxford?
>
> He said, "Yes, we do...."
>
> I then put to him my second question.
>
> 2. "Am I to understand that you, who with the Dean and Chapter have appointed me a teacher in a great department of Science, will consider me a mischievous and dangerous member of society, when I endeavour to do my duty in my office?"
>
> Dr. Pusey, who, whatever may be considered his faults by those who did not know him, was a strong, true man, and endowed with a sense of humour, threw himself back in his chair and laughed aloud.[61]

The two men were thereafter firm friends; Acland became, indeed, Pusey's own physician. More important, Acland could count on Pusey's support in the university, and that was crucial. It was not too hard to introduce more science in the curriculum, for the outside world clearly wanted this move, and the other scientists at Oxford still had some position in the community. The humanists handled the matter by throwing out Acland's plan of requiring some science for all students, and merely made it possible for students to specialize in science if any should have such an odd desire. The good students (that is, those who chose classics) were not to be subjected to any such educational frills. Here we see one of the beginnings of that divorce of science and the humanities that modern authors are so vocal about, and it is amusing to note that it was the humanists who forced it, over the protests of the scientists. Classical studies had improved their standing in the previous half-century, and their advocates had no intention of diluting their newly won standards. Instead, they pushed science into a corner and then went on, in the second half of the century, to consolidate their position. Benjamin Jowett, always suspicious of science, represents the triumph of the new Greek enthusiasts over the scientists, both at the university and in the nation.

Having won concessions in the curriculum, the scientists found it more difficult to get money for adequate laboratory space. A bitterly contested sequence of votes ensued before all of the appropriations for the building of the Oxford Museum were finally passed. At every stage there was determined opposition, even after construction had begun. Toward the end, a grant for gas-pipes for lighting the central court was carried in Oxford's Convocation by a slim majority. Then a grant for providing burners for the pipes was lost by an equally close vote. Acland was the masterly general of the forces of science in this contest, but the decisive factor was that Pusey threw the full weight of the Tractarian vote to Acland, and against the opposition of classicists, purse-savers, and old-fashioned churchmen.[62]

So it was because of Pusey that John Ruskin's dream building arose in the Parks. More precisely, it was because Pusey brought religion, in this struggle, back into its old alliance with science and material progress. Ironically, the first formal use to which the building was put was to act as setting for the meeting which produced a classic Darwinian scene: the encounter between Thomas Huxley and Bishop Wilberforce at the British Association meeting in 1860.

Nevertheless, Pusey never wavered in his conversion, and always insisted that science, even evolutionary science, was a worthy pursuit, although like many others he could "only hope that, in days which I shall not see, God may raise up some naturalists who may, in His hands, destroy the belief in our apedom." He expressed his own position in 1855 in an interesting, if not novel, image: "All the sciences move like planets round the sun of God's truth."[63] Thus, in order to express his belief in the centrality of religious truth, the religious leader had to make the prior assumption of the truth of Copernican theory. Rhetorically, at least, astronomy was the truest truth there was.

15

Even though I am expecting it, it is something of a shock when I open a Victorian book on architecture and encounter a passage such as the following:

> Lastly, before leaving the subject of structural deceits, I would remind the architect who thinks that I am unnecessarily and narrowly limiting his resources or his art, that the highest greatness and the highest wisdom are shown, the first by a noble submission to, and the second by a thoughtful providence for, certain voluntarily admitted restraints.... The example most apposite to our present subject is the structure of the bones of animals. No reason can be given, I believe, why the system of the higher animals should not have been made capable, as that of the *Infusoria* is, of secreting flint, instead of phosphate of lime, or more naturally still, carbon, so framing the bones of adamant [diamond] at once.
>
> ...But the architecture of animals *here,* is appointed by God to be a marble architecture, not a flint nor adamant architecture; and all manner of expedients are adopted to attain the utmost degree of strength and size possible under that great limitation. The jaw of the ichthyosaurus is pieced and riveted, the leg of the megatherium is a foot thick, and the head of the myodon has a double skull; we, in our wisdom, should doubtless, have given the lizard a steel jaw, and the myodon a cast-iron headpiece, and forgotten the great principle to which all creation bears witness, that order and system are nobler things than power.

Architecture, paleontology, and God working together—*that* is real Victorian thought!

The passage is from "The Lamp of Truth" in John Ruskin's *Seven Lamps of Architecture.*[64] As an undergraduate at Oxford, Ruskin had found in Henry Acland, a medical student four years his senior, his first real friend. Acland delighted Ruskin "as a leopard or falcon would," by his pride and determination, as well as by his love of beauty. At the same time—in the late 1830's—Ruskin was absorbing the message of William Buckland's geological lectures thoroughly, and at Buckland's breakfast table met, among others, young Charles Darwin.[65]

It is not surprising, then, to find that in "The Lamp of Obedience" the priority of Law over Liberty is proved by reference to science. Of Liberty we can say: "The stars have it not; the earth has it not; the sea has it not; and we men have the mockery and semblance of it only for our heaviest punishment...if there be any one principle more widely than another confessed by every utterance, or more sternly than another imprinted on every atom, of the visible creation, that principle is not Liberty, but Law.... Exactly in proportion to the majesty of things in the scale of being, is the completeness of their obedience to the laws that are set over them. Gravitation is less quietly, less instantly obeyed by a grain of dust than it is by the sun and moon."[66]

Ruskin, to be sure, had been interested in mineralogy since child-hood, long before he became something of a *protégé* of Buckland. Even so, it is still rather startling to realize that one of the chief messages of *Modern Painters* was that the modern painter does not know how to see. The painter must make "an earnest, faithful, loving, study of nature as she is," and what nature *is* must be learned from, of all people, the geologist. It is the geologist who *does* know how to see. And he knows what to do with his observations. "Stones, in the thought of the peasant, lie as they do on his field, one is like another, and there is no connection between any of them. The geologist distinguishes, and in distinguishing connects them. Each becomes different from its fellow, but in differing from, assumes a relation to its fellow; they are no more each the repetition of the other—they are parts of a system, and each implies, and is connected with the existence of the rest." And so generalizations of the kind that geologists make are "right, true, and noble."[67] Geology, then, is normative of truth, and the painter must train himself to see truth as well as the geologist does.

One fallacy of modern literary historians is to believe that the leading Victorian intellectuals and poets learned only from earlier intellectuals and poets, not from their teachers or from scholars. Thus Humphry House has attributed Ruskin's demand for accurate natural history in painting to a misunderstanding of Wordsworth.[68] Actually it involved a correct understanding of scientists such as Buckland and Acland. George Landow, in his *Aesthetic and Critical Theories of John Ruskin,* is typical in finding something odd about Ruskin's scientific passages: "After he has been led through long disquisitions on slaty crystallines, compact coherents, and the angles of aiguilles, the bored or bewildered reader may glance at the title page to reassure himself that he is still pursuing a work about Turner." Landow decides that Ruskin's use of science is allegorical, and outmoded.[69]

Allegorical in the usual sense it was not: Ruskin felt an actual connection between the laws governing human activity and the laws of the rest of the world. Ruskin's "language of Types," derived from Evangelical sources as it may have been in his case, was similar in kind to Richard Owen's "archetypes" in anatomy or Edward Forbes's "genera-ideas" in paleontology; and theories of types had a considerable vogue among French chemists. The successful attempt to find a real relation between artistic activity, general moral laws, and the language of stones was being conducted in private by Charles Darwin. Until he published, however, type-thinking was one of the interesting new ways to tackle scientific problems. Perhaps the usage in general derived from a method of Biblical interpretation; if so, Landow's remarks might lead to an interesting new

17

idea in the history of science. My point is that the method was not "out-moded" at the time Ruskin was writing.

Thus there are serious limitations inherent in the approach suggested by Jerome Buckley in his study of "the Victorian concepts of time, history, progress, and decadence." Buckley asserts that the student of literary history or the history of ideas should "heed first and last... the shapers of opinion in their own generation... the most eloquent of spokesmen—above all, the poets, and then the novelists and essayists...."[70] This could be a way to avoid studying the thought-creators and devote oneself to the middle men of ideas, without knowing much about the power, sources, or originality of the ideas they were drawing on. A particular problem comes when the influence of writings by a Ruskin or a Matthew Arnold suddenly seems to wax or wane. This is caused by changes in the Truth-Complex itself, on which a Ruskin or Arnold or Huxley has little effect. Scientists such as Buckland, Adam Sedgwick, and Charles Lyell, scholars such as Connop Thirlwall, Robert Willis, and John Kemble maintained the Complex; and it was only disagreements among such persons which could bring it down.

By not being too familiar with the high-level intellectual complex of which Ruskin was a part, it is easy to arrive at the position of Herbert Sussman's *Victorians and the Machine* and speak of Ruskin's "romantic, antiscientific position." Ruskin was not opposed to science—he encouraged young artists to learn from it, and they can be found in England and America actually studying geology and natural history in the 1850's and 1860's because he told them to. He was opposed to one kind of science—science as mere mechanism, a world-picture of precise geometrical forms and forces, the theodolite as the *only* measuring device of God's gift to us in nature. But his scientific friends and teachers were also opposed to this kind of science. The answer was not, of course, "vitalistic assumptions" (as Sussman suggests)[71] but a view of studying each part of nature with respect, on its own terms—since those are God's terms. It is easy enough to find passages in almost any serious writer of the early 19th century which seem to be an attack on science in general. Usually, on close inspection, the passage turns out to be an attack on "mechanistic"—that is, 18th-century French—interpretations of science, an attack by someone who knows the more modern 19th-century scientific position.

Of course this position should not be confused with a quite different one, a refusal to admire machines and/or technological civilization. One could be in favor of science and against air pollution as easily then as now; one could consider a locomotive beautiful but its soot deposits ugly.

This digression into historiography may be of some use in understanding that Ruskin could admire the geologist but refused to admire the zoologist or botanist. In 1841 he read Louis Agassiz's great *Poissons fossiles,* and what he learned from it was "that Agassiz was a mere blockhead to have paid for all that good drawing of the ugly nasty things, and that it didn't matter a stale herring to any mortal whether they had any names or not."[72] The biologist, Ruskin thought, was only interested in naming and classifying, not in grasping the true complexity of things. His knowledge was therefore "ignoble" as contrasted to that of the artist.

> This is the difference between the mere botanist's knowledge of plants, and the great poet's or painter's knowledge of them. The one notes their distinctions for the sake of swelling his herbarium, the other, that he may render them vehicles of expression and emotion. The one counts the stamens, and affixes a name, and is content; the other observes each of its attributes as an element of expression, he seizes on its lines of grace or energy, rigidity or repose; notes the feebleness or the vigor, the serenity or tremulousness of its hues; observes its local habits, its love or fear of peculiar places; its nourishment or destruction by particular influences; he associates it in his mind with all the features of the situations it inhabits, and the ministering agencies necessary to its support.[73]

This prejudice against botanists, or any kind of mere classifiers, is an interesting feature of the early Victorian period. It is not, however, new. It is a reassertion of the main line of English natural history against the incursion of Linnaean emphasis on classification, and on artificial schemes of classification at that. In the English tradition, from Ray in the 17th century through Gilbert White in the 18th, scientists had wanted to see the organism in its own setting, on its own terms. We can go back still further. Berowne, in a mock-serious speech in *Love's Labor's Lost* (I,i, 89–93), says of astronomers:

> These earthly godfathers of heaven's lights,
> That give a name to every fixèd star
> Have no more profit of their shining nights
> Than those that walk and wot not what they are.
> Too much to know is to know nought but fame;
> And every godfather can give a name.

The greatest practitioner of Ruskin's desired technique, however, was Charles Darwin, in the books he wrote after the *Origin of Species.* Not that he learned from such as Ruskin: he was using this technique in his notebooks in 1837 when Ruskin was going to college.

This seems a good place to recall the passage in Wordsworth's *Excursion* to which Adam Sedgwick objected, and to which Wordsworth replied that he was objecting to mineralogy, not geology. His objection to "the wandering Herbalist" and to him "who with pocket-hammer smites the edge/Of luckless rock," was that such a person chips off a splinter "to resolve his doubts;/And, with that ready answer satisfied,/The substance classes by some barbarous name,/And hurries on.../And thinks himself enriched,/Wealthier, and doubtless wiser, than before."[74] That is, the sciences of classification give worthless, superficial knowledge; whereas true sciences should show wider and wider interrelationships, should lead to an understanding of the system of the world and therefore should lead eventually to God.

This was Ruskin's position also, and indeed that of almost anyone who was a part of the Truth-Complex. Tennyson, in *Maud*, Part Two, Section II, expresses the same idea: "See what a lovely shell,/Small and pure as a pearl/...What is it? a learned man/Could give it a clumsy name./Let him name it who can,/The beauty would be the same." This position was a little hard on zoologists and botanists, but at least it kept them looking for some natural system of classification which *would* lead to an understanding of the system of the world.

Now that we have Ruskin placed properly with respect to science and Oxford: *protégé* of Buckland, friend of Acland, chief architectural innovator of the Oxford Museum; and now that we know of the science-religion alliance (Acland-Pusey) that won a larger place for science teaching at Oxford; Ruskin's own view of university education can almost be deduced. (Of course he spoke of "our present European system of so-called education" when he only meant what was currently taught at Oxford; but that is a common failing.) He wanted not Latin and Greek, not training the mind. He wanted natural science: "For one man who is fitted for the study of words, fifty are fitted for the study of things, and were intended to have a perpetual, simple, and religious delight in watching the processes, or admiring the creatures, of the natural universe." He wanted religion. "I know that this is much to hope. That English ministers of religion should ever come to desire rather to make a youth acquainted with the powers of nature and of God, than with the powers of Greek particles...." And he wanted the study of human relations, that is, what at Cambridge was being called the "moral sciences."[75]

Note particularly, and again, that the enemy was not the religious structure of the old universities, against which political reformers agitated. The real enemy to science and religion alike was the Greek and Roman classics.

The contrast between good science and bad science is shown very nicely in the writings of Charles Dickens. Science is not an important subject for Dickens as a novelist, although his references to it are more common than are those of more conventional writers. Anthony Trollope, for example, scarcely ever uses a scientific word—once or twice he refers to atoms. Dickens' references are more up-to-date, and (after the early and not very successful "Mudfog Papers") he uses science carefully. It is a standard of truth with which to contrast the shortcomings of less worthy enterprises, those of social scientists, moral reformers, economists, and arithmeticians. He is constantly alert lest a false front of apparent science be applied inhumanely to reduce real human beings to abstract categories.

The Chimes of 1844 is an especially good example of what Dickens disliked. Precise numbers are bad: they are for knowing that the poorhouse is 28 miles away. Imprecise numbers are good: thanking someone "a thousand times!" is meritorious. Addition and subtraction and division, averaging, the application of Nautical Almanack principles to men and women: all are bad. Anyone who can count accurately above 3 is suspect. Even 1 is bad if it means 1 alone instead of 1 member of a real group of people—preferably 3 people. In short, The Chimes is an out-and-out attack on arithmetic. Numbers, after all, are a way to classify living beings in a completely emotionless system, even more logical and less related to life than Linnaean botany.

A better-known attack on logical classification comes in Hard Times of 1854. The "Ologies" were often assumed to stand for modern knowledge. Thus in the Southern Rose magazine for girls published in Charleston, South Carolina, in the 1830's, there appeared a humorous fictional account of a visit to a Female Seminary. One of the girls in the story asserted, "I'll tell you, if I had a daughter, she should not learn anything but Botany, Astronomy, Ornithology, all the Ologies you know and the accomplishments. She should not be puzzled with grammar, arithmetic, reading, writing and such old fashioned things. Oh yes, she might learn to write poetry and billet doux."[76]

Dickens, however, was not so amused. Hard Times is, among other things, an attack on the "Ologies" if they are taught without imagination and concrete observation.[77] As Mrs. Gradgrind says, "But there is something—not an Ology at all—that your father has missed, or forgotten, Louisa" (Book I, ch. 9). Hence the famous definition of a horse (Book I, ch. 2) is considered by Gradgrind to be true knowledge, contrasted to which actual acquaintance with the living beast is worthless. "'Now girl number twenty,' said Mr. Gradgrind, 'You know what a horse is!'" This fallacy of substituting logical classification for direct knowledge is just what we have seen Ruskin and Wordsworth objecting to.

The objection was not new; Francis Bacon made it also. And here is another example, of 1828:

> Let us suppose that a zoologist should attempt to give an account of some animal, a porcupine for instance, to people who had never seen it. The por-

cupine, he might say, is of the class mammalia, and the order glires. There are whiskers on its face; it is two feet long; it has four toes before, five behind, two foreteeth, and eight grinders. Its body is covered with hair and quills. And, when all this has been said, would any one of the auditors have formed a just idea of a porcupine? Would any two of them have formed the same idea? ... The vital principle eludes the finest instruments, and vanishes in the very instant in which its seat is touched."

The author is Thomas Macaulay, in his essay on the poetry of John Dryden.[78]

Scientists themselves objected to the procedure Dickens satirized. William Whewell was a practicing scientist as well as a teacher and a philosopher. The following section of his *Philosophy of the Inductive Sciences* of 1840 might seem almost to be the textbook for Dickens. Whewell is explaining that definitions are neither clear nor complete in natural history, so that scientists actually work with a Method of Types rather than with logical definitions. Natural history, therefore, "must be studied by inspection of the *objects* themselves and not by the reading of books only." And Whewell noted:

> It is a curious example of the influence of the belief in definitions, that elementary books have been written in which Natural History is taught in the way of question and answer, and consequently by means of words alone. In such a scheme, of course all objects are *defined:* and we may easily anticipate the value of the knowledge thus conveyed. Thus, "Iron is a well-known hard metal, of a darkish gray colour, and very elastic." "Copper is an orange-coloured metal, more sonorous than any other, and the most elastic of any except iron." This is to pervert the meaning of education, and to make it a business of mere words.[79]

Dickens was in agreement with the scientific thought of his time. Wisely, he placed the blame for the perversion of science not on a particular science but on a particular kind of adult, the Gradgrinds of this world—and even so it is because they don't know any better.

Later in *Hard Times* (Book I, ch. 15), Dickens uses another science—astronomy—as a model by reference to which he can rebuke theorists who do not bother with concrete observation:

> As if an astronomical observatory should be made without any windows, and the astronomer within should arrange the starry universe solely by pen, ink, and paper, so Mr. Gradgrind, in *his* Observatory (and there are many like it), had no need to cast an eye upon the teeming myriads of human beings around him, but could settle all their destinies on a slate.

The passage draws its power from the assumption that what the astronomer really does is valid. (Correct scientific procedure is given in Disraeli's *Lothair:* "As an astronomer surveys the starry heavens until his searching sight reaches the desired planet, so Lothair's scrutinizing vision wandered till his eye at length lighted on the wished-for orb.")[80] Gradgrind is condemned because he has perverted scientific method. Dickens, like Newman and Pusey, found the rhetorical power of the "astronomy = truth" equation too useful to be ignored.

In the contrast between good science, as exemplified by physical astronomy, and bad science, as exemplified by logic and formal classification, we have one of the meanings of such sets of terms as "modern" and "out-moded" or even "progressive" and "reactionary" in early Victorian times. We also have one of the contrasts between Cambridge and Oxford. Some historians have expressed surprise that, after John Henry Newman left Oxford in 1845 and the religious antagonisms stirred up by the Tractarians died down, Oxford's next interest was the study of John Stuart Mill's *Logic* (1843; 2nd ed. 1846). The implication has been that this movement was a swing towards modernity, that Mill's book represented a progressive, or even a scientific position.

Actually, in spite of Mill's efforts, the *Logic* had little relation to scientific thought or discovery, about which Mill knew little. Much of what science it did contain was taken from John Herschel's *Preliminary Discourse on the Study of Natural Philosophy* of 1831, sometimes essentially verbatim. Mill's own contributions to scientific matters did not always meet with Herschel's approval, when Mill asked Herschel to criticize the first edition in preparation for the second.[81] The *Logic* was not an immediate hit, like Charles Lyell's *Principles of Geology* or Macaulay's writings; it seems possible to me that the success of Mill's *Political Economy* led some of his readers back to the *Logic*.

At any rate, by 1872 the student could buy Stebbing's *Analysis of Mill's System of Logic* or Kellick's *Student's Handbook of Mill's System of Logic*, each for 3.6d., if he did not feel up to the 8th edition of the real thing at 25s. By way of comparison, I note that the student could have Macaulay's *Essays* in Library, Cabinet, Traveller's, People's or Student edition for 36s., 24s., 21s., 8s., or 6s.; or he could have one of sixteen essays printed separately at 1s. or 6d.[82]

To a university such as Oxford, accustomed to Aristotle, Locke, and Butler, Mill's *Logic* would appeal as fitting in with, although being more up-to-date than, what it had been studying all along. The contemporary work which, with all its faults, did deal with science as it actually existed and was practiced was Whewell's *Philosophy of the Inductive Sciences*. Unfortunately Whewell's work was *too* modern and wide-ranging to fit into any neat teaching category, even at Cambridge. Whewell's philosophy reflected modern German thought since Kant, and that was rather a lot for Englishmen to accept. Then, too, Whewell was definitely an Idealist, and many scientists did not want to admit to too much Idealism, even though their science was making a simple empiricism practically unthinkable. "Empiricism" was a bad word in the early 19th century, and everyone (including Mill) enjoyed denouncing it; but if we use it to denote agreement with Hume that "cause" is merely a word for the constant antecedent of the effect, then there continued to be English empiricists. Mill may be considered one of them.

Roughly speaking, Humean empiricism thus defined is a position which would deprive scientists of confidence in their intuitive working concepts. If scientists had believed in it, neither field theory nor the conservation of energy would have been developed. As it turned out, few scientists bothered to notice whether their ideas were acceptable to Hume. One who did was John Herschel; the result was that he attacked Hume (see below, Chapter Eight). Another was James Clerk Maxwell, who considered himself more of an Idealistic even than Whewell.[83] The continued appeal of Mill and his approach to philosophers has ensured that much of English philosophic thought since the mid-19th century has been divorced from modern science and the realities it deals with.

When Oxford did get around to German thought, in the later 19th century, with a sure fatality it chose Hegel. Of all the German thinkers, Hegel is the one who is hardest to make compatible with science.

John Stuart Mill was mentally closer to John Henry Newman than he was to the processes of thought and discovery of 19th-century science. It is not a new observation to note that Newman's argument as to how a religious doctrine is developed by challenge and battle sounds similar to parts of Mill's *On Liberty*.[84] But if Mill was old-fashioned and Oxford was mildly reactionary, Newman was not. His anti-science stance may even be considered radical. At any rate, it is nicely symbolic, I think, that Newman said farewell to his Oxford friends at the Radcliffe Observatory, and then went out, leaving the Anglican Church, and the telescopes, behind him.

How, then, did cultivated Victorians learn what science is about?

They read Charles Lyell's *Principles of Geology* and John Herschel's *Treatise on Astronomy* (later *Outlines of Astronomy*) and thought they got a pretty good idea, regardless of what philosophers said.[85] I think they were very sensible.

Notes to Chapter One

The first version of this essay was given at the *Victorian Studies* colloquium at Indiana University in March 1962, and was published in the *Journal of the History of Ideas*, XXV (1964), 487–502. This second version is considerably expanded. I am grateful to the members of my advanced seminar at the Smithsonian in 1976 for thorough criticisms.

1. Quoted in Augustus J. C. Hare, *Memorials of a Quiet Life,* 12th ed. (London, 1874), I, p. 339.

2. Thomas Chalmers, *Discourses on the Christian Revelation Viewed in Connection with the Modern Astronomy* (New York, 1842), *passim;* J. B. Mozley, *Eight Lectures on Miracles* (Oxford, 1865), pp. 49–50.

3. Cf. Max Jammer, *Concepts of Force* (New York, 1962), pp. 242–248; and my remarks in Chapter One. Jammer is concerned primarily with the history of force under its own name; my remarks on energy go further, perhaps, than his text authorizes.

4. George Foote, "Science and its Functions in Early Nineteenth Century England," *Osiris,* XI (1954), 438–454.

5. R. K. Webb, *The British Working Class Reader, 1790–1848* (London, 1955), pp. 77–82; see also David Layton, *Science for the People* (New York, 1973).

6. Thomas B. Macaulay, "History," in *The Lays of Ancient Rome and Miscellaneous Essays and Poems* (London, 1958), pp. 24–25. The date of the essay is 1828.

7. William Whewell, "Report on...Electricity, Magnetism, and Heat," *British Association for the Advancement of Science Report,* V (1835), 1; Whewell, "Address to the meeting," *Brit. Assoc. Report,* III (1833), xiii; Charles Daubeny, "Address to the meeting," *Brit. Assoc. Report,* VI (1836), xxiii.

8. J. W. Clark and T. M. Hughes, *The Life and Letters of the Reverend Adam Sedgwick* (Cambridge, 1890), I, 271–272.

9. *Lux Mundi,* ed. Charles Gore, 5th ed. (New York, n.d.), pp. 13–14, 42, 87, 157–158, 161. Gore's statement is on p. 266; see also p. 399 and most of Essay 3.

    Later church historians have accepted the claim of Gore and his fellows that they were actually up to date and liberal-minded about evolution. (Actually they had not quite caught up to the idea of evolution = progress, the old idea which Darwin knocked on the head.) S. C. Carpenter's widely read *Church and People, 1789–1889* (London, 1933) is so thoroughly Anglo-Catholic that few of its analyses are well balanced; but the same position is found in more recent books. See James Carpenter, *Gore: A Study in Liberal Catholic Thought* (London, 1960), pp. 35, 75, 87n (here an Aristotelian idea is equated to "the modern scientific category of evolution" by Gore), 127, 151–155 (where "progressive evolution" turns out to be the Catastrophism of the 1830's—that is, the position of one set of Darwin's opponents). See also Arthur M. Ramsey, *From Gore to Temple* (London, 1960), pp. 3, 4, 17 (again, Catastrophism).

    On the other hand, John Dillenberger, *Protestant Thought and Natural Science: A Historical Interpretation* (London, 1961), after shuddering at Darwin as a person (p. 270), asserts that no rapprochement was possible between Darwinism as such and Protestantism as such (p. 224) but rejoices that Protestantism won (pp. 265, 267), and notes that in existentialism, "science stands for unauthentic existence" (p. 267). Dillenberger's classic sentence, to me, is: "It is difficult, in any case, for non-scientists to understand or to comprehend what science is about" (p. 274). But if they are to make judgments about it, they should try.

10. Karl Pearson, *The Grammar of Science* (London, 1937), p. 22n.

11. "Brief Memoir of Dr. Chalmers," in Thomas Chalmers, *Miscellanies* (New York, 1847), p. xi.

12. Thomas de Quincey, *The Confessions of an English Opium-Eater* (London, 1961), p. 216.

13. Thomas Carlyle, *Sartor Resartus* (New York, 1896), Book I, Chapter 11, pp. 56–57.

14. Thomas Carlyle, *Past and Present* (London, 1897), Book III, Chapter 11, p. 197.

15. John S. Mill, "Bentham," in F. R. Leavis, ed., *Mill on Bentham and Coleridge* (London, 1959), pp. 76–77.

16. E. J. North, "Paviland Cave, the 'Red Lady,' the Deluge, and William Buckland," *Annals of Science,* V (1942), 103.

17. William Buckland, "On the Adaptation of the Structure of the Sloths to their peculiar Mode of Life," *Transactions of the Linnean Society of London,* XVII (1837), Part Four, 17–27; Benjamin Disraeli, *Lothair* (New York, 1870), Chapter 26, p. 83; J. B. Delair and W. A. S. Sarjeant, "The Earliest Discoveries of Dinosaurs," *Isis, 66* (1975), 5–25.

18. "Lines Written as a School Exercise" (1784–85), *The Poetical Works of Wordsworth,* ed. Thomas Hutchinson, new ed. revised by Ernest de Selincourt (London, 1950), p. 482. All texts except that of the *Prelude* are taken from this edition.

19. Wordsworth, "The Tables Turned" (1798) and "A Poet's Epitaph" (1800), *Poetical Works,* pp. 377, 380. Thus I oppose the position of B. Ifor Evans, *Literature and Science* (London, 1954), pp. 43–57, which to me does not sufficiently allow for Wordsworth's changes of opinion. The same is true of Joseph Warren Beach, *The Concept of Nature in Nineteenth-Century English Poetry* (New York, 1936), p. 197. Both men try to explain away the scorn for scientists expressed in such passages because, I believe, they want to make early Wordsworth consistent with later Wordsworth. Quite simply, he wasn't consistent.

20. Wordsworth, *Poetical Works,* p. 150.

21. *Wordsworth's Prelude,* ed. Ernest de Selincourt, 2nd ed. (Oxford, 1959); 1805–06: Book X, lines 861–930; 1850: Book XI, lines 325–333.

22. *Wordsworth's Prelude,* Book VI, lines 169, 127–128.

23. *As You Like It,* I, i, 149–150, II, vii, 56–57; Jonson, as quoted in Bernard Beckerman, *Shakespeare at the Globe* (New York, 1962), p. 28.

24. *Poetical Works,* p. 738.

25. Roger Sharrock, "The Chemist and the Poet: Sir Humphry Davy and the Preface to *Lyrical Ballads,* " *Notes and Records of the Royal Society,* XVII (1962), 63.

26. "Influence of Natural Objects," line 60, and "A slumber did my spirit seal," *Poetical Works,* pp. 70, 149.

27. *Poetical Works,* pp. 367–368 and 283–284.

28. On this, see also Sharrock, "Chemist and the Poet," pp. 74–75.

29. Ben R. Schneider, Jr., *Wordsworth's Cambridge Education* (Cambridge, 1957), pp. 40–41.

30. William Wordsworth to William Whewell, 12 March 1829, Whewell Papers, Trinity College; see also Wordsworth to Adam Sedgwick, 14 May 1834, in Clark and Hughes, *Sedgwick,* I, 371; and William Wordsworth to Christopher Wordsworth, 13 March 1829, *The Letters of William and Dorothy Wordsworth: The Later Years,* ed. Ernest de Selincourt (Oxford, 1939), I, 371, #861. De Selincourt's edition is not definitive on letters to scientists.

31. William to John Wordsworth, 5 December 1833, *Letters of Wordsworth,* II, 679–680, #1065.

32. Dorothy Wordsworth to Catherine Clarkson, 13 February 1821 and 27 March 1821, *Letters of Wordsworth,* I, 19 and 30, #660 and 664.

33. William Wordsworth to Edward Quillinan, 10 September 1830, *Letters of Wordsworth,* I, 509, #941.

34. Clark and Hughes, *Sedgwick,* I, 249.

35. Clark and Hughes, *Sedgwick,* II, 41.

36. Frederika Beatty, *William Wordsworth of Rydal Mount* (London, 1939), pp. 230–233.

37. *Prelude,* Book III, lines 56–59 (1805–06), lines 58–63 (1850).

38. Beatty, *Wordsworth,* pp. 227, 230. The original title was *Ecclesiastical Sketches;* the name was changed in 1837.

39. Amy Cruse, *The Englishman and his Books in the Early Nineteenth Century* (London, 1930), p. 38.

40. *Letters of Wordsworth,* II, 920, 936, #1250 and 1260.

41. J. H. and J. Parker, Oxford and London.

42. "Evening," *The Christian Year,* 2nd Am. ed. (Philadelphia, 1840), p. 16.

43. *Christian Year,* p. 71. Now Hymn 168 in *Hymns Ancient and Modern.*

44. "The Travellers," *Christian Year,* p. 27.

45. "Balaam," *Christian Year,* p. 127.

46. "The Burning Bush" and "Christ Waiting for the Cross," *Christian Year,* pp. 94, 99.

47. Wordsworth, *Poetical Works,* p. 180, lines 40–44.

48. "The Flowers of the Field," *Christian Year,* p. 195.

49. Newman to Pusey, April 21, 1858, in Henry Liddon, *Life of Edward Bouverie Pusey* (London, 1893), IV, 78.

50. John Henry Newman, *The Idea of a University* (London, 1899), pp. 435, 73, 61.

51. Newman, *University,* 432; "The Tamworth Reading Room," *Essays and Sketches,* ed. C. F. Harrold (New York, 1948), II, 185; *University,* pp. 220–225.

52. William Whewell, *History of the Inductive Sciences* (London, 1837) I, 399–404; *Astronomy and Physics Considered with Reference to Natural Theology* (London, 1833), Book III, Chapters V and VI.

53. H. L. Weatherby, "The Encircling Gloom: Newman's Departure from the Caroline Tradition," *Victorian Studies,* XII (1968), 57–80.

54. Newman, *University,* pp. 42, 27.

55. Owen Chadwick, *From Bossuet to Newman: The Idea of Doctrinal Development* (Cambridge, 1957), pp. 91–95.

56. John Henry Newman. *An Essay on the Development of Christian Doctrine,* ed. C. F. Harrold, new ed. (New York, 1949), pp. 28–29, 94, 111.

57. Newman, *Development,* pp. 113–114.

58. Newman, *Development,* p. 34; Whewell, *History, passim.*

59. On the subjects discussed in this section see Henry Acland and John Ruskin, *The Oxford Museum* (Oxford, 1859); F. Sherwood Taylor, "The Teaching of Science at Oxford in the Nineteenth Century," *Annals of Science,* VIII (1952); 82–112; J. B. Atlay, *Sir Henry Wentworth Acland* (London, 1903): W. Tuckwell, *Reminiscences of Oxford,* 2nd ed. (London, 1907).

60. Henry W. Acland, *Oxford and Modern Medicine* (Oxford and London, 1890), p. 20.

61. Liddon, *Pusey,* IV, 331. The Lee's Readers at the time were entitled to £140 a year from Lee's endowment.

62. Liddon, *Pusey,* IV, 332. On the whole subject, see H. M. and K. D. Vernon, *The History of the Oxford Museum* (Oxford, 1909) Acland and Ruskin, *Oxford Museum.*

63. Liddon, *Pusey,* IV, 336; III, 409.

64. *Works of John Ruskin,* edition de luxe (London and New York, n.d.), I, 40–41.

65. Joan Evans, *John Ruskin* (New York, 1954), pp. 54, 57.

66. *Works of Ruskin,* I, 184–186.

67. *Works of Ruskin,* I, xliii, xxxvi.

68. Humphry House, "Man and Nature: Some Artists' Views," *Ideas and Beliefs of the Victorians* (New York, 1966), p. 226.

69. George Landow, *The Aesthetic and Critical Theories of John Ruskin* (Princeton, New Jersey, 1971), pp. 34, 330–334.

70. Jerome H. Buckley, *The Triumph of Time: A Study of the Victorian Concepts of Time, History, Progress, and Decadence* (Cambridge, Mass., 1966), p. ix.

71. Herbert L. Sussman, *Victorians and the Machine* (Cambridge, Mass., 1968), p. 77 for both phrases.

72. Evans, *Ruskin,* p. 81.

73. *Works of Ruskin,* I, xxxv–xxxvi.

74. Wordsworth, *Poetical Works,* p. 616, lines 161, 178–179, 182–189.

75. John Ruskin, *The Stones of Venice* (New York, 1885), III, Appendix 7, pp. 215–218.

76. *Southern Rose,* October 15, 1836, as quoted in Thomas C. Johnson, Jr., *Scientific Interests in the Old South* (New York, 1936), p. 115.

77. John Holloway, "Hard Times: A History and a Criticism," in *Dickens and the Twentieth Century,* ed. John Gross and Gabriel Pearson (London, 1962), pp. 159–160, 162 points out that Dickens was consciously concerned to attack arithmetic in this novel. Holloway does not trace this back to *The Chimes.*

    With a novelist, the only satisfactory procedure for the historian is to do as the literary critics do and go carefully through *all* of their published writings, or at least the bulk of them. If we are to profit from and enter into discussion with such scholars as Professor Holloway, we must bring ourselves up to their level. Remarks on the basis of reading one book or one essay are usually misleading.

78. Macaulay, *Lays and Essays,* p. 43.

79. William Whewell, *The Philosophy of the Inductive Sciences* (London, 1840), II, 519–520.

80. Disraeli, *Lothair,* Chapter 30, p. 110.

81. Herschel-Mill Correspondence, John Herschel Papers (Royal Society Library, London).

82. Longman's advertisement at the end of Mill's *System of Logic,* 8th ed., (London, 1872), II.

83. Lewis Campbell and William Garnett, *The Life of James Clerk Maxwell* (London, 1882), p. 261.

84. Cf. Newman, *Development,* pp. 36–37.

85. To place these scientific works, as well as the more overtly philosophical treatments of science and its methods, it is desirable to be aware of the late 18th-century British scene. Here are just a sample of the excellent recent treatments available: Arnold Thackray, *Atoms and Powers* (Cambridge, Mass., 1970); Robert E. Schofield, *Mechanism and Materialism* (Princeton, 1970); J. B. Morrell, "Professors Robison and Playfair, and the *Theophobia Gallica,*" *Notes and Records of the Royal Society of London, 26* (1971), 43–63; G. N. Cantor, "Henry Brougham and the Scottish Methodological Tradition," *Studies in History and Philosophy of Science, 2* (1971), 69–89; Arthur Donovan, "British Chemistry and the Concept of Science in the Eighteenth Century," *Albion, 7* (1975), 131–144.

Chapter Two

The Cambridge Network

The late Victorians popularized several ideas which have tended to obscure what was actually going on in intellectual matters in the early part of the 19th century. One of these is the notion—derived from Charles Lyell, spread by Thomas Huxley, and widely repeated ever since—that, whenever science and religion came into contact, some degree of scientific excellence was sacrificed, if only because the scientists themselves believed in the religious ideas. A second is the judgment that the typical Broad Churchman was a man like Dean Arthur Stanley, a "passive peaceable Protestant" always seeking compromise. And a third is the acceptance of Leslie Stephen's description of an arid "Cambridge rationalism" not only as enlightening (which it is) but also as characteristic of Cambridge as a whole.[1]

These and other similar misconceptions could be maintained because the later Victorians whom Noel Annan calls an intellectual "aristocracy,"[2] and which I characterize as a "self-reviewing circle," were not a group continuous from the leading figures of the earlier period. Such physical descendants as did remain—notably Matthew Arnold and Leslie Stephen—played quite different roles in the new circle from those which their fathers had filled in the older, looser, grouping; thus it was Stanley, not Matthew Arnold, who was the heir to Thomas Arnold's ideological position. The founders of the new grouping selected their mythic figures with an eye to current usefulness rather than with strict attention to the history of the earlier generation. This was to be expected. One could not expect Thomas Huxley to emphasize the great abilities of the geologist Adam Sedgwick when it was just such a reputation which supported "the old Adam" in his attack on Darwin's theories.

In order to show the inadequacy of the three conceptions given above, and to relocate the center of intellectual events in early Victorian England, I will sketch the coming together of those men who were mentors not only of Darwin but also of Stanley, of Tennyson, of Maurice, of Lord Kelvin, and of James Clerk Maxwell. Many of the men have heretofore been dealt with in various specialties: in religion as Broad Church leaders, in literature as propagandists for the Romantic poets, in histori-

29

ography as "Liberal Anglicans," in education as university reformers, and in the history of cliques as related to the Cambridge "Apostles."[3] Robert Robson, Martin Rudwick, and I have agreed that these are all *one* set of people with the same ideals, operating in all of these fields and in science as well.[4] In trying to indicate the relationships among so many individuals, I have not also tried to make the length of my treatment of each one a measure of his importance. Indeed, the more important the individual (John Herschel or Thomas Arnold), the less I have tried to give an adequate summary here.

For convenience it is desirable to have a label. Fortunately, Thomas Arnold has furnished one. He suggested that there was a "Cambridge Movement" stemming from Coleridge, which balanced what he considered to be the reactionary and un-Christian Oxford Movement led by Pusey and Newman.[5] There was indeed such a set of activities; Arnold, only secondarily interested in science, was only partly aware of its sources and power. Since it involved no tight-knit group of personal devotees such as initially surrounded Pusey and Newman, it may, by an amendment of Arnold's phrase, more properly be called the "Cambridge Network."

The grouping was a loose convergence of scientists, historians, dons, and other scholars, with a common acceptance of accuracy, intelligence, and novelty. It was made up of persons each of whom knew many but not all of the others intimately. Face-to-face contacts were sometimes regular, as with dons at the same college; sometimes often, as with leading members of the council of a scientific society; and sometimes periodic, as at christenings, Christmas celebrations, yearly terms of residence as a cathedral canon, and meetings of the British Association for the Advancement of Science. The chief agency of continual contact in such a network was the personal letter, of which many thousands still remain unread by historians.

Such networks have, perhaps, been common in Western history since the time of Erasmus. This one may properly be called the *Cambridge* Network, not only in opposition to the Oxford Movement, but also because its members set the pace for Cambridge University during the period in which it was transformed into a modern scholarly institution. Indeed, it was primarily Network members who made the title of Professor into one of modern prestige.

Let me emphasize, however, that a Cambridge affiliation is not a necessary qualification for membership. This would only be true if the standards of the Network had been learned *at* Cambridge. In some cases this is correct, but it is more often the case that it was these men who brought their standards *to* Cambridge, as well as to other English institutions. And there were, of course, good men educated at Cambridge who were not part of the Network. Thomas Arnold, Oxford trained, one of the

Oriel group around Richard Whateley, would be a problem for the terminology were it not that he suggested it. He sent young Matthew to Oxford, after much hesitation, only so that he could study that clear thinker, Aristotle; for, as Richard Trench remarked in objecting to his own appointment as Examining Chaplain for the Bishop of Oxford, "A man who has not read Aristotle is hardly fit to examine Oxford men on anything."[6] It is interesting to note that John Stuart Mill's godfather left £500 in his will for the express purpose of sending him to Cambridge; but James Mill refused to allow it.[7] So John's exposure to the new ideas had to wait until Maurice and Sterling, having learned them at Cambridge, brought them back to London and into John's circle.

There were women in the Network as well as men, notably the celebrated science writer Mary Somerville.[8] And there were many self-generated geniuses in the early 19th century; it is interesting to see how many of them, having modernized themselves on their own, were then drawn toward the Network because they saw in it a group which had the power to maintain high intellectual standards. Two of these are mentioned below: Francis Baily, who became a central figure in the London branch of the Network; and Michael Faraday, who is best classified as a friend rather than a member of the Network.

The Cambridge educational system emphasized "applied mathematics" (that is, physical science treated mathematically) but also rewarded classical merit. Thus Thomas Macaulay had a completely non-mathematical mind, and complained to his mother in 1818, "All my perceptions of elegance and beauty gone.... My classics must be Woodhouse, and my amusements summing an infinite series.... your most miserable and mathematical son." His problem was that he "regarded every successive mathematical proposition as an open question" suitable for debate. As a result he placed nowhere on the Tripos of 1822. Nevertheless, he won a fellowship at Trinity in 1824 because of his classical abilities; and this paid him more (about £300 a year) than he made writing for the *Edinburgh Review* in the late 1820's.[9]

At Cambridge, then, as in the Network generally, several kinds of excellence were recognized, but no difference in acceptable level among them, and no ultimate divergence in goals. This unitary ideal may have helped to prepare the ground for that self-confident pugnacity in the later Victorians which bothers some people today. After all, in 1843 it was perfectly normal for the methods of a historian to be compared to those of an anatomist, as in George Grote's praise of Barthold Niebuhr: "When we conceive his wonderful ingenuity in combining scattered facts,... his power of recomposing the ancient world by just deduction from small fragments of history, like the inferences of Cuvier from the bones of fossil

31

animals...he seems to stand alone even among so many distinguished countrymen and contemporaries."[10] Why, then, later in the century, should not a good anatomist like Thomas Huxley judge Jesuit theology, if standards of knowledge are invariant?

If one listens to the ex-undergraduate complaints of Wordsworth or Byron or James Stephen, it would be easy to overlook the sporadic intellectual vigor of Cambridge in the decades just before and around 1800, its Unitarian radicals and classical scholars, its lone enthusiasts for French mathematics and German scholarship and applied science.[11] It is true that the university or "Senate House" final examination for ordinary students as late as 1822 was at what would today be considered secondary school level. But we should try to evaluate the level rather in terms of what the university accomplished with its students with respect to contemporary cultural levels. The description given by Latham Wainewright in 1815 of *The Literary and Scientific Pursuits which are encouraged and enforced in the University of Cambridge* is useful. Students entered often with no knowledge of mathematics at all, so Wainewright was rightly proud that college lecturers could take them through trigonometry, geometry, algebra, fluxions, and natural philosophy, up to the point that some of them could "study with pleasure" Newton's *Principia*, "a work which on its first appearance in the world baffled the comprehension of some of the most renowned mathematicians of Europe." College lectures also treated Locke's *Essay Concerning Human Understanding*—which to Wainewright was an example of the application of Bacon's principles to the subject of metaphysics—as well as the responses to it by Reid, Beattie, and Dugald Stewart, and some elementary works on logic. (Hartley was considered too difficult for undergraduates.) The Greek New Testament was taught with critical textual remarks, so that students knew of the shortcomings of the King James translation. William Paley's relatively recent *Principles of Moral and Political Philosophy* represented the moral sciences; it was distinguished by "originality of remark and impartiality of judgment."[12]

As one who has tried teaching Locke's *Essay* to modern undergraduates, I can testify that these were not negligible teaching endeavours. Nevertheless, the curriculum was not modern, and there were individuals in several subjects who wanted to introduce new and up-to-date material. After 1800 Robert Woodhouse in mathematics tried to spread a taste for the methods of continental analysis. Herbert Marsh did the same for German biblical scholarship from 1793, when his translation of Michaelis's *Introduction to the New Testament* appeared; he later began a course of lectures.[13] Nor did this interest hurt his career; he held the well-paying professorship of divinity and then became Bishop of Peterborough. Isaac

Milner as Jacksonian Professor of Natural Experimental Philosophy did much, from 1783 to 1792, to demonstrate electrical, chemical, and mechanical devices and experiments; and the tradition was continued by William Farish from 1795. Farish lost the election for the Jacksonian chair and had to be contented with the chemistry one; but as the Jacksonian winner, F. J. H. Wollaston, wanted to lecture on chemistry, Farish lectured on machines and mechanism. "It seems fair," says the historian of Cambridge engineering, "to claim this as the first real course in mechanical engineering in any British university."[14]

Another of these individual enthusiasts was Edward D. Clarke, who as private tutor to one John Cripps took him on a Grand Tour which rather got out of hand. It was intended to last six or seven months, but it went on for over three-and-a-half years, from May 1799 to November 1802. They covered Norway, Sweden, Russia (including Cossacks), Greece, and Turkey, with a side trip to Galilee, as well as a return from Constantinople through Vienna and Paris. The travellers returned with over 150 packing cases of antiquities, manuscripts, and minerals, including the Eleusinian statue of "Ceres" for the University—which promptly gave Clarke an honorary doctorate.

Clarke became senior tutor of Jesus College, but, in order to be married, he entered the Church so that he could be awarded a vicarage. He then started lectures on mineralogy at his own expense in 1807. They were successful, and the university created the title of professor of mineralogy for him in 1808. The professorship had no salary attached; but Clarke was advanced in the Church to a rectory. He took up the chemical analysis of minerals assiduously, published a number of papers, and when he was dying (in 1822) worried his friends by concerning himself with the syllabus and experiments for his lectures rather than with proper devotions for the state of his soul.[15]

To change a society, the efforts and originality of a group of persons is required. The lone enthusiasts must make or find converts in the younger generation. For this reason, it is not possible to begin a continuous narrative of the scholarly revival of England until we reach the fortuitous conjunction of three undergraduates at Cambridge in the years 1811 to 1813. Or, at least the conjunction of this particular trio—John Herschel, Charles Babbage, George Peacock—was fortuitous; but where else would three young Englishmen who wanted to reform the world through mathematics meet more readily than at Cambridge?

John was the only son of William Herschel, the Hanoverian oboist who, after fleeing to his sovereign's less war-torn territory (England), had become the greatest astronomer in the world. John, however, had one advantage over the sons of other famous men. His father had seen more of

God's universe than any other human being ever had, but, in a post-New-tonian era, he did not know advanced mathematics. At Cambridge John met several students who, in the spirit of the 18th century, developed such a passion for mathematics that they eagerly studied Lagrange, Laplace and other such modern greats, even though they were not relevant to passing examinations. Among these were Babbage and Peacock. Full of disgust at the low state of the ordinary mathematical studies, these three intimates resolved to convert the university from Newtonian mathematics to the new French methods of analysis—methods which James Ivory, Edinburgh classmate of John Leslie and from 1804 professor of mathematics at the Royal Military College, was already using with complete mastery and with praise from the Continent.

The campaign of the trio rapidly succeeded. Much of the academic in-fighting fell to Peacock, who remained as a lecturer at Trinity College after Herschel and Babbage had gone off to London to see what they should do with their lives (for it was by no means obvious in 1813 that a gentleman should adopt science as a *career,* if he were not in the Church[16]). Peacock was able to convince his fellow-examiners on the Tripos examination to include questions involving the new methods, and the turning point in the campaign (which was helped along by frequent visits by the other two,[17] and by William Whewell) is sometimes dated from Peacock's holding this position of examiner ("moderator") in 1818-19. The alliance was strengthened by the adherence of the younger Whewell, who wrote new textbooks for mechanics, and by that of Peacock's student George Airy, the first great success of the new system, who did the same in optics and astronomy.

The work of these mathematical reformers was not merely that of teachers, translators, and publicists. On the contrary, one can see the basic originality of 19th-century English mathematics and physics in the way in which these men appropriated continental approaches. In a lengthy article, Elaine Koppelman has traced the rise of abstract algebra, and its associated symbolic logic, to the papers of Herschel and Babbage on the calculus of operations. This is surely correct.[18] I am not enough of a mathematician to decide which man's contribution was the more important, so I will only report that in 1816 it was Herschel from whom Peacock demanded a book on the new algebra. The book was looked forward to by his friends, but Herschel never quite got around to producing it, becoming interested in physical optics instead. After a lapse of some years it was Peacock who produced the desired algebra in 1830. Peacock was quite abstract in his mathematical tastes, and so he was the spiritual grandfather of such later Cambridge mathematicians as Arthur Cayley, who infuriated George Airy by posing Tripos questions which had no scientific

34

applicability at all.

In physical science, however, the main thrust was just the opposite. It was of the utmost importance that analysis was brought fully into English science by men who were firmly committed to a physical understanding of the processes described by mathematical symbols. Herschel and Babbage, and Whewell and Airy, deliberately reversed the Lagrangian impulse to reduce dynamics to mathematics. Whewell's textbook on mechanics of 1819, for example, stated firmly that there was no intention of substituting "the language of analysis for that of facts" in what is "really a physical and experimental science."[19] One of the clearest statements of this tradition occurs in James Clerk Maxwell's great *Treatise on Electricity and Magnetism.* Maxwell wrote:

> The aim of Lagrange was to bring dynamics under the power of the calculus.... Our aim, on the other hand, is to cultivate our dynamical ideas. We therefore avail ourselves of the labours of the mathematicians, and retranslate their result from the language of the calculus into the language of dynamics, so that our words may call up the mental images, not of some algebraical process, but of some property of moving bodies.[20]

Still a third tendency among the reformers was a continuing interest in physical devices in mathematics. Herschel and Whewell were principal agents in introducing graphical presentations and solutions as important tools in solving physical problems. Babbage developed his "engines," which we might call digital computers. Herschel devised a machine for solving one kind of transcendental equation.[21] We need not be surprised to find, a bit later, William Thomson developing his harmonic analyzer and his tide predictor (analog devices) to deal with Whewell's subject (the tides), or Maxwell working out ideas in electromagnetism in terms of interacting gears. Babbage, and Robert Willis after him, tried to work out methods of expressing the actions of machinery in mathematical symbols. Babbage needed such a "mechanical rotation" to work out engineering plans for his own engines. Willis, successor to William Farish as Jacksonian Professor, in his *Principles of Mechanism* of 1841 was interested in the more general problem of developing what Robert Hooke had dreamed of in the 17th century, "a Mechanical Algebra, for solving any Probleme in Mechanicks, as easily and certainly as any [in] geometrick by Algebra."[22]

These five men—Herschel, Babbage, Peacock, Airy, and Whewell— may be considered as one node of the Network. They converted the lone efforts of Robert Woodhouse into a modernization of the principal study of Cambridge University, and the isolated work of James Ivory into the beginning of a new era in British physical science. In so doing, they had to break through the prestige barrier behind which Englishmen cherished their filiation with Newton. They put students into contact with the best of international science, even though Newton was associated with Anglican theology and Frenchmen were associated with dangerous thoughts. By the 1840's, the new methods were even slipping into Oxford, through the teaching of the notable Evangelical preacher, and reader in experimental philosophy, the Reverend Robert Walker.[23] I believe that Bartholomew Lloyd introduced them at Trinity College, Dublin, in the 1820's.

By 1830 Herschel was a well-known figure not only in astronomy but in science as a whole. In the following quarter-century he became so famous that one has to go back to Newton for a parallel. Humboldt had been more famous, but Humboldt did not practice physical astronomy. "To be a scientist" meant "to be like John Herschel" in St. Petersburg as well as at Harvard College in Massachusetts. There is so much to be said about him that I shall not try to summarize it here.

Babbage is still remembered as the "scientific gadfly" of the period, and the site of his house at 1 Dorset Street, has been given a commemorative plaque by the Marylebone Society.[24] Babbage was interested in so many things, and outraged by so many things, that he has gained a reputation for being somewhat eccentric. To me, he seems to have had a fairly stable personal temperament as compared to John Herschel (whom he helped out of more than one scrape), David Brewster, Adam Sedgwick, and James South; he simply did not suffer indignities gladly, was not completely moderate in his professional conduct. His famous calculating engines suffered from his own ability, for he abandoned the partially completed "difference engine," which was to calculate mathematical tables, in order to work out plans for the far more sophisticated "analytical engine." The analytical engine, in his words, had as its objects: "to execute *by machinery*

1. All the operations of Arithmetic.
2. All the operations of Analysis.
3. To print any or all the calculated results."[25]

In the meantime, the government had nothing usable for the thousands of pounds it had given him. Babbage's contemporaries understood his engines well enough; but it was not clear who would use them; and it was fairly clear after fourteen years that Babbage would never push them to completion over the obstacles that contemporary engineering and engineers presented.[26]

To Peacock, more than anyone else, was due the re-founding of the Cambridge Observatory, which was then made important as a scientific center by Airy. It is Peacock, therefore, that one finds, in contemporary accounts, discussing details of the work with William Simms of the famous London instrument-making firm of Troughton and Simms. Some of Peacock's stimulation came from a visit to Italy, where he found observatories and practices much more interesting than those in England. In mathematics, Peacock's *Treatise on Algebra* of 1830 introduced the idea of algebra as "general reasoning by symbolical language," and the principle of formal invariance (or, as he called it, "the principle of the permanence of equivalent forms"). The *Treatise* in one of its two forms became the introduction to advanced thought on the subject for the next quarter-century. Herschel (if he is the author of Peacock's obituary in the Royal Society's *Proceedings*) much preferred the *Treatise* of 1842–45, which Peacock denied was merely a second edition of the first.[27]

It was George Boole who went on to develop what is recognizably mathematical logic. The self-generated Boole was the son of a Lincoln cobbler; his father was much interested in mathematics and in making optical instruments to observe the works of God. Young Boole, after learning Latin and Greek, had mastered Newton's *Principia* and at least read Lagrange's *Mécanique analytique* by age 19; by age 28, in 1844, he had won a Royal Society medal. The philosophic passages of his classic *Laws of Thought* of 1854 show that he was in general sympathy with the ideas of the Network, especially those of Whewell and of Peacock's best student, Augustus De Morgan—who admired his work in exchange—and was not very much impressed by John Stuart Mill's *Logic*.[28]

Some people considered being an Anglican dean as a sinecure for a scholar; thus in 1839 C. C. Bunsen proposed to his friend Bishop Edward Stanley of Norwich that they should try to get Thomas Arnold a deanery so that he would have time for writing and research.[29] This was not Peacock's way; and after 1839 as Dean of Ely Cathedral he transformed it into a model for the whole Anglican Church, both in careful physical restoration and in reform of its liturgical services. Persons with as diverse backgrounds as the liberal Montalembert and the High Church layman William Gladstone spoke enthusiastically of his accomplishment. Peacock, be it noted, was neither a Puseyite nor a Ritualist but was a firm, if moderate, liberal. The work done by such persons in revitalizing church services has probably been undervalued by historians, who have tended to notice more extreme modifiers of received practices. There were many clergymen, I imagine, who believed with Coleridge that the imperfections of Anglican liturgy justified improving it rather than abandoning it.[30]

Peacock produced a book of *Observations of the Statutes of the University of Cambridge* (1841) which included such topics as the rise (1230's) and the fall (1540's) of the "master of glomery" and his glomerells or gremials—that is, the master of grammar and his scholars. Such researches were not intended to add to the case for adhering to the old regulations; and Peacock played an active role in the parliamentary reform of Cambridge in the 1850's. He was also a leading member of Royal Society committees, and found time to write a biography of Thomas Young. He was a widely active, and quietly influential, early Victorian intellectual.

A useful personal friend of Peacock was Thomas Spring Rice, a Trinity graduate of 1811 who was secretary of the Treasury in Grey's Whig ministry, 1830–34; and Chancellor of the Exchequer in Melbourne's Whig ministry, 1835–39. He was M. P. for the town of Cambridge from 1832 until he was made Lord Monteagle in 1839 and retired from active politics.

Peacock's student George Airy revived university lectures on experimental philosophy and also made the new Cambridge Observatory one of the pace-setting scientific centers of England. Then in 1835 he became a political issue. The government had rather hoped to get John Herschel as Astronomer Royal;[31] but, when that failed and he gave his emphatic recommendation to Airy, both parties wanted the credit for hiring him. Whig and Conservative spokesmen explained to the House of Commons why they had, or had not, been successful: what was happening, as a matter of fact, was that Airy was squeezing the politicians into raising the salary of the position.[32] There were not too many young intellectuals in England whose salary would be the subject of a Commons debate.

After the Whigs had met his conditions, Airy carried on the campaign of his seniors in the Astronomical Society of London. Not himself one of the great individual observers but very inventive in practical devices of every kind,[33] he turned Greenwich into a cooperative, almost factory-like, enterprise possessing all the characteristics of precision and thoroughness with which the great German astronomers of the period were transforming positional astronomy. In addition, he was an all-purpose technical adviser often called on by the government.

Modern romantic notions of the cool impersonality of modern civilization have little to do, it may be, with how machines work, or with the impetuous natures of the men who invent them. These notions are perhaps more largely due to the personalities of men like Airy. Raised as an Evangelical, he became a *protégé* of the great anti-slavery worker Thomas Clarkson. Trained as a mathematician, he took as his task the reduction into order of his predecessors' acute visions, and calmly circumvented any interference with the routine carrying out of the observatory's duties. It is typical that he wanted a system for the photographic self-registration of magnetic and meteorological instruments so that his assistants would not have to stay up for twenty-four hours in a row, as was called for by the current observational scheme. This was not because he sympathized with his assistants; he simply considered it an improper interference with regular living habits. He felt the same way about changing methods in science; as he wrote to Edward Sabine concerning magnetic observations, "Change is an evil, but more particularly so when it is to affect habits which experience has shown to be effective."[34]

Perhaps such persons were more responsible than "Benthamites" or an ideal of "Benthamite efficiency" in reducing Britain's chaotic ways of getting things done into some kind of order. Such persons need not be as unsympathetic as Airy. We learn from Eugene Ferguson that Henry Maudslay's machine shop became the training-ground for leading machine-tool builders of the generation after 1820 as much because of his "innate love of truth and accuracy" as because of the precision of the technical devices he used. "'It was a pleasure to see him handle a tool of any kind, but he was *quite splendid* with an eighteen-inch file.'"[35]

Airy's reputation, perhaps unjustly, has never recovered from his apparently unenthusiastic attitude towards J. C. Adam's prediction of the location of the planet Neptune, an attitude which (coupled with that of Airy's successor Challis at the Cambridge Observatory) allowed the Frenchman LeVerrier to gain the priority. It seems clear that contemporary Cambridge opinion was based as much on chagrin that one of the university's own men had been forestalled as it was on judicious consideration of the duties of the Astronomer Royal.[36] But then judicious action was not what made Network scientists great; and it is hard to out-argue the judgment of one of Airy's colleagues who was a greater scientist than he was: Adam Sedgwick simply concluded, *"Curse their narcotic souls."*[37] Airy was one of the men through whom British science went to the front in international precision, although there were ways of being just as precise without losing something which Airy didn't even realize was there.

The fifth member of the mathematical reform group, William Whewell, had the most diverse career of all, although he never left Trinity College. Negligent of polished southern manners, fond of debate, always competent, Whewell was an archetypical product of Lancashire: he was an Englishman whom Macauley could not out-talk.[38] He became a theorist on crystallography, an expert investigator of the tides, a suggestive writer on medieval architecture, the great historian and philosopher of science of his times, the reviver of moral philosophy at Cambridge, and, as Master of Trinity, a powerful university figure who helped to force "moral sciences" (that is, social sciences generally—his own favorite was international law) into the Tripos. As with John Herschel, there is too much to be said about Whewell for me to summarize it here.

Around these five men, Herschel, Babbage, Peacock, Airy, and Whewell, several other men had grouped themselves. Adam Sedgwick had been rescued from the dreary fate of undergraduate instruction in mathematics by being elected professor of geology in 1818. He gradually turned himself into one of the world's best geologists, learning from persons he ran into on field trips, such as William Smith and (especially) W. D. Conybeare. In the 1820's he went on "many a lusty excursion" with William Wordsworth, who "delighted me (admidst the dry and sometimes almost sterile details of my own study) with the outpourings of his manly sense, and with the beauteous and healthy images which were ever starting up within his mind during his communion with nature, and were embodied, at the moment, in his own majestic and glowing language." In the Lake country Sedgwick also had friends in Thomas Clarkson and Robert Southey.[39]

Sedgwick was a worthwhile teacher. One of his assistants observed, "His aim was essentially synthetic. He tried to establish habits of thinking about the larger phenomena of nature.... [His lectures were] directed to concentrate attention on the principles on which facts hang together." His friend Joseph Romilly noted in his diary one of Sedgwick's field lecture days in 1835. Sedgwick covered about forty miles with 60 or 70 students on horseback, and gave five lectures during the day, one on the drainage of the Fens being given at the top of Ely Cathedral.[40]

It is interesting to note that Sedgwick lost money by resigning as an assistant tutor of Trinity to take the geological chair. But it paid in the long run. As I add up his income by the end of the 1830's, disregarding student lecture fees (out of which lecture expenses were to be paid), he received £100 a year as Woodwardian Professor, £100 for a second set of lectures each year, £600 as a prebendary of Norwich, and about £500 as a Senior Fellow of Trinity, for a total of £1300 plus rooms and commons. This was comparable to the salary of the head of a government office or a good bishopric. (Of course the fellowship income varied from year to year with the college's income.)

Sedgwick was an active liberal in politics. His *Discourse on the Studies of the University* of 1833, with its attacks on Locke and Paley, was of considerable importance in defining the new romantic-liberal-scientific stance; and Sedgwick kept up his interest by perusing writings of such younger men as Arthur Stanley, F. D. Maurice, and Leopold von Ranke, as well as those of his old friend Julius Hare.[41] He was made Secretary to the Chancellor of the University, Prince Albert, in 1847; this was almost purely an honorary post, but it made him into a friend at Windsor Castle. In 1863 the Queen had a long talk with him, in which "I believe I was the first person, out of her own family, to whom she fully opened her heart, and told of her sorrows.... 'He had the greatest regard for you,' she said, 'and that is why I had a strong desire to talk with you without reserve.'"[42]

If Sedgwick was the most important geologist of the group, a later recruit was William Hopkins, who became the teacher of many of the

"Cambridge School" physicists—Stokes, Tait, William Thomson, Clerk Maxwell, and others. As a private tutor from 1827, Hopkins was spectacularly successful. He trained seventeen Senior Wranglers and a total of around forty men who won one of the first three places on the Tripos. He and his family were very intimate with Sedgwick, who got him involved in geology. He used the principles of mechanics to explain the details of geological phenomena—a process for which he coined the term "physical geology"—and became one of the important theoretical spokesmen in geology in the 1840's and 1850's.

Whewell's friend the "Rev. R. Jones," as he usually appears in contemporary citations, was the member of the group who devoted himself to political economy, but he did not publish enough to overwhelm the Ricardians, as his friends hoped he would. I mention Richard Jones's role in the founding of statistical enquiries in Chapter Eight. He was a leading advocate of what I call a "comparative-historical method" in social studies; and his criticism of classical economists was basically that they "have confined the observations on which they founded their reasoning, to the small surface of the earth's surface by which they were immediately surrounded."[43] Rent, for example, is not the same thing in India as in England, because the Indian *ryot,* like the Irish peasant, is neither a laborer nor a farmer on the English model. Even John Stuart Mill accepted this amendment to Ricardian law.[44] It is not surprising that, after a short period at King's College, London, Jones found his proper place at the East India College, Haileybury, as Malthus's successor.

John Lubbock, who preferred original work to examination study, graduated only at the top of the middling good students in 1825 (first senior optime), joined his father as partner in the banking firm of Lubbock and Co., London, and devoted himself to various researches in physical astronomy and statistical methods. His best known scientific enterprise, undertaken with the encouragement of the Society for the Diffusion of Useful Knowledge, was a revision of London tide-tables. This met with much support. He interested his teacher Whewell in the subject, and their labors grew into an international enterprise, involving not only scientists but also foreign governments in the collection of data. A recent scholar, Margaret Deacon, has observed that in "discussing the general implications of new information and ideas and suggesting and organizing projects, [Whewell] became more prominent than Lubbock and came near to eclipsing him altogether in some people's minds."[45] I seem to have been one of the people she especially has in mind, so I hereby recant, and say that Lubbock was important in tidal studies and in a good many other ways, and deserves more study by scholars.

In 1832 Adam Sedgwick, Connop Thirlwall, and others talked Lubbock into running for M. P. for Cambridge University; William Wordsworth even wrote a (very poor) epigram for use in the contest; but there was not enough other support, and he withdrew after ten days.[46] He was important in the Royal Society's inner group from about 1830 for many years. He moved to Kent in 1840, where his family intermingled with the Charles Darwins and intermarried with the John Herschels—Kent was a locus of Network men behind only London and Cambridge, after the coming of the railroads.

I mentioned Robert Willis above as a theorist of mechanical engineering. In response to Whewell's *Architectural Notes on German Churches* of 1830, which used a comparative-historical approach to show that the pointed arch was not the defining characteristic of the Gothic style, Willis produced his more important *Remarks on the Architecture of the Middle Ages* of 1835. His knowledge of mechanics led him to refute Whewell's notion that the pointed arch derived from the mechanical requirements of vaulting, and, in general, to scrutinize closely all overly-simple structural explanations of architectural novelties. He became one of the most active and learned investigators of medieval architecture in England. His approach was, it seems, too technical for John Ruskin to follow. Ruskin, whose *Seven Lamps of Architecture* was partially indebted to Willis's work, became friendly with Whewell in the 1840's, and Whewell eventually brought Willis and Ruskin together.[47]

In 1819, through the initiative of E. D. Clarke, a joint idea of Adam Sedgwick and his close friend and hiking companion, the botanist John Henslow, materialized as the Cambridge Philosophical Society.[48] Its list of secretaries reads almost like a roll call of the resident members of the Network. Seven of the nine holders of that position between 1819 and 1850 were Sedgwick, Henslow, Peacock, Whewell, Willis, Hopkins, and William Hallowes Miller (Whewell's successor as professor of mineralogy and England's best theorist on the subject[49]). Nor is there any doubt as to where the initial intellectual power was. Of the papers of 1820 printed as Volume I, Part I, of the *Transactions,* eight out of a total of eleven were by Herschel, Babbage, Whewell, Sedgwick, and Clarke.

In later years the Society became famous as a vehicle in the early development of many of the "Cambridge School" physicists. It became so mathematical that Sedgwick boasted of its progress by saying, "I give you my word of honour that I have not been able to understand a single paper that has been read before the Society during the last twenty years."[50]

In Edinburgh a young ex-law student with Network as well as Scottish support, James David Forbes, defeated the famous David Brewster for the chair of natural philosophy in 1833.[51] Forbes was particularly close to Whewell. His own work on heat and meteorology was closely related to the interests of Herschel, whose "actinometer" (for measuring solar radiation) Forbes experimented with extensively. His most famous researches, on glaciers, led to a spirited fight with John Tyndall in the 1850's, which J. S. Rawlinson in a modern article asserts was the origin of the antagonism between Scots (Forbes, W. Thomson, Tait) and London (Tyndall, Huxley) scientists in the 1860's which produced a famous priority debate as to who was the discoverer of the principle of conservation of energy.[52]

Forbes's own work, unfortunately, has been overshadowed by the contribution he made to science by encouraging James Clerk Maxwell, even lending the youth his private equipment for experiments. When, on Forbes's advice, Clerk Maxwell went to Cambridge, he was tutored by Hopkins, was made an "Apostle," and became interested in the religio-social ideals of F. D. Maurice. All in all, the advanced education of Clerk Maxwell, adding to the philosophic interests he had been exposed to at Edinburgh, was one of the major achievements of the Network.

When, on Forbes's advice, Maxwell sought a job as professor of natural philosophy at Marischal College (Aberdeen) two years after graduating from Cambridge, he found it was a Crown appointment. So he acted on the suggestions of George Stokes "by asking Alex. Herschel to get his father to tell Lord Aberdeen that my case was worth attending to. Two of my friends are writing to two friends of the Lord Advocate. So much for private influence." Alex Herschel, an undergraduate, had been helping the young Trinity assistant lecturer with some experiments. More to the point, Alex was John Herschel's son; John Herschel's daughter was married to Aberdeen's son; and Aberdeen was the late Prime Minister. I don't know whether private influence or public testimonials were the more effective, but John Herschel was pleased when Maxwell got the appointment.[53]

Augustus De Morgan, like many others, had despised the usual Cambridge methods of grind and examination (members of the Network often reserved their harshest criticisms for their own university), but his interest in original work had survived, perhaps thanks to Peacock's teaching. He went to the new University College in London to be professor of mathematics, and in his own work extended Peacock's emphasis on symbolic algebra. In London he was intimate with Herschel, Airy, and another member of the group and Fellow of Trinity, Richard Sheepshanks, who was especially interested in improving scientific instruments. These four men more or less ran the Astronomical Society of London under the direction of its co-founder Francis Baily.[54]

Baily, after going on an adventurous trip in the back country of North America in 1796–97 and being disappointed in love by an American girl, returned to England in 1798. He was too old for a commission in the Royal Engineers; the East India Company was one alternative; rivalling Mungo Park as an African explorer was another. Thwarted in all of these paths, he settled down as a stockbroker to make himself rich, was one of the founding fathers of the Geographical Society, and became the chief instigator in England of the search for new standards of accuracy in scientific measurement.[55]

The functioning of Baily's group in London can be seen in Doris Borthwick's account of the visit of Lt. Charles Wilkes, USN, in 1836 to obtain instruments for a U. S. naval exploring expedition. At Baily's dinner party for Wilkes were Babbage, De Morgan, Sheepshanks, Thomas Galloway, William Fitton, and Lt. Stratford of the *Nautical Almanac*. Most of these plus Airy were at the Astronomical Society's dinner a few days later. Baily taught Wilkes about pendulum experiments and designed some pendulums which Wilkes then had constructed. Sheepshanks introduced him to a chronometer-maker. He went to Cambridge with William Simms, who was checking some work for the observatory. There he met Peacock and Challis, and discussed the tides with Whewell.[56] About the only person of importance he missed meeting was Herschel, who at that time was in South Africa.

Baily's crusade for standards of accuracy became a matter of importance beyond science proper when the British standards of length and weight were destroyed in the Westminster fire of 1834. The task of supplying a new standard of length was entrusted to Baily by a committee made up largely of Network members.[57] Baily soon died, but his disciple Sheepshanks was one of the best of observers with the micrometer, and he essentially completed the work before his own death. Airy knew Sheepshank's methods well enough to be able to codify and publish his results.[58]

This series of events may be compared to the work of Herschel, Peacock, and Babbage in extending the mathematical reform ideas of Robert Woodhouse. Here a self-generated ideologist, Baily, found among the younger generation not only enough support to carry out William Pearson's earlier idea of founding a whole society, but also specific individuals to carry to completion his own particular projects, perhaps better than he could have done himself.

The scientific men of the armed forces, especially Edward Sabine, an expert on geodesy and terrestrial magnetism, and Francis Beaufort, hydrographer to the Navy from 1829, found friends and allies in the Network. This collaboration produced many results, the most striking of which was James Ross's great Antarctic expedition of 1839–43 to check Gauss's theory of terrestrial magnetism, including his prediction of the location of the South Magnetic Pole. It was not in the usual sense an exploring expedition; Ross's ships were a mobile base for scientific instruments to be observed in exact correlation with those of fixed observatories which were established in the southern hemisphere. The striking feature of this expedition, to me, was that Herschel, Sabine, and Beaufort were able to force it through against the opposition of the powerful permanent secretary to the Admiralty, John Barrow. Barrow, who ran the Royal Geographical Society, was a great force in geographical exploration, but he wanted to concentrate on Arctic expeditions. Only Captain John Washington, of Barrow's circle, was noteworthy in his support of the Ross expedition, so far as I know.[59]

Another interesting development was the encouragement of naval officers to observe all sorts of scientific things. At one end of the publishing scale, we find in 1832 a twelve-page pamphlet by W. J. Broderip, "printed for the Hydrographic Office, Admiralty," on *Hints for Collecting Animals and Their Products,* with notes on how to catch, skin, and preserve various types. At the other end, we have an early Victorian classic, *A Manual of Scientific Enquiry; Prepared for the Use of Her Majesty's Navy* (London, 1849). This is a 488-page book edited by John Herschel and published by John Murray, with chapters by Herschel, Airy, Sabine, Whewell, Owen, Hooker, Prichard, Beechey, De la Beche, Darwin, and others. Sedgwick had been scheduled to write the geological chapter, but he became sick and recommended Darwin, writing to Herschel that "he knows practically what is wanted."[60] Sedgwick often became sick when there was writing to be done.

Around John Herschel clustered the hopes of the scientific men who wanted to wrest control of the Royal Society from noble lords and socialite physicians. He finally agreed to run for president in 1830 when he found that the friends to whom he had entrusted the task of denying his candidacy had instead enthusiastically joined the campaign to draft him. His three campaign managers, William Fitton, Charles Babbage, and Francis Beaufort, suggest three major clusters of scientists in the Society: well-established senior men, young Turks, and scientists from government departments. The revolt, that is, was not one of young malcontents; Robert Brown, the heir to Joseph Banks's scientific estate, was as much a supporter as was the young sportsman-turned-geologist, Roderick Murchison.[61] Loss of the election by a small margin did not end the revolt. By following committee actions, one can trace the defeat of the autocratic system of rule which Joseph Banks had passed on to Humphry Davy by the liberal followers of William Wollaston, Davy's sometime-collaborator and sometime-rival of the 1820's.[62] One reason was that, as Brown would not continue to support the policies that Banks had stood for, so Michael Faraday would not support those of Davy. Faraday was, indeed, very much an admirer of John Herschel.[63]

When William Vernon Harcourt announced the plan for a British Association for the Advancement of Science in 1831, Herschel was reluctant to support it. The original prospectus struck him as advocating the departmentalization and bureaucratization of science. He could never agree to anything which tended to divide "science" into "sciences."[64] Perhaps because of Herschel's opposition, his Royal Society supporters did not flock *en masse* to the founding meeting of the Association. However, the opposition faded after the first meeting, and Network names are prominent in the first decades of the Association. The goals of the Association arrived at at the first meeting were rather different from what Brewster had envisaged. The goals became still more different in a few years. Brewster blamed William Whewell and his Cambridge associates,[65] but I rather suspect that they were operating so as to take into account Herschel's ideas. That is the way Herschel's influence worked in the second quarter of the century. At any rate, there is room for more clarification of the subject; my Chapters Five and Six are only indicative, not definitive.

A second node of the Network may be considered from the standpoint of Julius Hare and his German scholarship. Julius was one of the sons of Georgiana Hare-Naylor, an international-minded bluestocking who introduced Julius to the best of Germanism by the direct method of taking the lad to Weimar in 1804. There they were welcomed by the Duchess and met the great men, Goethe, Schiller, and others.[66]

The eldest of the sons, Francis, left Oxford in contempt at its low and provincial scholarly standards. The second son, Augustus, stayed on to become a close friend of young Thomas Arnold. Julius went instead to Trinity College, Cambridge, in 1812, along with his Charterhouse friend, Connop Thirlwall. Both of these youths already knew far more of modern German thought than their teachers, who (especially Sedgwick) welcomed them as intellectual equals. An especially close undergraduate friend was T. Worsley, later to be Master of Downing. Herbert Marsh had been trying to spread a taste for German biblical criticism, but without much success. In contrast, Hare and Thirlwall as Trinity teachers in the 1820's, under Whewell as tutor, were the center of the developing cult of the great German historian Barthold Niebuhr. Because of their translation, Niebuhr reported, the second edition of his *Roman History* sold more widely in England than in Germany.[67]

The brief flurry of fashionable interest in German literature stirred up by Madame de Stael and exploited by Thomas Carlyle in the 1820's had died down by the end of that decade, as Carlyle glumly learned from his publishers.[68] William Hamilton's articles in the *Edinburgh Review* from 1829 were devoted to formal philosophy. But interest in all things German radiated out from Trinity with an increasing intensity. In part this was a function of Julius Hare's impulsive and enthusiastic nature: "Julius never loved or hated by halves. It was perhaps the very openness and demonstrativeness of character which rendered him peculiarly interesting to his acquaintances, and which made it impossible for him to pass unnoticed. He was often loved, frequently detested, but never ignored."[69]

The German interest even penetrated into Oxford in the 1830's, when an enterprising publisher there put out a student's cram-book, so strong was the need "to Niebuhrize" on examinations, as the term was.[70] For Oxford men the chief Niebuhrian was Thomas Arnold, who on Julius Hare's advice had learned German for the express purpose of studying Niebuhr, and had praised him in articles in the *Quarterly Review*. For Arnold, Coleridge, and Niebuhr were his twin modern masters, and "when these two authorities clashed, as they did occasionally, his thoughts were in a whirl."[71] Arnold's own *History of Rome* was avowedly based on that of Niebuhr, as was (in method) Thirlwall's *History of Greece*.

Unlike most of the men I discuss, Arnold did not think highly of natural science as an element of culture; the emphasis, he thought, should be on Christian and moral and political philosophy. He emphasized especially the need for social amelioration, given the state of England in the 1830's; and he wanted mass statistical analyses of living conditions. (One of his biographers is not surprised that Christian Socialism arose among people whom he influenced.) But at least Arnold was not worried about science, since all knowledge is for the glory of God. In his family relations it was represented by the Oxford geologist William Buckland, whose brother had married his sister; and he was on good terms with Buckland. What he disliked were the "Oxford Malignants;" he did not invent the term as a name for his article on the Oxford Movement in the *Edinburgh Review* in 1836, but he was not opposed to it. It is interesting to discover that the Prime Minister, Melbourne, was for a time wavering between Hampden and Arnold as professor of divinity at Oxford; so that the Hampden controversy, about which Arnold wrote the review, might have been the Arnold controversy.[72]

Hare and his friends stimulated interest in German thought generally as well as in German scholarship, in philosophy as well as in philology. This interest was not limited to the formal thinkers, Kant, Schelling, Schleiermacher, and others, but equally to the poetic thinkers, such as Goethe and Schiller—not that they themselves made quite these distinctions. Thus Hare on a trip to Germany in 1832 reported that he saw "Schelling, who, now that Goethe and Niebuhr are gone, is without a rival the first man of the age."[73] It is worth remembering that German thought was brought to England by Coleridge and his admirer Hare as

Kant-Schelling-Goethe in the service of Romanticism, and not as any isolated one of the three. Coleridge, indeed, in recent times has been criticized for not understanding Kant properly in contexts in which he was deliberately producing his own improvements on Kant, sometimes paralleling those of Schelling. (William Whewell has been subjected to the same kind of criticism.)

Hare's own contribution to philosophy was the book of "fragments" he wrote with his brother Augustus called *Guesses at Truth* (1827), the collection of fragments being a German romantic form which appealed also to Carlyle and Emerson. Duncan Forbes has called particular attention to the 1848 edition as a landmark in what he calls "liberal Anglican" philosophy of history, with its non-Hegelian idea of the gradual historical unfolding of all the intellectual and moral faculties of man, but not in a straight line of unbroken progress, rather with perturbations and hesitations like the path of a planet.[74] Comparing this to the Catastrophist ideas of Sedgwick and Whewell in geology and cosmography would make an interesting essay in itself; they are clearly related.

Of course the spread of Germanism brought a reaction, and the *Daily News* in its article on Prince Albert in 1854 managed to link a political to an intellectual attack. It spoke of him as "one who has breathed from childhood the air of Courts tainted by the imaginative servility of Goethe—who has been indoctrinated in early manhood in the stationary or retrograde principles of the school of Niebuhr and Savigny."[75] Niebuhr was indeed quite a conservative, but his more liberal admirers, somewhat ingenuously, dismissed much of that part of his thought as caused by local German conditions. Hare insisted that Niebuhr was merely an admirer of England, a supporter of Burke against the *contrat social*.[76]

At Trinity College in the 1820's the undergraduates met a harmonious and sociable set of Fellows who were spreading a heady combination of historical scholarship, German Idealism, and Romantic poetry, along with the best of modern science, and all this in a Christian context. One response was the flowering of the small discussion group which became known as the "Apostles." Led, in the first years of its Trinity phase by Frederick Denison Maurice, the group in the 1820's also included John Sterling, who was immortalized in Carlyle's biography; Richard Trench, who was to be Maurice's colleague at King's College London, William Buckland's successor as Dean of Westminister, and Richard Whateley's successor as Archbishop of Dublin; John Kemble, who became one of England's Anglo-Saxon experts in the exciting developmental years of linguistics; and Alfred Tennyson.

The historian of the group says that, "Niebuhr for them was a god, who for a lengthy period formed all their sentiments." Intense Trinity undergraduates had their religious crises not over the Real Presence or the Apostolic Succession, but over the application of Niebuhr's anti-mythical methods to the Bible and to Christian tradition generally. They worried not over early church councils but over natural science, natural theology, and Coleridge's distrust of natural theology—far too much so, Trench thought. But Monckton Milnes enjoyed it, writing, "We have had some capital debates.... We attacked Paley last night."[77] Thus when Noel Annan in his *Leslie Stephen* laments that there were no fierce partisan discussions at Cambridge to force prospective clerics to think out their Christianity, as he believes the Oxford Movement benefited Oxford by doing, he has overlooked the "Babel heights of 'Apostolic' dispute," which started earlier and perhaps reached more fundamental levels than Oxford debates.[78]

At least we can say that the Apostolic problems reached a wider audience. Tennyson was a rather lax Apostle in discussions and in writing papers,[79] but through his poetry some of the Apostles' doubts and wrestlings became the possession of a whole generation of middle-class Victorians. Tennyson was not really one who "brooded over popular science with vague uncritical reflectiveness," as Owen Chadwick has it;[80] he was one who worried about Charles Lyell's *Principles of Geology,* William Whewell's *Astronomy and General Physics,* and other issue-provoking works of the 1830's. Tennyson was one of Whewell's favorite pupils, and there are even verbal echoes of Whewell's *Astronomy* in Tennyson's *In Memoriam.* To be sure, these occur in contexts where Tennyson (as a good pupil will) wanted to contradict his tutor.

Hare's love of the poetry of Coleridge and Wordsworth was enthusiastically approved by these younger men and by Coleridge's son Derwent and his friends.[81] The results were not only noticeable in the lives of individuals such as J. S. Mill but were even statistically significant. Mill met Maurice and Sterling in 1828, and was introduced by them to Coleridge's thought, which opened up for him a whole hemisphere of ideas which Bentham had never taught. And it was only after propaganda from Cambridge, along with that from Oxford (coming from J. T. Coleridge's set and from John Keble, the professor of poetry), that the sales of Wordsworth's works became satisfactorily large in the 1830's.[82] One of the propaganda vehicles of Romanticism was in London. Some of the Apostles took over editorial control of the new *Athenaeum* review in 1828. It was run for most of its first two years by Maurice, Sterling, and their "band of Platonico-Wordsworthian-Coleridgean-anti-Utilitarians," as Trench called them. They were concerned "to fight the dragon of materialism," and were able to carry their admiration of Wordsworth and Coleridge to a wider audience. From 1830, under Charles Dilke, the *Athenaeum* shed the burden of Maurice's particular brand of social Romanticism, but it continued to praise Wordsworth as the greatest of contemporary poets until 1844, when it began to rate Tennyson as more modern.[83] To me, the *Athenaeum's* major role in early Victorian culture can be characterized by saying that it was the journal to which John Herschel naturally wrote a letter on any subject which aroused him, whether it be J. C. Adam's role in the discovery of Neptune or a proposed new musical scale.

Wordsworth himself, whose brother Christopher was Master of Trinity from 1820 (that is, during the period when all of this activity was going on), had already begun to modify his earlier unorthodoxy, and Cambridge was not responsible for his eventual High Church position. Nevertheless contact with the Fellows of Trinity, especially Sedgwick, weakened his suspicion of scientists expressed so neatly in his *Excursion* of 1814.[84] He was thus prepared for his great friendship with the Irish mathematical physicist William Rowan Hamilton. Hamilton was admired by the men of the Network who could understand his scientific work, but by no one more than by Wordsworth, who could not. This friendship produced the immortal tribute to Newton ("voyaging through strange seas of thought, alone") composed in the late 1830's and added to the second version of the *Prelude*.[85] The early Wordsworth had been far from seeing that the great scientist is a romantic quester just as is the great poet. In the 1820's and 1830's, Wordsworth's home at Rydal Mount became a summer center to which were attracted Hares, Arnolds (who built a house less than a mile away), Stanleys, Cambridge scientists, and even a group of Trinity students taken by their tutor on a walking-reading tour during the long vacation.[86]

The activities of the younger men such as Kemble and Trench in philology are discussed in Hans Aarsleff's *The Study of Language in England*. The subject is so interesting and important that I won't attempt to go into it except to note Aarsleff's account of the formation of the Philological Society of London in 1842. He considers this to be the beginning of a new stage in language study in England. The organizing meeting was held at the rooms of the Statistical Society, with Thirlwall in the chair. "A detailed examination of the list of original members, their education and earlier history," says Aarsleff, "will show a very close connection with Cambridge, and especially with Trinity College," the Apostles Club, John Kemble, and Richard Trench. (Thirlwall remained as president until 1868; Cambridge men were preponderant on the Council.) William Whewell gave them an account of the earlier Etymological Society at Cambridge around 1832, to which some of their members had belonged. Julius Hare had been the "'great workman'" then, with his and Thirlwall's short-lived journal, the *Philological Museum* of 1832–33.[87]

By putting together scattered contemporary references, we can get some idea as to the center of the group of Fellows at Cambridge. Whewell, Sedgwick, Thirlwall, and Sheepshanks were prominent "Northern Lights" (persons from the north of England). The four leaders of the movement for Catholic relief in 1829 were given in a ballad as Whewell, Sedgwick, Peacock, and Romilly. Four lion cubs were named by the students Whewell, Sedgwick, Peacock, and Simeon (I don't include Simeon in the Network; students often make mistakes). Macaulay took his sisters to Cambridge in 1831 and they were made much of by Whewell, Sedgwick, Thirlwall, and Airy; they found Oxford "terribly dull" by comparison. Arthur Stanley visited in 1834 after his first term at Oxford and met "the concentrated intellect of Cambridge" "except Sedgwick." He thought them less polished than Oxford Dons but with much less restraint among themselves; "they seem very happy all of them together." He also noted, "The lectures here are as good as those at Oxford are bad." In Joseph Romilly's diary, after his relatives and his walking companion John Lodge, the most-cited friends are Whewell (80 times), Peacock (74), and Sedgwick (70). The next most-mentioned person is William Carus (51); but several of these entries are notices of Carus's sermons at Holy Trinity Church. Finally, I note that in 1831, one of the largest local committees of the Society for the Diffusion of Useful Knowledge was at Cambridge. Four of the ten members were Sedgwick, Peacock, Thirlwall, and Henslow.[88]

Inevitably the Cambridge array moved apart as the individual members achieved individual status and patronage. Hare was coerced by family pressure into taking over its rectory of Herstmonceux (East Sussex) in 1832. He was not a very successful rural rector. His nephew reports that, "sick people in the parish used to say, 'The Archdeacon he do come to us, and he do set by the bed and hold our hands, and he do growl a little, but he do zay nawt.'"[89] He was, however, a good Archdeacon of Lewes, and he admired beyond all measure his fellow Archdeacon of Chichester, Henry Manning. His later years were saddened by Manning's conversion to Roman Catholicism.[90] This conversion was perhaps the greatest shock that the Anglican Church had to undergo in the nineteenth century, precisely because Manning was not a combative intellectual like Newman and his followers, but was, to his contemporaries, a true pastor.

(However, it seems to me that Manning's conversion may have saved the Anglican Church from a great deal of trouble. Whigs were worried about "Papal Aggression" in England; but it seems to me that the Papacy had quite a struggle in defending itself from English aggression. Manning did not quite take over direction of the Holy See at the Vatican Council, but I imagine the Holy Father had some rather uneasy moments.)

It was at Herstmonceux, in the rectory which seemed to be all library, so crammed was it with German books—but with a copy of Wordsworth always near the door which opened into the garden[91]—that young Arthur Stanley found a refuge, and a second beloved mentor in his "Uncle Jule," after the death of Thomas Arnold. Arthur had an absolutely non-mathematical mind, so it is just as well that he attended Oxford; but his father, Edward Stanley, was a liberal-minded Cambridge man who was a faithful member of the Geological Society of London and had even published a few minor scientific articles. After he became Bishop of Norwich in 1837 he urged the clergy to spread secular knowledge as well as religious instruction, and did not hesitate to entertain Jenny Lind at the episcopal palace. In Norwich the family were close friends of Adam Sedgwick—"dear old Sedgy" in Arthur's letters—who was one of the cathedral canons and resided there during a short part of each year.[92] Arthur went to the British Association meeting in 1835 with his father, could not follow the discussions, and decided that the Bible taught geology in Genesis no more accurately than it taught astronomy in Joshua. He could not see why the Bible "should not have taught [men] their relations to God as their Creator without teaching them the history of ichthyosauri, with which they had no relations whatever." Thereafter, he left science in the good hands of his older friends like Sedgwick and Herschel.[93]

The Stanleys and the Hares intermarried freely, and Julius Hare was Arthur Stanley's uncle in more than one way. Marital links also existed between Hare and his pupil Frederick Maurice; Hare married Maurice's sister, and Hare's half-sister was Maurice's second wife. Indeed, Hare and Maurice were involved with each other's relatives to such a point that little Augustus J. C. Hare felt that his childhood world was "over-run" with Maurices.[94]

Connop Thirlwall left Trinity in 1834 after a dispute with the Master, but as one of the most learned of liberal churchmen he was not neglected by Whig governments. Henry Brougham, the Lord Chancellor, found a rectory and a canonry available in late 1834; he wisely gave the rectory to Thirlwall and the canonry to Sedgwick. Melbourne made Thirlwall Bishop of St. David's in 1840; and he became one of the most influential clerics in the Anglican Church. By the 1860's, Thirlwall and Samuel Wilberforce of Oxford were the two most powerful bishops in the Church; a common question among their colleagues was, "What will the Bishop of St. David's think of what we are doing?" The differences between Thirlwall and Wilberforce defined the tension between the "Broad Church" and the "High Church" in the third quarter of the century.[95]

"Broad Church" was a label used by Arthur Clough, made public by

Arthur Stanley, and popularized by W. J. Conybeare (son of the geologist and Dean of Llandaff W. D. Conybeare) to describe people empathetic with Thomas Arnold. Conybeare noted that the Broad Church was not a compact body, had no newspaper or magazine or other rallying-point, and in contemporary terms was hardly a "party" at all.[96] These kinds of facts have led Owen Chadwick, in his recent *Victorian Church*, to overlook the kind of approach used in this essay and state, "The group was not a group but scattered individuals working toward similar ends."[97] Chadwick is handicapped by the tendency of British scholars to look primarily at Oxford men as the central figures in intellectual movements. Thus Chadwick finds, quite rightly, that Stanley at Oxford was a leading Broad Churchman but that his early intimate friend Benjamin Jowett was only on the fringes of the movement,[98] and that the Oxford scientist Baden Powell was really of no school at all. But this only shows that the Broad Church was not an Oxford group, although it was indebted to Whateley's earlier band of Oriel Fellows; and that Noel Annan's estimate of Jowett's central importance as a Broad Churchman is wrong.[99] By looking at the prominence of churchmen in the Church itself, and then seeing how they were interconnected by personal, family, and Cambridge links, we can see that the leadership of the Broad Church was an actually existing group. Its background has now been sketched, for the recognized clerical leaders after the death of Arnold were Hare, Maurice, Thirlwall, and Arthur Stanley.[100]

It is not clear to me whether Charles Dodgson at Oxford was a Broad Churchman. Is this a Broad Church poem?

> He thought he saw an Argument
> That proved he was the Pope:
> He looked again and saw it was
> A Bar of Mottled Soap.
> "A fact so dread," he faintly said,
> "Extinguishes all hope."[101]

If, then, there were Oxford dissidents who were only marginally related to the Broad Church, there were also students who were exposed to history at Cambridge but did not become members of the Network. One of Thirlwall's admirers was young Thomas Macaulay, even though Thirlwall's overall romantic approach—to education, to the study of Celtic remains, to the idea of progress—was quite different from what Macaulay came to stand for.[102] Macaulay held aloof from the Niebuhr craze when he was an undergraduate at Trinity, but still he urged the editor of the *Edinburgh Review* in 1830 to take notice of Niebuhr's *Roman History*, as the *Quarterly* already had done, saying, "The appearance of the book is really an era in the intellectual history of Europe." Macaulay's own career as a historian was inspired by Niebuhr, although in a rather backhanded way. He decided in 1835, as a result of his spare-time reading while Legislative Member of the Indian Council, that he could never match the art of Thucydides, but that he really did have much more historical judgment than Niebuhr.[103] He was especially concerned with the effectiveness, not the dignity, of style, saying that "to write what is not understood in its whole force, for fear of using some word which was unknown to Swift or Dryden, would be, I think, as absurd as to build an Observatory like that at Oxford, from which it is impossible to observe."[104]

The new kind of historical emphasis—scientific, comparative, social—if it did not much influence an old-fashioned historian like Macaulay, did, perhaps, have an influence in producing Henry Maine's *Ancient Law* of 1861. Maine's work impressed a contemporary like Leslie Stephen as being somehow related to Darwin's *Origin of Species*, even though it was not really evolutionary.[105] We can see, I think, that it was one kind of response to the position of the Network, whereas Darwin's was a reaction in a rather different direction. Maine, fittingly enough, wound up his career as Whewell Professor of International Law at Cambridge.

George Airy, as I have said, left Cambridge to become Astronomer Royal, and Peacock became Dean of Ely. John Henslow was given a crown living in Suffolk, and after 1841 was of great assistance to W. J. Hooker in developing Kew as a center of scientific botany. Whewell, Sedgwick, Hopkins, Willis, and Miller remained at Cambridge in the 1840's. Whewell became Master of Trinity and the most powerful Peelite Conservative in the university. Sedgwick was Vice-Master of Trinity, famous as a geologist, and the most vigorous liberal in the university. It is not surprising that when in the 1850's Parliament decided to reform the university, the majority on the Cambridge Commission was made up of Sedgwick, Peacock, Herschel, and Sir John Romilly. The opposition of the university was led by Whewell. The government's Cambridge Reform Bill of 1855 was too much in accordance with Whewell's ideas, so the four liberals (led by Peacock) attacked it in the Press, and a stronger bill was presented in 1856.[106] Through this as well as through other reforms, mid-century Cambridge was to a considerable extent what the men of the Network had been able to make it into.

It has been convenient to develop the Network around two nodes, one of science and the other of literary scholarship, although the composition of the Trinity High Table shows the artificiality of this dualism. It is also an inadequate representation of the general social situation. Sedgwick and Whewell were frequent guests of Julius Hare at Herstmonceux; so were Thirlwall and Baron Bunsen. Herschel's daughter Margaret was married in 1858, with Whewell coming down to Kent to perform the ceremony (he had married into the family of the groom). After the bridal pair were seen off, the hard core of the party descended on Arthur Stanley, then Canon at Canterbury, and continued the festivities in Stanley's house for another two days.[107]

Nor does the dualism of this presentation indicate the unity of ideals in the Network. Stanley did not have to worry about the threat of science to religion, for according to his scientific friends no such threat existed. The only danger was that science and religion might become divorced, and go their separate ways. In 1871 as Dean of Westminster he preached Herschel's funeral sermon, and in it stated the position of the entire Network on the true method of reconciliation of science and religion. One should, he said, trust "the grand and only character of Truth—its capability of coming unchanged out of every possible form of fair discussion." This statement was taken verbatim from Herschel's own *Preliminary Discourse on the Study of Natural Philosophy* of 1831. The revelations of geology, biology, and biblical criticism in the intervening forty years had not dissuaded the men of the Network from that position. This was natural enough: after all, they were responsible for much of the geology and biblical criticism. As for biology, it was Herschel's statement in 1836 that did much to rule out of serious consideration the attacks of bibliolaters on scientists who pondered the naturalistic origin of species.[108]

It is thus not surprising that young Charles Darwin was inspired by Herschel's *Discourse* as well as by Alexander Humboldt's *Personal Narrative*, or that he learned a good deal from Henslow, and from Sedgwick's practical instruction one summer. Darwin was made into a scientist by the voyage of the *Beagle*, and it is interesting to see how Network influence worked. This is how a student won a five-year travelling research fellowship (including basic food and lodging; no salary; expenses of dining at the Captain's table extra) in 1831. Captain Fitzroy of the *Beagle* was interested in physical science, chronometers and the like, and his instrument hut on board the *Beagle* eventually wound up as part of the magnetic observatory at Greenwich. He wanted a naturalist or geologist along to look over the land, investigate possible sources of metal ores, etc. He also wanted a social companion. So the Admiralty (that is, Herschel's friend Beaufort) asked Herschel's friend the mathematician Peacock at Cambridge to recommend someone; Peacock asked the advice of his Cambridge colleague Henslow; and Henslow recommended one of his favorite undergraduates.[109]

For Fitzroy, a high point of the voyage was a short stay at the Cape of Good Hope on the return to England; there he quizzed Herschel about chronometers. Later he took up one of Herschel's favorite sciences, meteorology, but was so much in awe of the great man that he apologized for bringing his own efforts to Herschel's attention.[110] As for Darwin, his letters to Henslow made him, even before he returned, eagerly awaited as a recruit to science by the men of the Cambridge Philosophical Society and then by those of the Geological Society of London. Sedgwick was a principal agent in promoting Darwin. Thus in November 1835 Henslow wrote to William Lonsdale at the Geological Society:

> I spent an hour or two with Sedgwick last Monday in looking over Darwin's letters and making memoranda for his report. He has taken the notes with him to Norwich with the intention of putting his matter together as quickly as possible. I am glad to find his remarks have excited so much interest for I have heard from two other persons on the subjects—I have made a few extracts from his letters which will be printed & distributed soon.[111]

Darwin's especial scientific hero, Charles Lyell, was a friend of, though not properly speaking a member of, the Network. In Darwin's case as in others, the Network helped provide the opportunities for research, the encouraging response, and the sheltered arenas for free discussion and debate, which made possible the free development of its pupils and heirs. Perhaps most important was the existence of an informed and interested audience of seniors and peers, prepared to be enthusiastic about new findings. Darwin never needed to feel intellectually isolated; he was always an important person within his scientific group, even though to the great world he was merely the author of a Humboldtian travel book. As late as 1856, this is how he appears in Macaulay's journal: "In the evening Darwin, a geologist and traveller, came to dinner."[112]

Another example of the uses of influence comes from 1866. Sedgwick did not like the actions of bureaucrats with respect to a small chapel of which he was senior trustee. He wrote a pamphlet about it, and the line of influence then went: to the Deanery of Westminster, to Windsor Castle, to the Archbishop of York, to the Prime Minister, to Parliament with the Ministry's support, to a new Bill passed in 1869.[113]

The belief expressed by Arthur Stanley in the Christian piety of free thought was not new, but was continuous from the eighteenth century. Georgiana Hare-Naylor wrote to her son Francis in 1797 discussing a famous preacher:

> He was for dividing *faith* and *reason*, while I am for uniting them; *true reason* must ever support *true faith*, since they both come from God.... Dr. Randolph, on the contrary, would have us *believe*, and not *inquire*. A Mohammedan, or a Pagan, can be advised to do no other, but the Christian surely has a nobler doctrine to teach.

Her son Julius agreed, and liked to point out that Coleridge had emphasized the superiority of Truth to the Church as such, without meaning to denigrate the Church in any way. As for his own German studies, Hare explained his devotion to them:

> Above all, to them I own my ability to believe in Christianity with a much more implicit and intelligent faith than I otherwise should have been able to have done; for without them I should only have saved myself from dreary suspicions, by a refusal to allow my heart to follow my head, and by a self-willed determination to believe whether my reason approved of my belief or not.[114]

Adam Sedgwick echoed Georgiana Hare-Naylor's opinion at a meeting in Norwich, in response to a fearful clergyman. Sedgwick took up a Bible and exclaimed, "Who is the greatest unbeliever? Is it not the man who, professing to hold that this book contains the Word of God, is afraid to look into the other volume, lest it should contradict it?"[115]

The Cambridge Network had its continental allies; one of the most prominent of these was Baron Christian Bunsen, disciple of Niebuhr, friend of Thomas Arnold, Egyptian historian, Prussian Minister at the Court of St. James. He put it all quite simply in a private journal in 1818: "Forasmuch as Thy truth is the truth, man need not, on the way to it, be terrified by any truth."[116]

When similar beliefs were expressed by Oxford authors in *Essays and Reviews* in 1860 (one of the essays was a review of Bunsen's work), it caused a stir not because such ideas were new in England. It caused a stir because the last citadel, Oxford, was shown to be safe no longer from freedom of thought.

Bunsen and Arnold first met in 1827 when Arnold visited Rome where Bunsen was Prussian Minister. They were friends from the first, wrote to each other regularly, and, according to Arnold's biographer, "seem to have been intoxicated with each other's minds." Bunsen obtained his appointment to England in 1841, and the Bunsens lived for two years at Herstmonceux near the Hares, until the difficulties of commuting without a convenient railroad wore them down.[117] Bunsen's scholarly goal, he wrote to his wife in 1838, was to complete a great history of Egypt and then to advance "into the region of history *before* history, and endeavour to establish the Keplerian laws, according to which the Eternal Spirit draws his orbit of eternity round the terrestrial life which we call time." Bunsen's learning and character impressed the English; and he appears in Charles Kingsley's *Alton Locke* as a hero of peace and civilization: "I am not ashamed to say," says Alton, "Chartist as I am, that I felt inclined to fall upon my knees, and own a master of God's own making."[118]

Bunsen summed up his principles of scholarship in one of his speeches:

1. The most unlimited liberty in the investigation of truth.
2. The ground for this to be, not the despair or mistrust of unbelief, but sound belief.
3. That as historical science is blind without natural, so, vice versa, natural philosophy is blind without philosophy of the mind and historical knowledge.[119]

The Network used this program, in and if necessary against a society that was increasingly Evangelical, to champion and advance truth in all departments—if indeed its members recognized "departments" of truth. They were careful to distinguish natural science from natural theology, natural theology from Revelation, and Revelation from biblical literalism. Unlike some other authors, they did not usually mix up geology and Christ in the same book (and are sometimes criticized for this by modern writers). John Herschel's writings on astronomy were Christian works, even though he did not fill them with constant references to God (as popular writers on astronomy often did). They were Christian because they were devoted to Truth, insofar as astronomy can establish Truth.

But departmentalization for its own sake they shunned. After all, no one of the actual phenomena of God's world is described exhaustively by one scientific specialty; each can be seen fully only if seen in its relation to all the others.[120]

Thus the contemporary scientist most admired by the Network was Michael Faraday, who in turn attributed some of his success as a discoverer to his reading of Herschel's *Discourse*.[121] Faraday began as a chemist, specialized in electromagnetism, and was eventually (perhaps always) looking for relations among all the basic powers of nature. After the deaths of the great English chemists of the first quarter of the century, John Dalton, Humphry Davy, and William Wollaston, there seems to have been much less intellectual excitement in chemistry in England for a number of years. One reason was that their great successor, Faraday, had gone off on a semi-Boscovichian search for the forces back of natural phenomena, and their interrelationships. Faraday's brilliantly productive quest was just the kind of unifying endeavour that the men of the Network liked. Even the least sympathetic member, George Airy, admitted: "I have always known Faraday as a mystic: I suppose however that a good large share of that character is indispensable for an advancing philosopher."[122]

In general, as I observe in Chapter Eight, there was a desire for massive amounts of factual material in the 1830's. But this was coupled with a desire for more and more widely applicable laws to interconnect the diverse phenomena, not merely laws of chemistry, or magnetism, or geology, but, for example, laws of electro-chemistry, of electromagnetism, of physical geology. This should dispose of the notion that a belief in "Baconian" empiricism was dominant. Rather, an interest in the detailed complexity of the actual may seem to suggest the influence of Romanticism. However, as "Romanticism" is not an independent entity or force which could "cause" anything to happen, I think we should reverse the formulation, and say that it was the development of this interest from the latter part of the eighteenth century onwards which led to overt doctrines which we recognize as Romantic.

Ardent admirers of Wordsworth and Coleridge as intuitive thinkers and not merely craftsmen in language, the men of the Network naturally rejected Utilitarianism. A sermon against Paley was almost a badge of membership among the clerical members at Cambridge (Paley's own university) in the 1820's and 1830's. The anti-Paley contribution of Adam Sedgwick amusingly brought forth an angry rebuttal from young J. S. Mill, in his first real venture into moral philosophy. Mill, writing in the *London Review* in 1835, was annoyed because Sedgwick had lumped Paley and Bentham together as Utilitarians and denounced them both. Mill made it clear that Paley's Utilitarianism didn't count, because Paley believed in God as well.[123] Thus we have this interesting tableau in May 1840 at Portman Square Hall. Thomas Carlyle is giving his series of lectures on Heroes. Caroline Fox (Quaker), whom John Sterling fell in love with, enters with her sister and sits next to Harriet Mill and Jane Carlyle. In front of them are Whewell (Broad Church) and Samuel Wilberforce (High Church). But John Mill is not there; he ceased attending after Carlyle attacked Bentham and Utility.[124]

The increasing amount of denunciation of Paley is a striking feature of the 1820's and 1830's. It serves to remind the historian that Utilitarianism was an old-fashioned, an 18th-century creed. One can find it even in Gilbert White's *Natural History of Selborne*. Indeed, Edmund Burke (like most practical political thinkers) was rather a Utilitarian himself. That is, if you remove God from the *Reflections on the Revolution in France* (admittedly rather a large excision), the doctrine can be collapsed under the stress of logical analysis into a Utilitarian justification of English self-satisfaction.

The 18th-century nature of Utilitarianism is why it received increasing support in the 19th-century political world. One does not need to look for specifically Benthamite influence everywhere; Utilitarianism was simply a good conservative way of being a reformer. The phenomenon is a fine example of cultural lag.

An interesting example of the contrast between the older rationalism of the Utilitarian radicals and the new interests of the Network is the argument between Mill and Whewell over the philosophy of science. A study of the debate, conducted in successive editions of Mill's *System of*

Logic and Whewell's *Philosophy of the Inductive Sciences,* shows that they were arguing at cross-purposes. Mill was seeking a logician's proof of the validity of induction, whereas Whewell was trying to formulate the processes by which scientific ideas advance. Like his Romantic friend Hare, Whewell was interested in the actualities around him, and what made them happen. Science was one of those actualities, so a Kantian approach was suitable: science exists, regardless of what logicians think, so let us try to understand how that is possible. He finally, in his third edition, renamed what he was writing about "the philosophy of discovery," leaving it up to the logicians to go on talking about "induction" as some other process of no particular interest to science, if they so chose. As Alvar Ellegärd has shown, the confusion about these two approaches continued into the 1860's and made the Darwinian debates even more confusing than they otherwise would have been.[125]

In religious matters, various Network men did not develop attitudes which we call "Broad Church" as a passive compromise among the novelties forced on them by others. Many of its members were themselves advocates of new theories and new methods. In history as in astronomy, they had a love of exact knowledge as well as clear thought.[126] The work of men such as Strauss in his sensational *Leben Jesu* was unacceptable not because it was novel but because it was of poor quality. Hare felt that Strauss was "deluded by an idea which he had taken from Hegel, and does not understand." (Hegel was well known to, but often denounced by, Network members—a character trait they shared with Mill and Leslie Stephen.)[127] A religious reviewer said, "Strict logic does not appear to be the strong point of Strauss' system." Bunsen commented, "Fortunately a negative tact for critical truth could never be better manifested...." And Sedgwick felt that Paley had gone over the question of the mythical nature of the New Testament long ago. The result was still the same: "If there be difficulties in revelation (as no doubt there are) there are ten thousand times greater difficulties in the transcendental metaphysics and mythicism of Strauss, and all the lesser fry of that school."[128]

British historians often teach the opinion which, for example, Standish Meacham uses in a review: he speaks of "the two enduring influences of nineteenth-century church history, the Evangelical revival and the Oxford Movement."[129] Actually, both of these have been overshadowed since the early 19th century by two other developments. The triumph of Broad Church scholarship, especially biblical scholarship, has been so overwhelming that modern historians don't even notice that it might not have happened. (It did not happen in several Christian countries.) Evangelicals and Tractarians have reacted, sometimes thoughtfully and sometimes frantically, to an intellectual revolution they could neither control nor shut out. And the modification of church services, by Gothicizers and Broad Churchmen and Ecclesiologists and Ritualists alike, has made the actual church-going experience rather different from what it was in the 18th century or under Evangelical or Tractarian auspices in the early 19th century. Whether these changes are for good or bad is a moot point; but the historian may accept the judgment of the Christians that a change in the experience itself, and not merely in ideas about it, is of major importance.

Another weakness of some historians is to believe that, except for the Evangelicals and the Tractarians, the 19th century was primarily a period of waning confidence in Christianity and of overt attacks on it. They do not like to admit what is obvious from contemporary sources: that such actions as Catholic emancipation, the opening of the universities to dissenters, the disestablishment of the Church of Ireland, the rejection of verbal inspiration of the Bible, and so forth—which are sometimes taken together as the "secularization" of society—were *actively desired by many Anglicans in order to implement the Christian message.* In direct opposition to the Tractarians, for example, Julius Hare wanted to "reanimate the Church, by giving back those functions to her members, which are now usurped by her ministers." Nor was this Broad Church position a stepping-stone to disbelief. As Conybeare observed in his essay on "Church Parties" in 1853, "On the contrary, the unbelievers of our age and country have come from the ranks of the Calvinists or the Romanists."[130]

In theology, Broad Churchmen did not (Maurice is an exception) concentrate on conventional theological treatises. This is because, I think, they realized that the central Christian problems of their day were in morality and anthropology, not in more metaphysical subjects. The problems were not, does God exist? but, is God evil? and, what is man that the universe is mindful of him? Strauss and Carlyle can hardly be compared with geology and paleontology as shifters of the Christian sub-strata on these questions. Charles Kingsley put it clearly in *Alton Locke* in 1850: "Poor suffering wretches! what is it to them to know that 'God is great' unless you can prove to them that God is also merciful? Did he indeed care for men at all? was what I longed to know; was all this misery and misrule around us his will—his stern and necessary law—his lazy connivance?"[131] These were not new questions. For Broad Churchmen they were central questions to answer at that time. The answers, they felt, should come from the fullest background of scientific, scholarly, and poetic knowledge and intuition.

It will be obvious by now that I have carefully—not by very much, but carefully—avoided identifying the Network with a particular approach in natural science, or a love of Niebuhrian history, or moderate reformism, or Romanticism, or the Broad Church, or Cambridge University. Scientists have traditionally recognized the existence of a "Cambridge School of Physics" in the mid-19th century. Theologians from the time of James Martineau have recognized a "Cambridge School of Theology" in the second half of the century (Lightfoot, Hort, Westcott, etc.) and have derived it from Coleridge through Hare and Maurice.[132] But I am not attributing such special groups solely to Network influence—not to speak of differences among themselves (Thirlwall could make even Stanley lose his calm, for example). Network members were not necessarily the most creative, or even the dominant individuals in all the phases of all the movements I have mentioned. What they did above all, I think, is to supply a modern intellectual matrix for the early Victorian period. They established or maintained—in London scientific circles, at Cambridge, in learned societies and on government advisory committees, at the Admiralty, in the Anglican Church—a common high standard of debate. Creative individuals found allies waiting to welcome them.

When young Charles Lyell published the first volume of his *Principles of Geology* in 1830, he was very nervous about his anti-Mosaic daring. He was startled to find himself a scientific hero, to hear his work praised not only by such friends as John Herschel and Roderick Murchison, but also by men such as Sedgwick and Whewell who rejected his scientific ideology.[133] James Stephen, with unorthodox ideas about Hell, found welcoming if not always agreeing spirits at Cambridge, chief among them Whewell; to be sure, he also found the usual antediluvians whose querulous voices can be heard at any university at any time.[134] When George Green, the miller of Nottingham, entered Cambridge as an undergraduate in 1833 at the age of forty after having already developed his own approach to physical science, he found an intellectual haven in the Cambridge Philosophical Society; and William Hopkins was in a position to pass on Green's work to the new generation of first-class mathematical minds.[135]

Even Plantagenet Palliser, Chancellor of the Exchequer in Trollope's *Phineas Redux,* was supported by the Network. His crochet was decimal coinage, with which De Morgan, Herschel, and others of the Network were connected, pro and con, in the 1840's and 1850's.[136] One of Herschel's desires as Master of the Mint was to ward off the danger of imitating the French, by bringing some kind of native British order into the odd collection of coins the Englishman had to use.

If the Network welcomed creative individuals with new ideas, it also exposed many individuals to ideas and approaches which they would not otherwise have encountered. This kind of importance in opening up the

English intellectual scene is indicated, in one case, by a statement of J. S. Mill. After coming into contact with Sterling and Maurice, Mill said, "The influences of European, that is to say, Continental thought, and especially those of the reaction of the nineteenth century against the eighteenth, were now streaming in upon me."[137]

As contrasted to the old-fashioned Utilitarianism of James Mill's generation, the backward-looking Oxford Movement, or the genius of individuals such as John Dalton and Michael Faraday, the Network promoted the modernization and internationalization of English science, scholarship, and religious thought. Its importance has not been seen so long as each member has been treated in one sub-specialty of history. For example, the basis of the new "physics" of mid-century is not to be found merely in the development from earlier stages of the same collection of subjects; there was a desire for a more basic unity. Physics also comes from a new mathematical approach imported from France, joined with a feeling for the reality of the processes described by symbols. Faraday's lines of force Lagrangianized, mathematized—that is the kind of combination on which the new theories were built. Was Faraday a chemist turned physicist—or was he a natural philosopher all along? John Herschel was a mathematician turned physical scientist turned astronomer (though he really liked chemistry best)—but this becomes too confusing, so perhaps none of these specialized classifications is satisfactory.

In a similar way, the revolution in Christian thought about the biblical message was accomplished as much by men labelled historians as by men labelled theologians. And it was the German breakthrough in *secular* history, climaxed by the work of Niebuhr, which was the foundation for much of the biblical history and criticism which followed.

In the period itself, these relations were seen, or sensed. General acceptance was symbolized at the Oxford commencement of 1839, when Wordsworth was given an honorary degree and was the great hit of the occasion. Also receiving degrees with him were Bunsen and Herschel.[138] It was fitting to have poetry, scholarship, and science honored together.

Each man of the Network saw his work not as an isolated bit of a specialty, but as part of an intellectual totality. We historians should see that totality as they did, in order to see the relevance of any individual's work. I suggest that the totality constitutes a progressive center of English thought of the period, with respect to which other intellectuals can objectively be classified as narrow or cosmopolitan, old-fashioned or modern, radical or reactionary. And by "objectively" I do mean without rancor or partisan pride—for Charles Darwin, one of my especial favorites, was by this test old-fashioned both in science and in the general intellectual position from which he derived the science. But that is another story.

63

Notes to Chapter Two

This essay began as a brief paper given at the Xth International Congress of the History of Science, 1962, and published in the *Proceedings* of the Congress, pp. 317–320 as "The Role of the Cambridge Movement in Early Nineteenth Century Science." A much expanded version appeared as "Scientists and Broad Churchmen: an Early Victorian Intellectual Network," *Journal of British Studies, 4* (1964), 65–88. This third version is expanded still more.

1. Charles Lyell, *Principles of Geology,* 2nd ed. (London, 1832–33), I, 2–13, 69–70; Thomas Huxley, "On the Reception of the *Origin of Species,*" in Charles Darwin, *Life and Letters,* ed. Francis Darwin (London, 1888), II, 179–204; Rowland Prothero, *The Life and Correspondence of Arthur Penrhyn Stanley* (New York, 1894), II, 173; Noel Annan, *Leslie Stephen* (London, 1951), pp. 140–148.

2. Noel Annan, "The Intellectual Aristocracy," in *Studies in Social History,* ed. J. H. Plumb (London, 1955), pp. 243–287. This is a more detailed treatment than is in his *Leslie Stephen* but not, perhaps, more interesting to read. Far broader than its title indicates is his brief but classic Hobhouse lecture, *The Curious Strength of Positivism in English Political Thought* (London, 1959).

3. See Charles R. Sanders, *Coleridge and the Broad Church Movement* (Durham, N. C., 1942); Duncan Forbes, *The Liberal Anglican Idea of History* (Cambridge, 1952); Hans Aarsleff, *The Study of Language in England, 1780–1860* (Princeton, N. J., 1967); D. A. Winstanley, *Early Victorian Cambridge* (Cambridge, 1940); Frances M. Brookfield, *The Cambridge Apostles* (New York, 1907).

4. Private communications.

5. Sanders, *Coleridge,* p. 113.

6. J. Bromley, *The Man of Ten Talents: A Portrait of Richard Chenevix Trench* (London, 1959), p. 86.

7. Michael Packe, *The Life of John Stuart Mill* (London, 1954), pp. 44–45.

8. See her Scientific Correspondence, in the Mary Somerville Papers (Somerville College, Oxford).

9. George O. Trevelyan, *Life and Letters of Lord Macaulay,* 2nd ed. (London, 1901), pp. 53, 60–61, 64, 101.

10. Quoted in A. K. Tuell, *John Sterling* (New York, 1941), p. 164.

11. Ben R. Schneider, Jr., *Wordsworth's Cambridge Education* (Cambridge, 1952) is the best study I have seen of the University's intellectual life at the end of the 18th century. There is a nice description of the Senate House examination in Thomas J. N. Hilken, *Engineering at Cambridge University, 1783–1965* (Cambridge, 1967), pp. 11–12; he describes examination reforms from 1822 on pp. 76–78. D. A. Winstanley's *Unreformed Cambridge* (Cambridge, 1935) and *Early Victorian Cambridge* (Cambridge, 1940) contain much indispensable information but tend to reinforce old prejudices on intellectual matters. See also G. M. Trevelyan, *Trinity College* (Cambridge, 1946). Sheldon Rothblatt's important publications are centered on a later period than I discuss.

12. Latham Wainewright, *The Literary and Scientific Pursuits which are encouraged and enforced in the University of Cambridge* (London, 1815), pp. 47, 61–67. On Paley, see D. L. LeMahieu, *The Mind of William Paley* (Lincoln, Nebraska, 1976).

13. See Herbert Marsh, *Course of Lectures Containing a Description and Systematic Arrangement of the Several Branches of Divinity* (Cambridge, 1809).
   Woodhouse was not merely a mathematician; see his "Some account of the tran-

sit instrument made by Mr. Dollond, and lately put up at the Cambridge Observatory," *Phil. Trans., 115* (1825), Part II, 418–428 and Plates XXVII and XXVIII.

14. Hilken, *Engineering at Cambridge,* pp. 36, 40–42. Farish succeeded to the Jacksonian chair in 1813. See Peter J. Booker, *A History of Engineering Drawing* (London, 1963), pp. 114–127: "Farish: Isometrical Perspective."

15. E. W. B. [Edward W. Brayley], "Biographical Sketch of the late Rev. Dr. Clarke," *Annals of Philosophy,* n. s. *8* (1824), 409–419, drawing on the longer account by William Otter, *Life and Remains of E. D. Clarke* (London, 1825) [I assume Brayley saw an advance copy].

16. *The Herschel Chronicle,* ed. Constance A. Lubbock (Cambridge, 1933), pp. 348–50.

17. Cf. John Herschel to Fearon Fallows, 8 December 1816 (John Herschel Papers, St. John's College, Cambridge).

18. Elaine Koppelman, "The Calculus of Operations and the Rise of Abstract Algebra," *Archive for History of Exact Sciences, 8* (1971), 155–242; see De Morgan's judgment of Peacock's work on p. 234. See also Lubos Nový, *Origins of Modern Algebra* (Leyden, 1973).

19. William Whewell, *An Elementary Treatise on Mechanics* (Cambridge, 1819), p. vii.

20. James Clerk Maxwell, *A Treatise on Electricity and Magnetism,* 3rd ed. of 1891 (New York, 1954), Part IV, ch. 5 (II, 199–200).

21. John Herschel, "Description of a Machine for resolving by Inspection certain important forms of transcendental equations," *Trans. Camb. Phil. Soc.,* IV (1833), 425–440.

22. Another important member carried further the interest in practical mechanics which is also shown in some of Whewell's textbooks and is related to contemporary French work on the subject. See T. M. Charlton, "Contributions to the Science of Bridge-Building in the Nineteenth Century by Henry Moseley," *Notes and Records of the Royal Society of London, 30* (1976), 169–179, and, for a general setting for Moseley and his family, J. L. Heilbron, *H. G. J. Moseley* (Berkeley, 1974), 1–9.

23. W. Tuckwell, *Reminiscences of Oxford,* 2nd ed. (London, 1907), p. 43.

24. *Post-Mercury,* 25 April 1969, p. 5. The house itself no longer stands, so the Greater London Council has refused to provide one of its usual blue plaques.

25. Charles Babbage, "General Description of the Analytical Engine 4 May 1869," in his "Scribbling Book 10 Dec. 1868" (Babbage Manuscripts, Science Museum, London).

26. See Charles F. Mullett, "Charles Babbage, a Scientific Gadfly," *Scientific Monthly, 67* (1948), 361–371; and Philip and Emily Morrison, eds., *Charles Babbage and his Calculating Engines* (New York, 1961). Maboth Moseley, *Irascible Genius: A Life of Charles Babbage, Inventor* (London, 1964) has useful details of Babbage's life but is not, to me, very convincing on his career and personality.

27. George Peacock, *A Treatise on Algebra* (Cambridge, 1830), pp. v–xxxv, is a good short statement of his position; the quotations are on p. xvii and p. 1. See also his "Report on the recent progress and present State of certain branches of Analyses," *Brit. Assoc. Report, 3* (1833), 185–288, and his *A Treatise on Algebra* (Cambridge, 1842–45); *Cambridge History of English Literature* (Cambridge, 1916), XIV, 555; Anon., "Obituary of George Peacock," *Proc. Roy. Soc., 9* (1857–59), 536–543.

28. George Boole, *An Investigation of the Laws of Thought* (New York, 1854), pp. 239, 299, 403, 416n, 420 (sec. 9); Boole to William Whewell, 7 February 1854 (Whewell Papers, Trinity College, Cambridge); Boole's "Sonnet to the Number Three," quoted in William Kneale, "Boole and the Revival of Logic," *Mind, 57* (1948), 157; William and

Martha Kneale, *The Development of Logic* (Oxford, 1962), pp. 398, 404; Geoffrey Taylor, "George Boole, F.R.S., 1815–1864," *Notes and Records of the Royal Society, 12* (1956–57), 44–52; Augustus De Morgan, *A Budget of Paradoxes,* 2nd ed., ed. David Smith (Chicago, 1915), II, 80.

29. Frances Bunsen, *Memoir of Baron Bunsen* (London, 1868), I, 544–545.

30. Sanders, *Coleridge,* p. 88.

31. Herschel to Francis Beaufort, 5 March 1831, Beaufort to Herschel, 7 March 1831 (John Herschel Papers, Royal Society).

32. George Airy, *Autobiography,* ed. Wilfred Airy (Cambridge, 1896), p. 48.

33. Hilken, *Engineering at Cambridge,* p. 49.

34. Airy to Edward Sabine, 8 November 1841, in the file "Establishment of Magnetic Observations at Greenwich and in the Colonies—to 1848" (Airy Papers, Royal Greenwich Observatory, Herstmonceux).

35. Eugene S. Ferguson, "Metallurgical and Machine-Tool Developments," in *Technology in Western Civilization,* ed. Melvin Kranzberg and Caroll Pursell, Jr. (New York and London, 1967), pp. 278–280.

36. For a good account of the affair, see Morton Grosser, *The Discovery of Neptune* (Cambridge, Mass., 1962), which is not exhaustive as regards English manuscript sources. Robert Grant, *History of Physical Astronomy* (London, 1852), pp. 198–200, contains a spirited defense of Airy's conduct.

37. J. W. Clark and T. M. Hughes, *The Life and Letters of the Reverend Adam Sedgwick* (Cambridge, 1890), II, 288; the italics are Sedgwick's. Airy made significant contributions to medicine: see John R. Levene, "Sir George Biddell Airy, F.R.S. (1801–1892) and the Discovery and Correction of Astigmatism," *Notes and Records of the Royal Society of London, 21* (1966), 180–199; and "Sir G. B. Airy, F.R.S. (1801–1892) and the Symptomatology of Migraine," *Notes and Records of the Royal Society of London, 30* (1975), 15–23. "Astigmatism" is one of Whewell's coinages for his scientific friends.

38. Trevelyan, *Macaulay,* p. 480. Thomas Carlyle was a Scot.

39. Clark and Hughes, *Sedgwick,* I, 220–221, 248; II, 426, 509; the quotation, I, 249. On Northern scientists in general, see Robert Fox, *The Hard Progeny of the North* (Lancaster, England, 1976).

40. H. G. Seeley, quoted in Clark and Hughes, *Sedgwick,* II, 489; *Romilly's Cambridge Diary, 1832–42,* ed. J. P. T. Bury (Cambridge, 1967), p. 73.

41. Clark and Hughes, *Sedgwick,* II, 109–114, 295, 388.

42. Clark and Hughes, *Sedgwick,* II, 391.

43. R. Jones, *An Essay on the Distribution of Wealth* (London, 1831), p. xxii, as quoted in Eric Roll, *A History of Economic Thought* (London, 1961), p. 313. It was Jones who helped Charles Wheatstone to an academic job: see Brian Bowers, *Sir Charles Wheatstone* (London, 1975), p. 55.

44. Eric Stokes, *The Utilitarians and India* (Oxford, 1959), p. 135; this is an important book on both of its subjects. My feeling is that Ricardo would have accepted the amendment also.

45. Margaret Deacon, *Scientists and the Sea, 1650–1900* (London and New York, 1971), p. 257, text and note 29. See also J. R. Rossiter, "The History of Tidal Predictions in the United Kingdom before the Twentieth Century," and Wolfgang Matthäus, "On the

History of Recording Tide Gauges," *Proc. Roy. Soc. Edinburgh, Section B, 73* (1971–72), 13–23 and 25–34.

46. Clark and Hughes, *Sedgwick,* I, 398.

47. Joan Evans, *John Ruskin* (New York, 1954), pp. 141, 146, 165n; Ruskin's letters to Whewell (Whewell Papers, Trinity College, Cambridge). On Willis's life and career, see the fine sketch in Hilkin, *Engineering at Cambridge,* pp. 50–57.

48. Clark and Hughes, *Sedgwick,* I, 204–207; J. W. Clark, "The Founding and Early History of the Society," *Proc. Cam. Phil. Soc., 7* (1891), ii–vii.

49. On Whewell's and Miller's work (not to be confused with the chemist William Allen Miller of King's College, London) see Herbert Deas, "Crystallography and Crystallographers in England in the Nineteenth Century," *Centaurus, 6* (1959), 129–148.

50. Clark and Hughes, *Sedgwick,* I, 208.

51. See the "Testimonials in favor of James D. Forbes...for the Chair of Natural Philosophy in the University of Edinburgh" (Roderick Murchison Pamphlet Collection, IV, Geological Society Library); Forbes to Herschel, 1 December, 13 December 1832 (John Herschel Papers, Royal Society).

52. J. S. Rowlinson, "The Theory of Glaciers," *Notes and Records of the Royal Society, 26* (1971), 189–204; on Forbes's work in general, see J. C. Shairp, P. G. Tait, and A. Adams Reilly, *Life and Letters of James David Forbes* (London, 1873), Chapters XIV–XV; on his relation to Clerk Maxwell, see Lewis Campbell and William Garnett, *Life of James Clerk Maxwell* (London, 1884), esp. p. 57.

53. Maxwell to Stokes, 22 February 1856, *Memoir and Scientific Correspondence of the Late Sir George Gabriel Stokes,* ed. Joseph Larmor (Cambridge, 1907), II, 2; Herschel to Adam Sedgwick, 7 May 1956 (John Herschel Papers, Royal Society). See George Davie, *The Democratic Intellect* (Edinburgh, 1961), pp. 191–197, on Clerk Maxwell's debt to his Scots education.

54. Sophie De Morgan, *Memoir of Augustus De Morgan* (London, 1882), pp. 46–48; J. L. E. Dreyer and H. H. Turner, eds., *History of the Royal Astronomical Society* (London, 1923), pp. 3, 20, 26.

55. J. Bonnycastle to Baily, 20 December 1798; B. Edwards, Secretary of the African Association, to Baily, 28 May 1799 (De Morgan Papers, Royal Astronomical Society); Hugh Robert Mill, *The Record of the Royal Geographical Society, 1830–1930* (London, 1930), p. 25. For background, see D. B. Hermann, "Some aspects of positional astronomy from Bradley to Bessel," *Vistas in Astronomy, 20* (1976), 183–186.

56. Doris Esch Borthwick, "Outfitting the United States Exploring Expedition: Lieutenant Charles Wilkes' European Assignment, August–November, 1836," *Proc. Am. Philosophical Soc., 109* (1965), 159–172.

57. Airy, *Autobiography,* p. 134.

58. Dreyer and Turner, *Astronomical Society,* p. 118.

59. On Barrow and Washington, see Mill, *Geographical Society,* pp. 41–42; L. P. Kirwan, *A History of Polar Explorations* (London, 1959), p. 159. The work of Herschel, Sabine, and Beaufort is made clear in their correspondence in the Herschel and Sabine Papers at the Royal Society.

60. Sedgwick to Herschel, 2 January 1848, in Clark and Hughes, *Sedgwick,* II, 141. There is a copy of Broderip's pamphlet in the Roderick Murchison Pamphlet Collection, IV (Geological Survey Library).

61. Herschel to Fitton, 18 October 1830; Herschel to Babbage, 26 November 1830; Herschel to Beaufort, 26 November 1830 (John Herschel Papers, Royal Society); Mill, *Geographical Society,* p. 33.

62. A useful beginning is Henry Lyons, *The Royal Society, 1660–1940* (Cambridge, 1944), Chapter VII.

63. Faraday to Herschel, 21 August 1827, 10 November 1832 (John Herschel Papers, Royal Society).

64. Herschel to Vernon Harcourt, 5 September 1831; Herschel to William Whewell, 20 September 1831 (John Herschel Papers, Royal Society).

65. A. D. Orange, "The British Association for the Advancement of Science: the Provincial Background," *Science Studies, 1* (1971), 316 and note 10.

66. Augustus J. C. Hare, *Memorials of a Quiet Life,* 12th ed. (London, 1874–76), I, 149. There were 19 editions of this book.

67. Tuell, *John Sterling,* p. 162; Bunsen, *Bunsen,* I, 396.

68. Julian Symons, *Thomas Carlyle* (New York, 1952), pp. 105, 128.

69. Hare, *Memorials,* I, 191.

70. Tuell, *John Sterling,* pp. 162–163.

71. T. W. Bamford, *Thomas Arnold* (London, 1960), p. 195.

72. Bamford, *Arnold,* pp. 11, 98–99, 120, 195, 205.

73. Hare, *Memorials,* I, 458.

74. Forbes, *Anglican Idea of History,* pp. 64–65; see also the articles by Robert Preyer, "Victorian Wisdom Literature: Fragments and Maxims," *Victorian Studies, 6* (1963), 245–262, and "Julius Hare and Coleridgean Criticism," *Journal of Aesthetics and Art Criticism, 21* (1957), 449–460.

75. Quoted in Kingsley Martin, *The Crown and the Establishment* (London, 1963), p. 36.

76. Julius Hare, *A Vindication of Niebuhr's History of Rome from the charges of the Quarterly Review,* with a postscript by Connop Thirlwall (Cambridge, 1829), pp. 21, 27 [a copy of this pamphlet is in the Roderick Murchison Pamphlet Collection, VI, Geological Survey Library, inscribed to Francis Hare from the Author]. See also Bunsen, *Bunsen,* I, 340, 394–396.

77. Brookfield, *Apostles,* pp. 8, 333, 12. The name of the group was doubly satirical. "The Apostles" in regular Cambridge terminology were the twelve men who received the lowest passing marks on the Tripos examination. See [John Wright] *Alma Mater* (London, 1827), I, 3.

78. Annan, *Leslie Stephen,* p. 45; Brookfield, *Apostles,* p. 332.

79. Brookfield, *Apostles,* p. 10.

80. Owen Chadwick, *The Victorian Church. Part I* (New York, 1966), p. 567.

81. Amy Cruse, *The Englishman and his books in the early nineteenth century* (London, 1930), p. 53.

82. Packe, *Mill,* p. 83; *The Letters of William and Dorothy Wordsworth: The Later Years,* ed. Ernest de Selincourt (Oxford, 1939), II, 934–936, #1260.

83. Leslie A. Marchand, *The Athenaeum, A Mirror of Victorian Culture* (Chapel Hill, N. C., 1944), pp. 10–11, 15, 246–250, 264, 275.

84. Clark and Hughes, *Sedgwick,* I, 246–249; II, 41; Wordsworth to William Whewell, 4 December 1828, 12 March 1829 (Whewell Papers, Trinity College, Cambridge); *Later*

Letters of Wordsworth, II, 556–557, 679–680, #973 and 1065. The especially relevant section of the *Excursion* is Book III, lines 173–189.

85. Frederika Beatty, *William Wordsworth of Rydal Mount* (London, 1939), pp. 230–233.

86. Hare, *Memorials,* II, 273; III, 224; Augustus J. C. Hare, *The Years with Mother* (London, n.d.), pp. 40, 51; Bamford, *Arnold,* p. 56; Prothero, *Stanley,* I, 100; Clark and Hughes, *Sedgwick,* I, 431; *Later Letters of Wordsworth,* I, 508–510, #941; II, 627–628, 666, #1025 and 1055; III, 1204–05, #1527.

87. Aarsleff, *Study of Languages,* pp. 214, 216–221.

88. *Dictionary of National Biography,* article "Richard Sheepshanks;" Clark and Hughes, *Sedgwick,* I, 336, 469; Trevelyan, *Macaulay,* p. 129; Prothero, *Stanley,* I, 136–137; *Romilly's Cambridge Diary, 1832–42*—if my count is off by 2 or 3 entries, it would not affect the result; Charles Knight, *The Results of Machinery,* 3rd ed. (London, 1831), p. 2.

89. Hare, *Years with Mother,* p. 35.

90. Hare, *Memorials,* II, 332; III, 235; *Years with Mother,* p. xi.

91. Hare, *Memorials,* II, 87.

92. Prothero, *Stanley,* I, 3, 112–113, 181–185; II, 415; Hare, *Years with Mother,* p. 51; Clark and Hughes, *Sedgwick,* II, chapter 9; Edward Stanley, "Summary of Observations for Ten Years, from 1815 to 1824, containing the Mean of Thermometer and Barometer, number of Days of prevailing Winds, Snow, and Rain, Fall of Rain, &c at Alderley Rectory, near Knutsford, Cheshire," *Edin. Phil. Jour., 12* (1825), 300–304.

"Faithful" is defined as "paying dues willingly." In the Letter Books of the Geological Society, see E. Norwich (Edward Stanley) to Geological Society, 29 July 1837 (III, #107), 16 Oct. 1840 (V, #223), 7 March 1842 (VII, #61), 14 June 1842 (VII, #144), 14 February 1844 (VIII, #158); compare Bishop of St. Davids (Connop Thirlwall), 23 August 1845 (IX, #148) and 24 Sept. 1846 (IX, #414). Contrast to S. Wilberforce, 24 August 1845 (IX, #149), quibbling over whether he owes resident or non-resident dues; and to the resignations of G. B. Airy, 14 June 1837 (III, #89), George Poulett Scrope, 8 May 1841 (VI, #236), and W. J. Hooker, 20 April 1846 (IX, #285) who stated that he gave his original ten guineas just to encourage a new society, but he never received its publications, went to its meetings, or got any good out of it.

93. Prothero, *Stanley,* I, 150.

94. Hare, *Years with Mother,* p. 66.

95. Clark and Hughes, *Sedgwick,* I, 432–433; John Connop Thirlwall, Jr., *Connop Thirlwall* (London, 1936), pp. 177, 179, 191 (the quotation), 245–246.

96. W. J. Conybeare, "Church Parties," in *Essays Ecclesiastical and Social,* 5th ed. (London, 1855), p. 147.

97. Chadwick, *Victorian Church. Part One,* p. 545.

98. *Cf.* Prothero, *Stanley,* I, 212, 326–327, 390, 473–475; II, 31–34, 41–43.

99. Annan, *Leslie Stephen,* pp. 134–136.

100. Thirlwall, *Thirlwall,* pp. 179, 245.

101. Quoted in Florence B. Lennon, *Victoria Through the Looking-Glass: The Life of Lewis Carroll* (New York, 1945), p. 254. See also p. 74, on Jowett's salary: "In an earlier age of mathematics J would have been referred to rectangular axes, and divided into two unequal parts—a process of arbitrary elimination which is now considered not strictly legitimate."

102. Forbes, *Anglican Idea of History,* p. 117.

103. Trevelyan, *Macaulay,* pp. 142, 312, 316–317, 319.

104. Trevelyan, *Macaulay,* p. 416.

105. *Dictionary of National Biography,* article "Henry Maine," by Leslie Stephen.

106. Clark and Hughes, *Sedgwick,* II, 226, 292–293.

107. Hare, *Memorials,* II, 226; Herschel's diary for 20–26 January 1858 (John Herschel Papers, Royal Society).

108. Prothero, *Stanley,* II, 317; Herschel, *Discourse,* p. 10; Forbes, *Anglican Idea of History,* pp. 100–101, 172; Walter F. Cannon, "The Impact of Uniformitarianism," *Proc. Am. Philosophical Soc., 105* (1961), 301–302. See also Connop Thirlwall, "Present State of Relations between Science and Literature," in his *Remains: Literary and Theological,* ed. Perowne (London, 1877), III, pp. 284–311.

109. Darwin, *Life and Letters,* I, 191–193; Nora Barlow, "Robert Fitzroy and Charles Darwin," *Cornhill Magazine,* n.s. *72* (1932), 497, 505; Report of the Astronomer Royal to the Board of Visitors, 5 June 1841 (Airy Papers, Royal Greenwich Observatory, Herstmonceux); Beaufort to Fitzroy, 1 Sept. 1831, in *Darwin's Journal,* ed. Gavin de Beer (London, 1959), p. 6 note 10. See the interesting articles in the *British Journal for the History of Science:* Jacob Gruber, "Who was the *Beagle's* naturalist?", *4* (1968–69), 266–282, and Harold Burstyn, "If Darwin wasn't the *Beagle's* Naturalist, why was he on Board?" *8* (1975), 62–69.

110. See, for example, Fitzroy to Lady Herschel, 29 June 1836; Fitzroy to Herschel, 21 February 1841, 8 April 1858 (John Herschel Papers, Royal Society).

111. Henslow to the Geological Society (this usually meant Lonsdale, the assistant secretary), 28 November 1835, in Letter Books of the Geological Society, Book II, #19 (Geological Society Library).

112. Trevelyan, *Macaulay,* p. 633.

113. Clark and Hughes, *Sedgwick,* II, 437–438.

114. Hare, *Memorials,* I, 132, 195; Sanders, *Coleridge and the Broad Church Movement,* p. 138.

115. Clark and Hughes, *Sedgwick,* II, 582.

116. Bunsen, *Bunsen,* I, 140.

117. Bunsen, *Bunsen,* I, 315–316; Bamford, *Arnold,* p. 196; Hare, *Years with Mother,* pp. 36–37.

118. Bunsen, *Bunsen,* I, 481; Charles Kingsley, *Alton Locke, Tailor and Poet* (New York, 1886), p. 228.

119. Bunsen, *Bunsen,* I, 506.

120. Herschel, *Discourse,* p. 259.

121. Faraday to Herschel, 10 November 1832; Thomas Andrews to Herschel, 11 February 1871, reporting remarks Faraday made to Andrews (John Herschel Papers, Royal Society).

122. L. Pearce Williams, "Boscovich and the British Chemists," *Roger Joseph Boscovich,* ed. L. L. Whyte (London, 1961), pp. 153–167, and "The Physical Sciences in the First Half of the Nineteenth Century," *History of Science, 1* (1962), 7, 10; Airy to Herschel, 20 March 1857 (John Herschel Papers, Royal Society).

123. Adam Sedgwick, *A Discourse on the Studies of the University* (Cambridge, 1833); John S.

Mill, "Professor Sedgwick's Discourse," *Dissertations and Discussions* (Boston, 1864), I, 121–185; and see Packe, *Mill,* pp. 197–198.

124.   Packe, *Mill,* p. 264.

125.   Isaac Todhunter, *William Whewell* (London, 1876), I, 129–130; II, 187, 298, 416–417; Alvar Ellegärd, *Darwin and the General Reader* (Göteborg, 1958), pp. 220 ff. Ellegärd's analysis of the contending positions is, however, not satisfactory to me.

126.   Forbes, *Anglican Idea of History,* p. 95; see also his interesting comments on teaching the classics as classical history, p. 114–115, which I can neither confirm nor deny by my own knowledge.

127.   Karl Britton, *John Stuart Mill* (London, 1953), p. 35; Annan, *Leslie Stephen,* p. 52n.

128.   Hare, *Memorials,* II, 186; "Recent Latitudinarian Theology," *Christian Remembrancer,* n.s. *38* (1859), 406 [I have no idea what the author's affiliation was]; Bunsen, *Bunsen,* I, 427; Clark and Hughes, *Sedgwick,* II, 300.

129.   Standish Meacham, in *Victorian Studies, 10* (1967), 441.

130.   Conybeare, *Essays,* p. 144 (on Hare), p. 155.

131.   Kingsley, *Alton Locke,* p. 182.

132.   Thirlwall, *Thirlwall,* p. 188; Arthur Ramsay, *F. D. Maurice and the Conflicts of Modern Theology* (Cambridge, 1951), p. 105. Thirlwall (the author) includes Wordsworth and Coleridge, Thirlwall and Hare, Sterling and Tennyson, Kingsley and Maurice, Colenso, Lightfoot, Vaughan, Williams, Alford, and Donaldson.

133.   Charles Lyell, *Life Letters and Journals,* ed. K. Lyell (London, 1881), I, 270, 312, 355, 359; Todhunter, *Whewell,* I, 57; II, 108–109; [William Whewell], "Lyell-*Principles of Geology,*" *British Critic, 9* (1831), 195; Adam Sedgwick, "Address of the President (1831)," *Proc. Geological Soc. London, 1* (1826–33), 302–303.

134.   Stephen-Whewell Correspondence (Whewell Papers, Trinity College, Cambridge).

135.   H. Gwynedd Green, "A Biography of George Green," *Studies and Essays in the History of Science and Learning Offered in Homage to George Sarton* (New York, 1946), pp. 552–593.

136.   De Morgan, *De Morgan,* pp. 235–255.

137.   Packe, *Mill,* p. 86.

138.   Beatty, *Wordsworth of Rydal Mount,* p. 228.

Chapter Three

Humboldtian Science

Probably most of us have some vague idea of what modern historians mean when they characterize much of science in the early and mid-19th century as "Baconian." This "Baconian" notion is so widespread that it cannot be attributed to any one person; there is more missionary work to be done than can be accomplished by attacking one or a few books or articles. There are even historians who want *British* science in this period to be "Baconian," although according to others Baconianism is supposed to be natural for the colonies, not for the mother country.

I am not going to try to define Baconianism in a neat analytic way, since authors use it rather loosely. It means something like the collection of facts, lots of facts, in all sorts of places, and on queer applied subjects; the absence of an analytical theory or of sophisticated mathematical tools; the belief that a hypothesis will emerge somehow from the accumulation of facts; and so forth, and so on. It means an activity very suitable for a new country where collecting new rocks and species is easy, whereas learning advanced mathematics is hard.

I may note that Baconianism advocates seem to have a rather odd notion of natural history methods. They seem to envisage a "collector" as one who wanders into a jungle and grabs now this species, now that one, or perhaps just everything in sight. But actually one collects according to plans and categories; it is by reference to the pre-determined categories that one can recognize a genuine novelty when one encounters it. Collecting at its best is done for or against a set of theories; and great collectors like Alfred Russell Wallace are often also great hypothesizers.

Baconianism, it should be noted, is a set of beliefs in the minds of 20th-century historians which they associate with the philosophy of Francis Bacon. Quite erroneously, some of them then proceed to find scientists in the early 19th century who said that they admired Bacon, and display them as evidence of a wide-spread belief in Baconianism. I say "erroneously" because, rightly or wrongly, Bacon was often taught in the early 19th century not as an advocate of Baconianism but as advocating a method of analysis, generalization, and deduction: a rather modern-sounding combination. This teaching radiated, as one might expect, from

Cambridge. It is expressed admirably in John Herschel's *Preliminary Discourse on the Study of Natural Philosophy* of 1831, which, indeed, had an engraving of a bust of Bacon on one of its two title pages. I am not concerned to determine whether Herschel represented the historical Bacon correctly. The points that are relevant are that the *Discourse* was read by some scientists, and that in this position as in others in the book Herschel seems to have been representing the general opinion of his professional set. He saved particular novelties of his own for later publications. The *Discourse* represented many things, but not John Herschel's personal philosophy of science.[1]

We can go further, with the help of a line from an ex-Trinity College undergraduate, Lord Byron: "Was it not so, great Locke? and greater Bacon?"[2] Locke and Newton were sometimes taught as the two great appliers of Bacon's method, the one in the moral sciences and the other in the physical sciences. Many scientists of the early 19th century meant by "Baconian": that approach to natural philosophy which, as contrasted to other approaches such as those of Aristotle, Descartes, and perhaps even Galileo, had led to Isaac Newton's *Principia.* There were, to be sure, other people who thought that Bacon's philosophy called for facts and more facts (which in a sense it does), especially if they felt antagonistic to some current theory. These were not always scientists; some of them were clergymen or review writers.

If, then, a scientist is found approving of Bacon, it does not at all follow that he was a believer in what modern historians call Baconianism; it only shows that he was in agreement with a then-common interpretation of what Bacon stood for.

Rather than to worry any more about Baconianism (since it is an idea in the minds of modern historians, I would have to study *them* instead of the 19th century), I propose instead to sketch what Alexander Humboldt stood for in science, and suggest that the activities of many scientists in the early 19th century match up to Humboldt's schemes fairly precisely, rather than to this vague Baconianism fairly vaguely. What I know about British and American science suggests that the match is possible in a number of cases. Obvious candidates for consideration in America are Joseph Henry at the Smithsonian, New York's James Hall, William Rogers, Asa Gray, Matthew Maury, J. N. Nicollet, the Long expedition, the Wilkes expedition, and various state geological surveys. I particularly recommend Nathan Reingold's article on Alexander Dallas Bache, the longtime director of the U. S. Coast Survey. Reingold's description seems to me to call out for the phrase "Humboldtian science" to characterize Bache's interests.[3] I will mention some British candidates later in this essay. I will not assert that two works by Stanley Guralnick support my position but rather that they may be studied to see if they can be fitted into it: see his "Sources of Misconception on the Role of Science in the Nineteenth-Century American College," *Isis, 65* (1974), 352–366, and *Science and the Ante-Bellum American College* (Philadelphia, 1975).

We must, of course, distinguish what a scientist himself wanted to do from the arguments which convinced a public body to support his activities. Any state might want to exploit its mineral resources; any national government might want to explore its western lands; but these factors do not explain why Hall and Long went about their jobs the way they did. It is the *unnecessary* science attached to useful activities which shows an extra influence at work.

Alexander Humboldt (he disliked his inherited "von," but his publisher would not drop it from his books)[4] was careful not to call himself an explorer; he was a scientific traveller. He measured accurately what explorers had reported inaccurately. The Humboldtian travelling party was characterized by its care for and resort to the latest instruments of measurement, if only a two-inch snuff-box sextant by Troughton (Humboldt's own favorite for canoe trips).[5] Humboldt himself set the example. Here is a list of the measuring devices he carried with him on his American travels from 1797 to 1804. Even this was not, by his standards, a complete collection: some of his meteorological devices, a chronometer by Arnold, and several achromatic telescopes never caught up with him.

Chronometer by Berthoud, once possessed by Borda.

Demi-chronometer by Seuffert.

Three-foot achromatic telescope by Dolland (for determining longitude by observing eclipses of the satellites of Jupiter.)

Smaller telescope by Caroché, which could be fixed to the trunk of a tree.

Lunette d'epreuve with micrometer, by Kohler, for measuring small distant angles.

Ten-inch sextant by Ramsden.

Two-inch snuff-box sextant by Troughton.

Twelve-inch reflecting and repeating circle by LeNoir.

Theodolite by Hurter.

Artificial horizon by Caroché.

One-foot quadrant by Bird.

Graphometer by Ramsden.

Twelve-inch dipping needle, on the principles of Borda and LeNoir, "of the most perfect execution," from the French board of longitude.

Variation compass by LeNoir.

Twelve-inch magnetic needle, suspended to an untwisted thread, for horary variations of the magnetic variation and for intensity changes with latitude.

Saussure magnetometer by Paul [he did not like this on trying it out].[6]

Invariable pendulum by Megnie.

2 barometers by Ramsden.

2 barometrical apparatuses to find the mean height of the barometer.

Several thermometers by Paul, Ramsden, Megnie, and Fortin.

2 hygrometers, of Saussure and Deluc.

2 electrometers, to measure the electricity of the atmosphere [on board ship, this had increased to 6 electrometers].[7]

Cyanometer by Paul, colored conformably to that of Saussure, for comparing the blueness of the sky to a fixed standard.

Fontana eudiometer, for the quantity of oxygen in the atmosphere.

Phosphoric eudiometer by Reboul.

Apparatus by Paul to determine the degree at which water boils at different altitudes.

Thermometrical lead by Dumotier.

Two aeromotors of Nicholson and Dolland.

Compound microscope of Hofmann; standard meter by LeNoir; land-surveyor's chain; assay-balance; rain gauge; tubes of absorption for carbonic acid and oxygen; some Haüy's electroscopical apparatus; vases to measure the evaporation of liquids; artificial horizon of mercury; small Leyden phials; galvanic apparatus; reagents for experiments in the composition of mineral waters; "and a great number of small tools necessary for travellers to repair such instruments as might be deranged from the frequent falls of the beasts of burden."[8]

Thus the complete Humboldtian traveller, in order to make satisfactory observations, should be able to cope with everything from the revolutions of the satellites of Jupiter to the carelessness of clumsy donkeys.

For short trips away from the main base, Humboldt contented himself with a sextant, artificial horizon, dip needle, device for magnetic variation, thermometer, barometer, Saussure hygrometer, and sometimes the invariable pendulum.[9]

Alexander Humboldt, then, from about 1800 to about 1840 successfully directed the attention of many European scientists, especially the younger ones, to a complex of interests for which there has never been a completely satisfactory phrase; Nathan Reingold and I have discussed the problem for some time, without reaching a really satisfactory conclusion. This is not too surprising, as Humboldt himself considered current terms as rather vague.[10] "Physical geography" was a phrase much used in England, and by Matthew F. Maury in his *Physical Geography of the Sea* of

1855; but "geography" conveys a rather limited and superficial meaning now. "Physique de la terre," one of Humboldt's attempts, does not sound as though it covers the geographical distribution of plants and animals. Our modern "geophysics" introduces the same objection; "physics" is a much more limited subject than was Humboldt's "physique." Suppose we consider the fields which Cleveland Abbe's dream institution of the 1860's, as described by Reingold, was to cover: an astronomical observatory training young men in "'practical astronomy'," with research programs in the tides, meteorology, terrestrial magnetism, the temperature within the surface of the earth, geodesy, earthquakes (especially as they affect the stability of instruments), gravitation, and so on. We would have to call this, not "geophysics," but "astronomy and geophysics;" and since Abbe was not interested in natural history, we would still have a term for only half of Humboldt's emphases.

For such complexes as Abbe proposed, Reingold has introduced a good historical approach by omitting the modern idea of "geophysics" as distinct from "astronomy," and calling the whole complex "applied astronomy." This enables him to point out that something like this complex has a tradition going back as far as 1660, and that it was the basis of "a significant percentage of the labors of scientists of the period 1660–1860."[11] It had distinguished support; there are many more topics treated in Newton's *Principia* than is usually indicated in standard presentations. This applied astronomy complex, then, long ante-dated the larger Humboldtian complex, of which it became a part, but with somewhat different emphases than it had had in the time of Newton.

We have, in short, no term available which includes astronomy and the physics of the earth and the biology of the earth all viewed from a geographical standpoint, with the goal of discovering quantitative mathematical connections and interrelationships—"laws," if you prefer, although they may be charts or graphs. Hence I call it "Humboldtian science" in spite of the obvious objections that some of its parts (like applied astronomy) and some of its own characteristic emphases (like the accuracy of portable instruments) were major concerns before Humboldt appeared on the scene. If Humboldt was a revolutionary (as I think perhaps he was), it was not in inventing all the parts of Humboldtian science. It was in elevating the whole complex into the major concern of professional science for some forty years or so.

William Goetzmann, in his excellent *Army Exploration in the American West,* is well before me in asserting a predominant Humboldtian influence on American scientists; indeed the phrase "Humboldtian science" is his, except that he spells it "Humboldtean." In his narrative Humboldt, if not so important as Oregon and the Oregon Trail, is about as common as St.

Louis. Both are fixed points of departure for the American West. But the use of the same phrase and the same man should not mislead anyone into believing that we are really talking about the same thing.

When Goetzmann connects the Humboldtian spirit to Romanticism, I can go along with him, in part. But his explanation shows, to me at any rate, that he has not penetrated to the reasons *why* Humboldt wanted people to collect much and varied data, or even to the manner in which Humboldt wanted them to collect it. Goetzmann represents Humboldt's as an "essentially cosmic approach," vague and unfocused as compared to the classical, involving necessarily an interest in "cosmology or idealistic metaphysics," calling for "all these pieces of data of infinite variety and exotic appeal." "Everything was relevant, everything was important. The trivial and the obvious had to be observed and recorded, though Fremont carried this to extreme when he climbed what he thought was the highest peak in the Wind River Mountains and commented chiefly upon the habits of a bumble bee he found buzzing there."[12]

This is a fairly normal account of one modern view of what Romanticism was like; but it is hard to reconcile it with Humboldt's insistence upon measurement, and upon more accurate measurement at that. Goetzmann has, perhaps, been misled by believing that the product of Humboldt's old age, his *Cosmos,* was typical of his message; at least it is the *Cosmos,* in its 1845–61 English translation, which Goetzmann especially cites.[13] It was not typical; it disappointed a number of his followers. And it came late enough not to influence anyone that I shall discuss.

If we take Fremont and his bumble bee as an example, we can see some of the difference between the "usual" romantic and the Humboldtian. The usual romantic, I take it, should be excited by climbing the highest mountain. The Humboldtian knows that the performance of plants and animals under extremes of geographical conditions is a key to their ability to adapt and hence to their geographical distribution generally. Was the effect of mountains on organisms living there due to cold, rainfall, excess solar radiation, oxygen content of the atmosphere, or what? Any romantic can climb a mountain; the Humboldtian wanted to use its height as a laboratory for observing extreme conditions of existence.

Goetzmann's "Humboldtean spirit" sounds too much like Baconianism with a Romantic justification to fit the Humboldt I know. His important work in stressing the influence of Humboldt on American scientific expeditions in the West cannot have its deserved effect so long as the nature of that influence is misunderstood. It is important to see also that the mathematical scientists in the civilized East were equally Humboldtian, not in spite of, but because of, their mathematics.

Thus Reingold comments on Alexander Bache that he was not opposed to natural history as such, but he did little of it. "The principal difference between what Bache would or would not touch was the presence or absence of mathematical theories providing structure to the operations and to which data were referred."[14] Bache was more of a purist than the master; Humboldt went into natural history with only the *hope* of eventually getting mathematical laws. This emphasis produced some notable results. The men who collected many species of butterflies were encouraged to worry about the laws which regulated their distribution; and one at least, Alfred Wallace, worried about the dynamic forces which might produce the laws. Not all Humboldtian travellers theorized about their observations, of course. The long-standing distinction in natural history between observer-collector and theorist-back-home was too well-established, and too useful, to be overcome. But the compleat Humboldtian was the man like Wallace who thought about what he collected.

With this kind of orientation, it would then be possible to re-study the Western expeditions in detail to see why they did just what they did: what equipment they used, how satisfactory it was, what measurements and observations they made in detail, why these rather than others, what the results were and how accurate they were, what the balance between physical and biological and anthropological observations was and why. In short, it would be possible to study the expeditions as the functioning scientific enterprises many of them were. Goetzmann's story is interesting but tantalizing, like *Hamlet* with the Prince of Denmark off-stage most of the time.

It is not really surprising that most Americans I have talked to are like Goetzmann in thinking of Humboldt as a sort of vague Romantic idealist (that is, if they think of him at all). A number of Germanic philosophers have an intellectual and emotional investment in their own attachment to Humboldt's philosophy. In order to rescue a great scientist from his Romantic admirers, I shall proceed, unrepentant at being inspired by a German engineer, and by a French historian. After all, Humboldt chose to live his adult life in Paris, and to write in French.

One of Humboldt's early and basic interests was geology. He had studied at Freiberg for a year and then had been a mining administrator for four years before he became financially independent at his mother's death. Especially, he wanted to locate a dynamical explanation for the origin of mountain-chains, given the strange fact that in Europe they are oriented not at random but generally in a northeast-southwest line. However Humboldt was in revolt against the overly simple and overly universal explanations for which today, possibly with some error, we often condemn Werner and Buffon and such theorists (Humboldt rather admired Werner[15]). If mountain-chains often run in this direction, there are many exceptions. The dynamical law which explains them should explain the exceptions, and should also explain the inclination of the strata, and even the magnetic axes of an individual mountain where such exists.[16]

Humboldt's great trip to South America, 1799–1804 (he originally wanted to go around the world, but never made it) was in some degree motivated by a desire to see new mountains in a new setting; this might stimulate new ideas as to causes. His accumulated measurements eventually enabled him to disprove the original idea, but he felt that he could prove something more interesting: that mountain-chains were formed by the same forces that upheaved and fractured the adjoining strata.[17] Thus the hypothesis of a set of mammoth elevating forces in the interior of the earth was much more plausible than that of a universal ocean or other Neptunist agency.

In short, in this as in other fields Humboldt wished to promote worldwide observations not because of the intrinsic interest of a great many varied facts, but because he wanted general theories, and would not settle for the overly simple ones of the past based on generalizing merely local observations. "It is by isolating facts, that travellers, on every other account responsible, have given birth to so many false ideas."[18] Thus, for example, he opposed extending European names of formations in geology to American formations supposedly analogous.

Humboldt was quite clear that much of what was "known" wasn't true. "One of the most important services, that modern chemistry has rendered to physiology, is it's having taught us, that we are still ignorant of what illusory experiments on the chemical composition and salubrity of the atmosphere had led us to admit fifteen years ago."[19] This general position sometimes led him to place more emphasis on disproving an 18th-century theory than on establishing a new one. He often ostentatiously drew attention to the fact that he was limiting himself to verifiable regularities. But it is a characteristic of his presentation that often, when Humboldt had completed a denunciation of hypothetical thinking and a praise of accurate and repeated measurement, he immediately followed it by an unprovable hypothesis of his own.[20]

Humboldt's creed was measurement; and having cleared his conscience by repeating this creed for the reader, he then relaxed and enjoyed his profane love, speculation. As he said, "Some parts of physics and geology are merely conjectural; and it might be said, that science would lose much of its attraction, if we endeavoured to confine this conjectural part within too narrow limits."[21]

Some historians have not noticed this feature of Humboldt's rhetoric. They have concluded that Humboldt could not, for example, have influenced Darwin because Humboldt considered the origin of species to be an insoluble problem. But Humboldt declared all sorts of subjects to be beyond the limits of science; and then proceeded to entice the reader with speculations or challenges. What he meant was: without more detailed first-hand observations, you don't have a case.

The geological quest for the origin of mountain-chains, then, is the kind of thing Humboldt liked. Because he himself failed to find the precise laws and dynamical causes he was seeking in subject after subject, some persons did not realize that that was the point of the whole enterprise: Humboldt had nothing but scorn for what he called "the sterile accumulation of insulated facts."[22] It took, however, one or two or three generations of scientific disciples before the Humboldtian approach paid off, and sometimes it did not do so even then.

Thus in the 1830's the Humboldtian science which seemingly had succeeded, that is, in which an analytic law was produced, was the study of terrestrial magnetism. To promote this study, Humboldt had originally planned to make measurements all the way around the world in his 1799 voyage, although the fortunes of war determined that he never got beyond America.[23] Among the suppliers of analysis were Hansteen, and then Gauss. This successful endeavour became an inspiration, and supplied a model. It seemed in the 1830's that such complex studies as meteorology and the movement of the tides might yield in another ten or fifteen years to the same kind of concerted effort which had done so well in magnetism. In these subjects, unfortunately, no new Gausses arose.

Now Humboldt preached that all matters which vary with geographical position should be measured, in their variations. One of these is the effect of sunlight; and one of the origins of modern astrophysics is Humboldt's attempt to measure the power of the sun in different locations and at different heights.[24] Humboldt did not do it quite right; it took John Herschel with his actinometer, then Pouillet with his pyrheliometer, and finally Langley with his bolometer to carry out Humboldt's desires. The linkage was complex enough that I wonder if Langley even knew that the man at the other end was Humboldt.

In astronomy, I am toying with the notion that the interest of some astronomers in not merely locating double stars but also in measuring their rotations accurately to show that these are explicable by dynamical law is related to Humboldtian influence. But one who speculates about influences in astronomy in this period without knowing where to place Bessel is a rash person indeed. No such Humboldtian explanation is exactly demanded for John Herschel; his interest can always be explained away as "filial" piety; but it might help for others. John was, in other areas at any rate, influenced by what Humboldt had done with William Herschel's ideas, as well as directly. Humboldt's "astrometer" for measuring the relative intensity of starlight is a case in point. It was a device inspired by William Herschel's work, and it in turn stimulated John Herschel.[25]

Humboldt had been trained as an administrator of state economic matters before he escaped by going into geology, and his continued interest in political economy is obvious in his more popular writings. I have not gained the impression (though others may differ) that this interest led him in the direction of Malthusian population problems, or even into a profound comparison of the geographical distribution of humans to that of other organisms. Rather, his interest in quantitative measurement led him to suggest a formal analogy between any two means and how they are calculated. For example, a mean of population statistics is similar to a mean of temperature measurements. Moreover, we can use the methods learned in astronomy to handle our figures in political economy. Particularly useful are the probable results obtained by using different astronomical methods for determining position: "We should employ numerical elements after having well discussed them and determined the limits of the error. We might almost compare the different degrees of probability which statistical results furnish in the Ottoman Empire, in Spanish or Portuguese America, in France or in Prussia, with those geographical positions which are founded either on lunar eclipses, on the distances of the moon from the sun, or on the occultation of the stars."[26] Humboldt's intellectual heir in this subject, therefore, would seem to have been Quetelet.

I find it interesting to speculate as to who, in my own field of British science, carried out Humboldt's plans most thoroughly under direct Humboldt influence. In his work on terrestrial magnetism Edward Sabine fits these twin criteria; but his geodetic work, in which he first made a major reputation, was in the tradition of applied astronomy ante-dating Humboldt. George Gabriel Stokes, the theorist of geodesy and geodetic instruments, is an interesting possibility. This would explain why anyone in 1849 should care about the hydrodynamics of a viscous fluid as it affects a cylinder oscillating in the fluid—it was for a gravity pendulum swinging in the air.[27] Abstract mathematical analysis to deal with sources of error in a precision instrument in order to determine the exact shape of the earth: an excellent example of Humboldtian science. John Herschel's giant scheme for worldwide meteorological observations was directly Humboldtian; it was modelled after the terrestrial magnetism network of observers; and it is, much enlarged, still with us. That is, today's coordinated meteorological observations are continuous with the coordinated scheme Herschel set up. But Herschel, as I said, was perhaps a Herschelian first, and a Humboldtian second.

A comparison of subjects treated shows that one of the men greatly indebted to Humboldt was that speculative young geologist, Charles Lyell. It is interesting to see why Lyell was a follower, but only a partial follower, of Humboldt. He had come to know and admire Humboldt during his stay in Paris in 1823.[28] But the insistence on measurement was simply the wrong emphasis if one wished to be a great geological discoverer at that time. The new British way of doing geology made Humboldt's methods unduly restrictive, however useful they may have been earlier.

Humboldt's emphasis on measurement had enabled him to reject the grandiose speculations of the past. The new way developed in Britain, however (with Cuvier's blessing), was that of carefully reconstructing the local history of a limited area. The identification of fossils played a large but not exclusive role in this version of archeology; the exact measurement of anything played a quite limited role for most geologists. (Lyell indeed was one of the exceptions; his Eocene, Miocene, Pliocene periods were differentiated on the basis of percentages of extinct to living shellfish species.) This British approach to geology permitted very nice debates about the quantity and intensity of this or that, as for example between Uniformitarians and Catastrophists. Since neither group had much quantified data, such debates could go on happily for decades; and they did.[29]

Humboldt's catastrophism, if it should be called that (the terminology is questionable outside of Britain), was less restrictive. It was his way of denying the slow processes of a Wernerian universal ocean and on insisting on the study of earthquakes, volcanoes, and processes of widespread elevation and fracture of strata. Even the distribution of some plants, he suggested, might be related to widespread subsidence in recent times, which had left such isolated remains as Portugal, the Atlas Mountains, the Azores, and the Canaries.[30] Thus, it is well to emphasize, catastrophism was not primarily a Biblical prejudice. For Humboldt, it was a way of substituting processes which could be investigated for hypothetical universal oceans, and other such myths. As for particular geological forces, Humboldt noted that one school of geologists explained everything by violent commotions (such as are related to earthquakes and volcanoes), while the other "has recourse to small powers, which produce their effect almost insensibly by progressive action." He himself refused to become involved in such vague discussions when they involved "the origin of things." He would confine himself to "the small number of facts hitherto well observed." Among these, quite clearly, was the cumulative effect of water running down a gentle, uniform slope.[31]

Best of all for the young geologist, perhaps, was Humboldt's insistence on the importance of the "geography of plants" for geology, and vice versa. The problems of neither could be solved without reference to the other. In his brief essay of 1807, Humboldt pointed out that, by studying fossil plant distribution, one should be able to trace the ancient connection of continents, the primordial history of the globe. Plant migration was assisted by winds, currents, birds, now by men, all of which are related to geological geography. But one particular classic problem, the existence of tropical fossil plants in the far north, was probably not explicable by simple migration, Humboldt thought. There was probably a different climate at some previous time. An astronomical cause of the magnitude of, say, the shifting of the earth's axis is unlikely. Can the geologist provide a local source of heat? Or was there a variation in the intensity of the solar rays? Or a local perturbation of the earth's orbit?[32]

Humboldt observed that this subject was not yet really a science, only a name. That he did not consider it to be mere botany is indicated by his credits: Delambre, Prony, Decandolle, but above all others, Biot. Delambre and Prony provided height measurements and calculations, Decandolle provided material on the geography of plants in the Alps. But why was Humboldt so appreciative of the advice and "sagacity" of the *physicien* Biot? I wish I knew.[33]

Let me briefly summarize my impressions on going through just two of the works of Humboldt which became available in English for a young geologist to read, if he wanted to learn more about Humboldt's approach. Humboldt was originally scheduled to do the essay on the geography of plants for the *Dictionnaire des Sciences Naturelles,* but it was turned over to Decandolle, and Humboldt did the article on "Geognosy" instead. This was made available as a separate book in English in 1823, *A Geognostical Essay on the Superposition of Rocks in Both Hemispheres.* After reading British geologists of the same period, I found myself unable to become very interested in this treatment, even the theoretical part (most of the book is a long description of where which rock is found).

Humboldt in the *Geognostical Essay* emphasized comparative cross-sections of geological structure, citing William Buckland's table of parallel formations of 1821. He insisted that in different countries one found equivalent formations, not identical formations, and praised the use of geographical place names for formations: e.g., the lias of Dorsetshire, the sandstone of Nebra. He even tried to work out both a geometric and an algebraic method of showing the superposition of formations.[34]

Humboldt noted the use of fossils in showing the equivalence of various strata, but pointed out that fossils are absent or rare in the older formations. Then he gave a long list of questions the fossil geologist might investigate—not a particularly novel or original list, as far as I can see, unless being rather anti-Cuvier counts. That is, he rather liked the idea of "a progressive development of vegetable and animal life on the globe," and saw no reason to believe that the destruction of genera and species was correlated with the change from one formation to another. On the whole, his position was one of caution: do not divide up formations merely on the basis of fossils, if all other evidence indicates that they are a unity.[35]

Not very exciting. If "geognesy" (dating from Werner) had not already succumbed to "geology," this kind of treatment would probably have accomplished it.

In contrast, Volume VI, Part 2 of the *Personal Narrative,* translated at least by 1826, is entitled "Sketch of a Geognostic View of South America" and is much livelier. Even so, one can see what Darwin meant when he said that Lyell's manner of treating geology was far superior to that of any other author he had read.[36] Humboldt's sketch does not tell the person reading it what he should do to construct his own geology—except to use a barometer and a sextant. This approach produced a good deal of topography, and a lot less discussion of strata, than the British were accustomed to. It did get one accustomed to thinking of gigantic elevating forces; and Humboldt's assertion that there had been such a large-scale system of forces for several very large regions of South America, of several thousand square leagues each (a league was about 3 miles), was perhaps the high point of Humboldt's geological theorizing. He believed this was demonstrated by measurements on the direction of the strata.

What he believed to be *not* demonstrable (but he kept on insisting on it anyway) was that major geological features must somehow be related to the action of general forces in the interior of the globe. All ideas of what the relation is, he thought, are as yet vague conjectures.[37]

Ironically, perhaps Humboldt for all of his caution was reaching for too large results too soon. It was by being more localized and particular that British geology was advancing. Humboldt did not realize, perhaps, how far the men of the Geological Society of London, by ostentatiously renouncing all vast schemes and continent-wide explanations, were succeeding in building up a set of intermediate theories and dynamical explanations which were to make geology—the science Humboldt loved first, and despaired of most—the foundation for the solution of many of the problems he raised.

It seems fair, as a first approximation, to say that Charles Lyell's involvement of geology with the geographical distribution of plants and animals, both currently and in the geological past, was an actualization of the science which Humboldt had called for in 1807. If this seems too specific an influence to postulate—for the work of Cuvier and Brogniart, the men of the Geological Society of London, and Lyell's own teacher, William Buckland, lay in between—at any rate we can say that Lyell could accept Humboldt's proposed interweaving of geology with biology. He could use Humboldt's essay on isothermal lines as the basis for his own celebrated theory of climate change, which explained the tropical plants within the Arctic Circle which Humboldt and others had worried about.[38] He could use statistical methods for his Eocene-Miocene-Pliocene divisions. But there is much in Lyell which does not seem to me to be Humboldtian.

My search for the best British Humboldtian was thus stalled until work by Dr. Sandra Herbert supplied me with the key to make all of the other pieces fit together.[39] Humboldt's successor in Britain was exactly the young man who said he was inspired by Humboldt's *Personal Narrative;* who set off to South America and covered that part of the continent which Humboldt had missed; who brought back geological specimens including fossil monsters, as Humboldt and Bonpland had been the first (so Humboldt asserted) to do; who published his own personal narrative in imitation of Humboldt, and who tried to furnish a continent-wide dynamical theory for the mountains and strata of South America. It was young Charles Darwin, of course.

In his *Autobiography,* Darwin speaks of the impact that reading Humboldt had on him in much more intense terms than he uses for any other book:

> During my last year at Cambridge, I read with care and profound interest Humboldt's *Personal Narrative.* This work and Sir J. Herschel's *Introduction to the Study of Natural Philosophy* stirred up in me a burning zeal to add even the most humble contribution to the noble structure of Natural Science. No one or a dozen other books influenced me nearly so much as these two. I copied out from Humboldt long passages about Teneriffe and read them aloud on one of the above-mentioned [walking excursions].... Some of the party declared they would endeavour to go there; but I think that they were only half in earnest. I was, however, quite in earnest, and got an introduction to a merchant in London to enquire about ships.[40]

Then came the opportunity to go around the world on the *Beagle.* Darwin's copy of the Williams' translation of Volumes I–II of Humboldt's *Personal Narrative* (3rd ed., 1822), is inscribed, "J. S. Henslow to his friend C. Darwin on his departure from England upon a voyage round the World 21 Septr 1831."[41] It is not clear to me that Darwin had read all of the *Personal Narrative* at this time—the Williams translation runs to about 3,754 pages— or that Henslow gave him the entire multi-volumed set. However, Dr. Herbert informs me that the *Beagle's* library was substantial, and included the standard travel works for the places it visited; so in one way or another the complete *Personal Narrative* was probably on board.

It was also a Humboldtian idea, or something like one, which made a position available for Darwin. Humboldt observed at the beginning of the *Personal Narrative* that purely marine expeditions could not advance the progress of geology and other parts of physical science as well as travels in the interior. "It is not by sailing along a coast that we can discover the direction of chains of mountains and their geological constitution, the climate of each zone, and its influence on the forms and habits of organized beings." He was also very emphatic that a geologist should be consulted before investing time and money in mining operations. Captain Fitzroy of the *Beagle,* on his first voyage to South America, had come to the same conclusions—including the investigations of useful ores. Hence he took a naturalist-geologist-mineralogist on his second voyage to *take care of the land,* while he and his officers did the applied astronomy part of the *Beagle's* work.[42]

Darwin's imitation of Humboldt went much deeper than accepting specific scientific theories. It affected his instinctive behaviour as a scientist. He found a pattern for argument: "Why should not...," "in order to explain... we must admit," "which seems to me not unlikely." This Humboldtian method of hypothesis is basic to, is indeed at the logical heart of, the *Origin of Species.*[43] Darwin also learned from Humboldt to consider plants as almost self-conscious migrants, as "colonists."[44] Darwin may even have learned from Humboldt how to be properly emotional. Humboldt did not dictate the choice as to whether Darwin should find the peaks of the Andes or the tropical rain forests more sublime (his own choice was the cataracts of the Orinoco) but he made it clear that the traveller should find *one* of these things sublime. That is one of the reasons for going travelling. Hence Darwin wrote to John Henslow at Cambridge from Rio de Janeiro in 1832: "Here I first saw a Tropical forest in all its sublime grandeur....I formerly admired Humboldt, I now almost adore him; he alone gives any notion of the feelings which are raised in the mind of first entering the Tropics."[45]

On another emotional point, it is my feeling that Darwin's denunciation of slavery, while normal for an Englishman of the period, was moderate compared to Humboldt's continual condemnation not merely of those who continued the institution but almost of Western Civilization as a whole, which showed its intrinsic evil by having introduced it.[46]

It is, however, most obviously Humboldt's emphasis on the geographical distribution of plants and animals (he was more of a plant man, but made some observations on animals, which were in theory just as much a part of his system) that was transmitted to Darwin. Here we must be careful to consider the ways in which it reached Darwin. On the one hand, simple geographical distribution was not Humboldt's point. His approach included not only the relation of plants to each other, but also to their soil, air, sunlight, latitude, height above sea level, and so forth. This is often overlooked because Humboldt's astronomical, trigonometrical, and barometrical observations were published in separate volumes; they are not very important in the *Personal Narrative,* or at least they are not presented strikingly there. But if it is the *Personal Narrative* which Darwin read, then this simpler version of, as it were, straight naturalist's geographical distribution is what we would expect to find in Darwin. Note, however, that even at this simpler level geographical distribution as the more advanced biological science, to which the sciences of classification were preliminary and useful—much as mineralogy may be useful to the geologist.[47] Hence in approaching natural history from the standpoint of geographical distribution, one would not think of oneself as a humble worker compared to true scientists such as the anatomists and the physiologists. On the contrary, one would think of oneself as the true problem-solver who used the results of these more specialized workers.

The other way in which geographical distribution reached Darwin, was, of course, through Lyell's *Principles of Geology,* and it would take a far more detailed study than I have made to decide whether Darwin in each individual case was influenced more by Humboldt directly or more by Humboldt in Lyell.

With these warnings, I want to draw attention to a few striking passages in Humboldt. The first set of these concerns an apparently negative formulation: Humboldt asserted in the *Personal Narrative* that "the causes of the distribution of species are among the number of mysteries, which natural philosophy cannot reach," since this would require "the investigation of the origin of beings."[48] Put positively, this means that, if you want to explain geographical distribution, you must explain the origin of beings.

We remember that Humboldt said elsewhere, "In vain would reason forbid man to form hypotheses on the origin of things, he is not the less tormented with these insoluble problems of the distribution of beings." Add Humboldt's assertions that climate and other gross environmental factors are not a sufficient explanation for geographical distribution, his references to the history of nature, and his preference for developmental speculations.[49] Then the *right grouping of ideas was available in English translation* from 1821 on, for anyone who believed that the origin of beings was a mystery that natural philosophy *could* reach. And I doubt that I am the first person to realize that, when Humboldt made such statements about the impossibility of proving this or that, he didn't really mean it; he only really meant that methods used in the past were not good enough to solve such problems.

However, if Humboldt's negative formulation may have kept Charles Lyell from trying too hard to solve the problem, there was no need for Darwin to have to worry about it. His other hero, John Herschel, in his monumental letter to Lyell in 1836, opened by discussing "that mystery of mysteries the replacement of extinct species by others"—a phrase which struck Darwin so forcibly that he used it in the second sentence of the *Origin of Species:* "That mystery of mysteries, as it has been called by one of our greatest philosophers." Herschel's point was that "the origination of species" would be found to be "a natural in contradistinction to a miraculous process," that is, one suitable for scientific inquiry. But, Herschel said, "we perceive no indications of any process actually in progress which is likely to issue in such a result." Here, then, was a direct challenge to any enterprising young naturalist with Uniformitarian leanings, from England's best-known scientist: can you find the process actually in progress?

The dates are very suggestive: Herschel's letter was being discussed in Geological Society circles in 1837 when Darwin came up to London, visited Lyell, gave his first paper to the Society, and then moved to London from Cambridge, where he had spent several months after the *Beagle* came home in the autumn of 1836. Lyell used this first paragraph of the letter to support his own position; and Charles Babbage printed it, along with other extracts, in May 1837 in his *Ninth Bridgewater Treatise.* By May also, William Whewell's *History of the Inductive Sciences* was available, with its repetition of Whewell's earlier objection that Lyell's Uniformitarian geology depended on the gradual introduction of new species, yet gave no satisfactory evidence that this was true, and no indication of a process whereby it came about.

Darwin began to form ideas on the subject in March, the same month in which he moved to London and saw much of Lyell, and in which Lyell wrote to Whewell explaining his position and citing Herschel's letter in his defense, so that he would be fairly represented in Whewell's *History.* Darwin opened his first notebook on the "Transmutation of Species" (his title) in July. We know that at this time he had not encountered Herschel's paragraph, because of his excitement when he did find it in Babbage's book in December 1838—that is, two months after he read Malthus and was seeing how all of his ideas and evidence fell into place. There is no specific evidence as to when Darwin first heard of Whewell's discussion of the "Transmutation of Species" in his *History;* Darwin was not reading it in detail, with copious notes, until October 1838, so far as I can demonstrate—that is, almost directly after reading Malthus. He did look at parts of the *History* briefly, at some earlier time. Until there is a better proposal, I am going to believe that Herschel's letter, followed by the pending publication of Whewell's *History,* was the direct cause for the origin of species being an available topic of debate and consideration in just those professional circles which Darwin joined in early 1837. From Lyell and other friends he may have learned both that it was a professionally valid problem and that it was one of the ones for which Uniformitarians in particular did not have a solution. The indirect or general cause was not so much Lyell's position (perhaps he would have been willing to let the matter remain unsolved) as it was Whewell's repeated claim that the lack of a solution was one of the good reasons for rejecting Uniformitarianism altogether.[50] Darwin probably opened his notebooks and used the term "transmutation" as a direct result of looking at Whewell's *History*.[50]

Given, then, either that Darwin could have ignored Humboldt's prohibition against investigating the origin of beings on his own, or at least that in his own circle in 1837 it was being ignored by the quasi-Humboldtians he knew and respected, let me describe two sections of Humboldt which seem relevant to Darwin's problems. For one of them, there is no evidence that I know of that Darwin was inspired by that particular section; but we can see that the kind of thing Humboldt discussed was the right frame for Darwin's thought. For the other one, there is evidence that he marked the passage for one purpose, and then used it in his transmutation speculations for another purpose. In this case also, it is easier to see that Darwin was a Humboldtian than it is to prove definitely what he took directly from a passage in Humboldt.

Darwin saw just how specialized geographical distribution could be during his visit to the Galapagos Islands; but this was no more specific than Humboldt's observations on insects on the Orinoco River, where each new tributary was characterized by a new species of gnat. Humboldt's was the more complex problem; the species varied not only from tributary to tributary, but also from hour to hour of the day, by altitude above sea level, and by distance from water.[51] By comparison, Darwin recognized his classic example in a simple case, where the geographical limits were so clearly and simply the stretches of water between the islands. What Humboldt was interested in proving was a negative: that climate alone is not a sufficient explanation. Darwin benefitted by being free of the old theories which Humboldt had swept away; he could speculate on what else could have caused the observed diversity of species.

Note, however, I have said only that the simplicity of the Galapagos Islands situation made it easy for Darwin to see and think about the problems, not that he concluded from it that geographical isolation was the key to, was necessary for, the formation of species. This would have been a one-variable solution in conflict with all good Humboldtian instincts, and it would have contradicted another passage in Humboldt, which could be one source of Darwin's jaguars. Humboldt insisted that, in discussing and distinguishing among varieties and among species, we should consider not merely observable structure (as the ordinary classifier might) but also living habits. "In all the mammifera, and particularly in the numerous family of the apes we ought, I believe, to fix our attention less on the passage from one colour to another in the individuals, than on their habit of separating themselves, and forming distinct bonds." One needed to live long in tropical countries, he observed, to be able to distinguish with certainty the varieties and species of jaguars.

Paul Barrett and Alain Corcos have noted that Darwin did indeed make a marginal note in his copy of Humboldt concerning the jaguars—not this point about the jaguars, but Humboldt's description on the previous page about the thousands of them which were killed for their skins. Darwin's marginal scribble was on such slaughter as a positive check to overpopulation.

In Darwin's second notebook on the transmutation of species in 1838, Humboldt's idea of habits as a key to species came out as: "Instinct goes before structure... hence aversion to generation, before great difficulty in propagation"; and as (in a page excised later for use in preparing his big book): "My definition of species has nothing to do with hybridity, is simply an instinctive impulse to keep separate, which no doubt [can] be overcome, but until it is these animals are distinct species." What suggests that these opinions of Darwin are related specifically to the passage in Humboldt, rather than to other authors expressing a similar idea, is that shortly after the first one in the notebook comes Darwin's interesting use of jaguars, as an example of how changed habits have led to changed structure. The jaguars are, we might say, a trial run for the celebrated swimming bear of the *Origin*, which was an "improvement" of Lyell rather than Humboldt:

> for instance, fish being excessively abundant & tempting to the Jaguar to use its feet much in swimming, & every development giving greater vigour to the parent tending to produce effect on offspring—but *whole* race of that species must take to that habitat.—All structures either direct effect of habit, or hereditary & combined effect of habit—perhaps in process of change—Are any men born with any peculiarity, or any race of plants.—Lamarck's willing absurd [because] not applicable to plants.

If I have correctly identified the passage in Humboldt which may have been one of Darwin's starting-points, then it is clear that it was, while very suggestive, only the starting-point; Darwin was determined to do *something* with Humboldt's suggestions. Of course it may seem that I am pushing the evidence too hard in this specific case. There are jaguars in the first transmutation notebook here and there—not very interesting ones, to me, but they serve to remind the historian that Darwin had been in South America himself. Perhaps Darwin's jaguars were primarily his own, and he noted them in Humboldt only in passing. The most I can really conclude is that, if Darwin was not thoroughly influenced by Humboldt, he somehow had or obtained a discrimination in observing remarkably like that of Humboldt.[52] He had, controlling his thought, a *topographical vision* of the world, its organisms, and its history, like that of Humboldt.

Humboldt was the complete professional—he did nothing but science (except when drafted by his sovereign, the King of Prussia)—and in Paris became, in effect, the opposite center to Cuvier. Cuvier represented official science and conservative politics, Humboldt represented what we might call the liberal spirit. At least this is how it seemed to a number of visiting Englishmen after 1815. This "liberalism," though, had nothing to do with opposition to government support of scientific projects. Humboldtian expeditions could be costly; and directly or indirectly Humboldt extracted more money from monarchs for science than perhaps any other man ever had done. Real monarchs were easier to handle, it would seem, than representative assemblies. The Elector of Hanover provided his University of Göttingen with new magnetic equipment several years before, as King in England, he was allowed to consent to the requests of his advisers to help Greenwich Observatory catch up.

Typically, it was not the old official academies of science alone, but also newer organizations which represented the Humboldtian spirit. Thus in September 1828 the German *Naturforscher,* or Society of Naturalists and Natural Philosophers, had its triumphantly successful meeting at Berlin, and it was particularly distinguished by having obtained Humboldt as its president for the year. He was a greater attraction than the scientific sessions, if we judge by attendance figures. There were about 460 members present; at one dinner some 850 people attended; but at Humboldt's *soirée* there were 1200 guests, including His Majesty the King of Prussia. In his opening speech to the Society, Humboldt demonstrated how he got along with monarchs, saying of the King:

> ...he modestly and in simple grandeur, adorns every year this royal city with new treasures of nature and of art [i.e., for the museums]; and what is of still greater value than the treasures themselves—what inspires every Prussian with youthful strength, and with an enthusiastic love for the ancient royal family—that he graciously attaches to himself every species of talent, and extends with confidence his royal protection to the free cultivators of the understanding.[53]

But what I am primarily concerned with are Humboldt's views on science, here stated as officially as possible. (How else can a free scientist make an "official" statement of his views better than as the president of his kind of scientific group?) "Knowledge of truth and performance of duty are the highest objects of morality," Humboldt said. "Emulation in mental struggles has called forth" the best of science, art, and the humanities. The chief object of this assembly is not to present innumerable memoirs. It is "to bring those personally together who are engaged in the same field of science. It is the immediate... interchange of ideas, whether they present themselves as facts, opinions, or doubts." This was the method used by the ancient Greeks. "The discovery of the truth without difference of opinion is unattainable, because the truth in its greatest extent can never be recognized by all, and at the same time." When the disputes of the learned are finished, so will be the continued advance of science.

We have properly linked together in one organization all the branches of science. "Important as it is not to break that link which embraces equally the investigation of organic and inorganic nature," still we have specific meetings for single branches as well as the general meetings for all. "For it is only in such contracted circles—it is only among men whom reciprocity of studies has brought together, that verbal discussions can take place."[54]

Humboldt is notoriously difficult to translate because of his easy style, constant undercurrent of wit, and precise use of technical terms; this translation of his speech by Charles Babbage is no more inspiring than most such, of the period or of the present. Nevertheless I have used this translation, and have taken my account of the meeting from Babbage's article, for a reason which should be clear to readers of Chapter Five. This is the article which appeared in the *Edinburgh Journal of Science* in 1829, as one of the events leading up to the founding of the British Association for the Advancement of Science. What you have read, therefore, are the opinions of Humboldt as they were presented to many of the participants in that affair. To me, at any rate, these opinions of one of the world's two leading professional scientists seem to correlate with, and probably helped to increase the power of, the opinions of one particular individual in that sequence of events—John Herschel.

One characteristic of some leading British scientists after 1800 is so obvious to many historians, and yet so unexplained, that I am tempted to classify it as Humboldtian; it fits in so well with his general approach of debunking the simple theories of the immediate past. Actually, I imagine, the attitude had native sources as well, which were reinforced by Humboldt. Expecially in philosophy it had Scots sources. The characteristic is that of taking a light, sophisticated view of hypotheses about fundamental entities. For example: atoms may well be; atoms may well not be; but the hypothesis is an interesting one to play around with. Or, in general, as John Herschel taught: "Hypotheses have often an eminent use: and a facility in framing them, if attended with an equal facility in laying them

aside when they have served their turn, is one of the most valuable qualities a philosopher can possess." It is safer, no doubt (said Herschel) to
confine oneself to laws and higher laws.[55]

Thus many chemists, taking the safe route, agreed to believe in the
Law of Definite Proportions, but not in the rather naive atoms John Dalton used to explain it. More sophisticated philosophers (including Herschel) rather liked atoms, but they sometimes liked Boscovich's atoms
even though such atoms did not lead to any definite numerial results.
They were rather like Humboldt, who yearned for a developmental explanation for the origin of things, stars as well as organisms, but kept remind-
ing his readers that there was not really enough evidence for such beliefs.
In the same way, Herschel was delighted by Prout's hypothesis that all
atomic weights were simple multiples of one smallest weight—"it opens up
views of such importance as to justify any degree of labour and pains in
the verification of the law"—but had to insist that chemical analysis was
simply not accurate enough to demonstrate the validity of the belief.[56]
This attitude, not of disbelief so much as sitting loose to the basic mechanisms proposed by their predecessors, was of considerable importance, I
imagine, in the initial acceptance of the Young-Fresnel wave theory of
light. If everyone around 1830 had really had to *believe* in an elastic solid
ether co-extensive with the universe, there might have been trouble.

Perhaps related to this increased sophistication with respect to hypothetical entities was a widespread feeling among Humboldtians that the
laws they sought could not be obtained as direct deductions from the established truths of the past. This was true even in a subject in which Newtonian law clearly applied: the tides. There were so many varied and
arbitrary factors involved—depth of water, width of channels, shape of the
land, etc.—that the application of Newtonian law gave only general guidance, not an accurately verifiable and predictive mathematical expression. In a still more complicated subject like meteorology, it was still more
obvious that basic laws of laboratory natural philosophy would not be
sufficient, by themselves, to enable one even to notice, much less predict,
all the variations which took place.

The Humboldtian was interested, therefore, in new approaches con-
formable to the complex subject itself. These would use the simpler laws
of the kind developed in the past, drawn from not one but several
branches of the old science. No real phenomenon is explainable merely by
chemistry or merely by mechanics or merely by anatomy. Humboldt's
"geography of plants" was such a superior real subject, using such simpler
studies as classification, vegetable physiology, climatology, topography,
etc. So was Herschel's meteorology; or, for that matter, so were "laws of
the lowest degree of generality" such as Well's theory of dew.[57]

It is possible (although this is only a speculation) that such a situation led to an increased interest, in England at any rate, in methods of discovering or formulating valid laws in science. If there was such an increase, it would help explain why more people were discussing Bacon than formerly (if there were more people doing this), and hence why modern historians have thought that the early 19th century was a period of "Baconianism." To start a discussion of how to arrive at scientific laws, an Englishman might naturally start with Bacon; but he might not stop there.

In what I have said so far, it may appear that telling a Humboldtian from a non-Humboldtian is a rather iffy matter. Actually, in most cases the reverse is true. Humboldtians are easy to spot, because of the two technical characteristics of Humboldtian science. One was a conceptual tool so common these days that we may forget that it was startling in its day. The other was a finicky fiddling with instruments which we also forget today; that is, scientists still spend a major part of their time doing it, but they don't always emphasize this in reporting their results.

Humboldt, the theorist, was quite clear that "observations are not really interesting, except when we can dispose their results in such a manner as to lead to general ideas."[58] Recording numerical results in tables of numbers, which Humboldt often did, was nevertheless not always such a manner. Geographically distributed data can best be displayed on a map. But not just any kind of map. Humboldt came to stand for what I call the "iso-map," a map showing lines of equal something: isothermals, isobars, isotherals, isodynamics, isogonics, or iso-anything you like: "these magnificent lines," as George Harvey called them in the *Encyclopaedia Metropolitana's* article on "Meteorology" in 1832.

Not all of these iso-maps led to something in particular, but the impetus they gave to the display of data in visual and then in graphical form led to all sorts of important things. It is difficult for us to remember just how novel graphs were in the 1820's, especially as real tools in science, not just illustrative pictures in a hypothetical book by Boscovich. If there is anything that makes it clear that Lagrange, the great exponent of pure analysis in mechanics, was not entirely successful even among his admirers, it is the development of graphs. Lagrange boasted that his mathematical methods made the use of diagrams unnecessary. But many of the early Lagrangians in England were convinced visualizers as well. They liked a physical representation of the mathematics they were using. And the line from tides to graphs of tides to harmonic analyzers and synthesizers—or, as we would say, to analog computers—is fairly clear even if not fairly simple.

If we add in straight mapping problems, we also can go fairly directly from geodesy to much of topology, that part that is related to Gauss. Humboldt's terrestrial magnetism interests led to the development of absolute measurements. And the line to cosmic rays is quite direct, though time-consuming. So the off-shoots in the mathematical-physical sciences from the study of geographically distributed data have been quite important indeed.

In short, if you find a 19th-century scientist mapping or graphing his data, chances are good that you have found a Humboldtian. Chances are good also that he will have spent considerable effort in mapping the wrong thing: maximum temperature instead of mean temperature, wind velocity instead of wind volume, latitude of plants instead of height above sea level. Indeed, any *one* set of coordinates is almost surely the wrong answer, a point Humboldt insisted on. A typical Humboldtian parameter—mean minimum temperature, let us say—varies not only with latitude and longitude, season, time of day, height above sea level; it also varies with geological epoch. And vegetation, for example, may be responsive not to temperature but to rate of change of temperature.[59]

If graphs and iso-maps were characteristic tools of the Humboldtians after about 1820, improving the accuracy of instruments and worrying about sources of error in measurement was a concern inherited from earlier times which the Humboldtians went a long way towards dealing with satisfactorily. I have not said, and do not believe, that Humboldt invented all of the pieces of Humboldtian science; even the iso-map goes back, in some sense which I have only looked at, to Edmond Halley. (Humboldt did, I think, put the pieces together rather nicely). Reading the *Personal Narrative* suggests to me that what I had already concluded for the history of geology is also true for the history of science as a whole: the most important now-neglected scientist of the latter 18th century was Horace Benedict de Saussure. Indeed, he may have been one of the three or four most important scientists of the latter 18th century, neglected or not.

But de Saussure is not the only one of importance to my story before the emergence of Humboldt; one is tempted to go back to Daniel Bernoulli's prize-winning essay for the Paris Academy of Sciences in 1743, "Mémoire sur la manière de construire les boussoles d'inclinaison."[60] Nevertheless, interest in a subject is not enough. It would seem that Humboldtian science, the science of measuring world-wide variables, *could not come into existence even as late as 1770 because the available instruments were still too crude.*

In the general interest in voyages, scientific expeditions, improvements in navigational instruments, etc., which make the 18th century such an enjoyable period for the historian of science, the stimulus of Cook's voyages in particular for the development of Humboldtian science seems clear. Humboldt himself was turned on to voyaging by George Forster, who went on Cook's second voyage. But the improvements in portable instruments are made too little of in ordinary histories, because the basic idea of the instrument was already in being.

One thing that emerges from the study of Humboldt and other travelling scientists after 1800 is their contempt for scientific "knowledge" based on measurements made 20 or 30 years earlier. Humboldt contemptuously noted that on the trip to South America, because of his instruments "we were surer of the position of the vessel than of that of the land to which we directed our course;" the French, Spanish, and English charts all gave different positions. Later he sarcastically noted two mountains which "wander on our maps" through 2½° of latitude. He spoke disparagingly of Chappe's 2½-foot quadrant used for transit-of-Venus measurements in 1769; a modern sextant of 5-inch radius, he thought, gave more accurate results.[61] The work of Ramsden and his successors, and Harrison and his successors, had made all previous measurements by travellers questionable. We usually hear of sextants and chronometers in the context of navigation; but sextants by Ramsden, or chronometers by Berthoud may have had a marginal impact on the history of commerce. In contrast, land travellers could re-locate the boundaries of empires, position mountain-chains, unravel river systems, with these devices whose extreme dimensions were 12 inches or less.

Even the most detailed instrument study can benefit from a knowledge of Humboldt's requirements. Thus we find out, from the standard history of the barometer, that Gay-Lussae was trying to improve the syphon barometer, but not why. The barometer was the device for determining height above sea level, one of Humboldt's prime variables. Fortin had assembled all of the improvements made on the mercury barometer—including Ramsden's decisive improvement of the reading scale—into one useful portable instrument. According to Humboldt, it was accurate in the torrid zones, where the barometer is very steady, to about 4 or 5 toises [26 to 32 feet]. This, he felt, was accurate enough to make feasible the building of an Atlantic-to-Pacific canal across Nicaragua. Without a Fortin barometer, Humboldt's own height measurements varied by as much as 13 or 14 toises [83 to 90 feet].

Clearly, improvement was still desirable. The next major problem seemed to be the effects of capillarity in the mercury barometer. Humboldt's friends in Paris, Arago and Gay-Lussac, tackled the job.[62]

It was, then, not the invention of totally new devices so much as it was the improvement of the barometer, thermometer, hygrometer, dip needle, sextant, and other known devices, which made a science of the whole world possible. Investigation of this topic would lead us to a very mixed world of experience, where ideas as to what to measure and ideas as to how to improve needle pivots were completely intertwined. We should remember, for example, that the French engineer Coulomb sent a memoir to the Paris Academy of Sciences on the best manner of making compass needles; those historians who connect this with his subsequent development of the torsion balance would seem to have probability on their side. One of the most intriguing inhabitants of this mixed world was Borda, explorer and instrument-maker, intriguing both in his achievements and in his possible shortcomings. Borda will take his proper place in history only when he is dealt with by someone who is willing to get his mind dirty with problems of how well a needle will dip, or if a barometer will break because of the expansion of its wood support, in non-European humidities.[63]

To be sure, humidity questions were ones which the British had faced in the 1760's, in connection with Mason and Dixon's geodetic work in the North American colonies; and in geodesy Borda's devices were perhaps less satisfactory than those of Ramsden or of an ex-British subject, David Rittenhouse, in the 1780's. Thus, by looking at problems related to instruments, we are led quite naturally back to those enterprises in applied astronomy which Nathan Reingold has emphasized.

In general, in the late 17th century and even until the mid-18th, a 2½-foot quadrant was considered accurate enough for most latitude determinations in the field. But the development of the zenith sector telescope made it obsolete. The zenith sector was a long, not especially powerful, telescope mounted so as to point directly overhead (at the zenith) and movable only along a rather small arc or sector away from the zenith. Originating probably from an idea of Robert Hooke, the zenith sector was made famous by James Bradley, who with it discovered stellar aberration, 1725–29. After mid-century, use of a quadrant in field work is a sign that the expedition was not quite of the first rank. Mason and Dixon had available a quadrant bought for New Jersey, but it was only accurate to 30 seconds, and they used their zenith sector by Bird instead.

If we take as our base Greenwich Observatory, then by reading the preface to the Reverend Nevil Maskelyne's *Astronomical Observations Made at the Royal Observatory* covering the years 1765 to 1774, and by browsing through it and his *Tables for Computing the Apparent Places of the Fixt Stars, and Reducing Observations of the Planets* (London, 1774), it is apparent that the Astronomer Royal was much concerned with locating and estimating the extent of instrumental error. Whether or not he succeeded is another matter; he had, as far as I can see, no general approach to or theory of the subject. His general estimate, however, if my browsing among his figures was adequate, seems plausible: "The place of any heavenly body may be always found by them [his instruments] within ten seconds of a degree, both in Longitude and Latitude, and generally much nearer."[64]

An actual problem which faced scientists for a hundred years or more can be stated in simple form for astronomy (and then applied to other sciences) something like this:

Given that Bradley's zenith sector by Graham, altered by Bird to Maskelyne's specifications,[65] and mounted on the wall of Greenwich Observatory, where it can be checked against the great 8-foot quadrant by Bird, has an accuracy within the limits given by Maskelyne for determining the latitude of Greenwich Observatory, what is its accuracy for determining the latitude of Lancaster, Thomas Penn's colony in North America?

The answer is: zero. A zenith sector mounted on the wall of Greenwich Observatory cannot determine the latitude of Lancaster at all. All of the genius of all of those men has been devoted to producing an instrument which is useless—for our purpose.

One solution, known to the French expedition to Lapland, is to procure a *portable* zenith sector, which can be used as a latitude instrument in Pennsylvania and can also be brought home and compared to a primary standard there (or even be used as the primary instrument there). This requires that the instrument not be affected by the voyage; and it requires a method for determining that it is not so affected. One finds, therefore, scientists carrying things from and to places: barometers from the Royal Society's barometer in London to somewhere else and then back to London; gravity pendulums to India, and back; chronometers all around the world and back to the starting-point. Not one chronometer—3 chronometers, or even (on the voyage of the *Beagle*) 24 chronometers.

Another solution is to build a zenith sector in Pennsylvania, and be knowledgeable enough both of theory and of practice to be reasonably sure that your primary determinations are accurate enough to be better than the errors which would be introduced by carrying a device across the ocean for direct comparison with the results obtained by the Royal Observatory's own instruments at Greenwich. If your device is itself portable, then you can measure the *difference* in latitude between, say, Natchez and Lancaster, very nicely.

Where there is a developed body of theory, you can try to check your accuracy by starting with the observations at Greenwich and deducing what should be observed, given your situation relative to Greenwich. The difference between that and what you do observe might then be an indication of inaccuracies in your instruments, or poor observing techniques, or both. What happened in the particular case I am interested in, however, that of Andrew Ellicott in Lancaster, is that by 1806 he was able to conclude that the major source of inaccuracies, for the longitude at any rate, was *in astronomical theory,* not in his instruments or techniques.[66]

It would be tempting to develop an approach to the history of physical science in America, starting with the standards maintained by Maskelyne at Greenwich and transferred to America by Charles Mason and

Jeremiah Dixon in their astronomical and geodetic enterprise of 1763–68. Fortunately, articles by Cope and Robinson give us an elaborate treatment of the Mason-Dixon affair and indicate its relation to the Royal Society and some of the other expeditions in which it was interested.[67] The line would then run through David Rittenhouse (primarily) to his disciple Andrew Ellicott, who by 1800 was the center of a loose network of like-minded men, such as Simeon de Witt, Andrew Porter, Robert Patterson, William Dunbar, and Jose de Ferrer.

Such a study would dispense with the conception of a "colonial" science which was backward and unsophisticated. Whatever may have been the situation earlier, Rittenhouse's zenith sector telescopes were apparently at least the equals of ones imported from Europe; and Rittenhouse realized the importance not only of temperature but also of humidity and barometric pressure for the accuracy of instruments, given the extremes of American weather. When I first announced, in my talk of 1966 on Rittenhouse, that his large zenith sector was in my opinion the most accurate instrument of any kind made in North America before 1840, I did it to stimulate a search for counter-examples. I have had no rival candidates proposed either by my Smithsonian colleagues or by anyone else. The sector was good to two seconds of arc in latitude measurements when used with great care by Ellicott, according to my estimate; five seconds was a more routine result. The District of Columbia (which Ellicott did the boundary survey for, using the sector) is, as a matter of fact, about two seconds (two hundred feet) off.[68] Ellicott improved on the methods of Mason and Dixon, improved one of the methods of calculation of Newton, and was as aware of and interested in dealing with error as was Maskelyne himself.[69] In short, these men were alert practitioners of applied astronomy, carefully adapting it to the American environment—in which not only extremes of humidity and pressure but also fires lit to drive off mosquitoes, and impertinent Indians leaning against a support, affected instrumental accuracy more than in England. Europeans had heretofore been much aware of the problems caused by temperature change; that was not enough, in the New World, with its great humidity changes.

The great distances in the New World led Rittenhouse (apparently) to one new device, the surveyor's compass adjustable for the variation in magnetic declination; this variation was quite noticeable along such a distance as that from one end of Pennsylvania to the other. This device is often called the "vernier" surveying compass; but the adjustment is not in the first instance for increased accuracy in reading off a single observation; although the compass may have a vernier scale as well.

What the Americans *were* backward in was making optical glass. The lenses were the part of his zenith sectors which Rittenhouse imported from Europe. Powerful achromatic telescopes were usually imported also. But the problem of producing optical glass has its own history, separate from "colonial" backwardness. By the 1820's, England herself was backward in optical glass; but this did not make her astronomy unsophisticated.

The myth of American lack of sophistication suited some British tastes after the American Revolution; it was expounded very well in Henry Brougham's review of Volume V of the *Transactions of the American Philosophical Society* for the new *Edinburgh Review* in 1803.[70] To be sure, Brougham's opening target was the radical English chemist, Joseph Priestley, whose papers in the *Transactions,* Brougham said, "are distinguished, if possible, by a more eminent want of scientific precision, than even his periodical volumes of chemical experiments." But he was equally scornful of American attempts at science; and noted by way of dismissal that "150 [pages] consist of journals kept by a person sent to make meteorological and astronomical observations, that is, notes of the weather, and of the geographical position of certain places in the territory of the United States."[71]

These 150 pages are, as a matter of fact, Andrew Ellicott's reports on his expedition to survey the American-Spanish border in 1796–1800, perhaps the best and probably the most ambitious enterprise involving the accurate use of instruments in North America during that period. It is interesting to note that the new astronomical circle by the great London craftsman Troughton turned out to be more trouble than it was worth; whereas Rittenhouse's two zenith sectors, in Ellicott's hands, were stellar performers, as was a 7-inch sextant by Ramsden.[72]

Brougham began his review by saying, "This volume, one of the very few that ever issue from the American press, will not, we apprehend, repay the labors of him who may be induced to wade through it." He ended with "the purpose of stating and exemplifying a most curious and unaccountable fact—the scarcity of all but mercantile and agricultural talents in the New World."[73] This is surely the voice of the hurt child, pretending that what he lost wasn't worth anything anyway. Brougham, of course, is notorious in the history of science for having scorned both William Herschel and Thomas Young. His view of American science shows about the same level of scientific awareness. Why it has been accepted by so many American historians is, I confess, not at all clear to me.

I am not, of course, saying that there was a large number of sophisticated scientists in the new United States, only that there was at least one group in the advanced school of applied astronomy related to Greenwich Observatory. A comparison of Ellicott's Spanish border expedition with the roughly contemporary trigonometrical survey of Roy in England would show, I think, that the differences were in the requirements of the task, and in the physical environment (Ellicott had to make some observations in swamps, for example), not in the scientific acuteness of the senior scientists involved. And there may have been other American scientists or groups of scientists at this level; I have investigated only the group which corresponded with Ellicott and published in the *Transactions* of the American Philosophical Society.

The men whose detailed reports I have looked at, from Maskelyne on, were clearly conscious of the problems of error and the probable limits of error. Just as clearly, they did not know what to do about it; and, to me, all of their estimates of accuracy are unconvincing. If we jump to the 1830's, I would guess that the basic problems had been solved, both in locating errors in the most refined astronomical circles, and in handling the theory of errors by Gauss's or someone else's techniques. One still worries about error, but one knows how to worry about it.

How this was done, I do not know. It would be a good subject for a detailed narrative involving both practice and theory. If, for example, Ellicott and others by 1800 or so could readily out-distance lunar theory with their observations, then we can see why English physical astronomers in 1810 or 1820 were still making their reputations with what might seem to an outsider as rather esoteric refinements of this part or that part of lunar theory.

That Gauss was involved in the search for a theory of error certainly makes it look like a Humboldtian affair. Gauss, with his magnetometers and theory of terrestrial magnetism and correlated network of observers, with his surveying heliotrope and his non-Euclidean geometry inspired by the mapping requirements of geodesy, was to Humboldtian science what James Clerk Maxwell was, later, to physics. He was the fully-formed professional.

At any rate, we can say that what Ellicott was doing was the kind of thing Humboldt appreciated. It should not surprise us to find that Humboldt cited Ellicott's work in his *Personal Narrative;* cited, indeed, just that work which Brougham dismissed so contemptuously in the *Edinburgh Review*. After Humboldt's visit to Washington, D. C., in 1804, and his discussions with Thomas Jefferson, among others, it is not surprising that he went to Lancaster, Pennsylvania, and spent some time with Ellicott. Nor is it surprising that young Meriwether Lewis visited Ellicott in 1803 as one of the preliminaries for his western expedition with William Clark and a contingent of soldiers.[74]

The Lewis and Clark expedition was a crude affair scientifically, by the best standards of the time. There might be many reasons for this. One, perhaps, was the scientific amateurishness of its sponsor, Thomas Jefferson. Certainly Jefferson's conception of science, while perhaps laudable in a politician, lagged well behind that of Rittenhouse and his disciples—it was full of the simplicities of 18th-century quasi-philosophic speculation of the kind Humboldt despised. Jefferson stood for what Humboldt was trying to abolish. Lewis and Clark did very well with the equipment provided. Jefferson did not (perhaps he could not) provide them with the best that scientific travellers and explorers could use.

On Rittenhouse's death, the American Philosophical Society chose Jefferson as his successor as president. It is probably only symbolic that American science, then at a peak, sagged slowly downwards during Jefferson's long tenure of the presidency until 1815.

After the War of 1812, physical science revived slowly, in part by a blending of modern French personnel and methods with modern British ones.[75] The old Rittenhouse men encountered the new applied astronomers on the commission to carry out the articles of the Treaty of Ghent (24 December 1814, ending the War of 1812) which pertained to determining the U.S.–Canadian border. Ellicott wrote to his wife from Albany, New York, on July 24, 1819:

> Since I came here, I have had much conversation with my old friend and astronomical companion S. DeWitt, surveyor gen[1] of this State, who is a man of science, and a good practical astronomer: he informs me, that he spent several days with Mr. Hastler and the British astronomers on the boundary last summer; but could not entirely comprehend the nature of their operations, and (between ourselves), he assured me that as far as he could comprehend them they appeared better calculated for expense than accuracy.

The next week he was in a position to see for himself, and wrote to his wife again, from Burlington, Vermont:

> M[r]. Hastler has taken away every instrument with which he was furnished by the government to determine the boundary; this he done contrary to the opinion of our Commissioner M[r]. Van Ness. By this conduct of M[r]. Hastler, we should have been without any instruments this season had I not fortunately brought on my own....
>
> As to our business I can say nothing at present, and candidly confess that I do not yet comprehend the method pursued by the British astronomer and M[r]. Hastler, it is different from anything I have yet seen or heard of, not more than one observation in ten can possibly be applied to the boundary—those that can are probably good, but their mode of calculation is laborious in the extreme.[76]

Ellicott was probably not the first, and certainly was not the last, American to complain of the methods and manners of Ferdinand Hassler, with his expensive and time-consuming new European standards of finicky accuracy. Nor was Hassler at the U.S. Coast Survey the only such extravagant scientist. The new scientists wanted to know a great deal; wanted to know it accurately; and were not hesitant about getting government to pay for their ventures, well beyond the limits of political or commercial applicability.

103

Even in boundary surveys, then, the new Humboldtians were not merely a continuation of the applied astronomy tradition of the 18th century. There was a new spirit involved, a kind of discontinuity which the non-meeting of Ellicott and Hassler in 1819 may be taken to symbolize.

As this example of applied astronomy shows, the concept "Humboldtian science" contradicts conventional ways of looking at the history of science as developments in discrete special subjects, each with its continuous comprehensible internal history. In such a traditional arrangement, Humboldt did not do a great deal, since interest in many of the special subjects which made up Humboldtian science was in existence before Humboldt. But then I am not very fond of the traditional arrangement. If pressed to analyze the new elements (but I really prefer to describe, as I have been doing), I would isolate four elements:

1. A new insistence on accuracy, not for just a few fixed instruments, but for all instruments and all observations.

2. A new mental sophistication, expressed as contempt for the easy theories of the past, or as taking lightly the theoretical mechanisms and entities of the past.

3. A new set of conceptual tools: isomaps, graphs, theory of errors.

4. An application of these not to laboratory isolates but to the immense variety of real phenomena, so as to produce laws dealing with the very complex interrelationships of the physical, the biological, and even the human.

Immense variety, interest in individual detail, complex inter-relationships, refusal to accept the abstractions of the past: all this may sound like Romanticism. I think it was; or rather, I think Romanticism was part of it. But remember that the Humboldtian did not renounce Newton's *Principia* in favor of the Queries to Newton's *Opticks,* did not look for a soul or vital force in the universe, and did look for mathematically precise laws. The Humboldtian was almost bursting with the desire to produce new abstract theories, but they must be theories that account for particular details, if possible numerically accurate details.

Here we have a hint that Charles Darwin was not, in the last analysis, a complete Humboldtian. His speculations had the plausible general logic and lack of exact quantitative verification characteristic of the 18th century, not of the early 19th century. Darwin in his formulative period (1837–42) almost never considered arithmetic; his experiments, in mid-career, were little ones, not designed to verify his whole theory but to see if one particular postulated mechanism did really exist and have some effect; and his great venture into a particulate theory of inheritance, "pangenesis," was a thorough failure.[77] Darwin's intuition did not have the drive for numbers and measurement that Humboldt's did; and it is probably just as well. Numbers might have been the wrong emphasis for natural history at that time, just as they were for geology at that time. The geographical scope and the complex interrelationships of the Humboldtian vision were enough for Darwin and, wisely, he simply ignored the part that would have been a burden to him.

To summarize my primary assertion in this chapter: the great new thing in professional science in the first half of the 19th century was Humboldtian science, the accurate, measured study of widespread but interconnected real phenomena in order to find a definite law and a dynamical cause. Compared to this, the study of nature in the laboratory or the perfection of differential equations was old-fashioned, was simple science concerned with easy variables. Insofar as you find scientists studying geographical distribution, terrestrial magnetism, meteorology, hydrology, ocean currents, the structures of mountain-chains and the orientation of strata, solar radiation; insofar as they are playing around with charts, maps, and graphs, hygrometers, dip needles, barometers, maximum and minimum thermometers; insofar as they they spend much of their time tinkering with their instruments and worrying about error: they are not Baconian, they are not backward, they are not colonial, they are not doing that merely because they are amateurs and calculus is too difficult. They are eagerly participating in the latest wave of international scientific activity: they are being cosmopolitan.

We can see why Humboldt has impressed a historian like Goetzmann as one for whom "everything was relevant, everything was important." This is not correct; only those things which could be observed accurately and become part of a general theory were important. But Humboldt saw so many areas for theory that it would take many scientists many years to cover them all with the accuracy Humboldt insisted on. As a program for one man, it was hopeless. As a program for ten or twenty years, it was gigantic. But as a program for the world—well, the hosts of scientists *were* forthcoming, and the expeditions *did* take place, and Gauss and Darwin and others *did* propose theories. Of course I am not saying that one man "caused" all of this activity; but he did nudge it in particular directions.

It was only toward the end of the 19th century, after physics and laboratory physiology had risen to their position of dogmatic self-assurance, that this kind of activity was in its turn seen as old-fashioned, that Gauss's theory of terrestrial magnetism was judged not to be a theory at all, and Darwin not a professional biologist. That is where the "Baconianism" advocates are, at the point of view of 1900. But that does not tell us what it was like in 1830, to someone who wanted to join the avant-garde.

Notes to Chapter Three

My first idea for the analysis of Humboldt came from hearing the paper of Dr. Hans Baumgärtel on Humboldt at the New Hampshire Conference on the History of Geology in September 1967, now published as "Alexander von Humboldt, Remarks on the Meaning of Hypothesis in his Geological Researches," in *Toward a History of Geology,* ed. C. Schneer (Cambridge, Mass., 1969), pp. 19–35. The first version of this essay was given to the seminar conducted by the Smithsonian Institution's History of Science and Technology Department in August 1969. The second version was a talk to the History of Science Society in December 1969. This third version is greatly expanded and incorporates ideas from my talk on David Rittenhouse to the History of Science Society in December 1966. I am indebted to a number of discussions with Dr. Nathan Reingold of the Smithsonian, and to encouragement from Professor Francis Haber of the University of Maryland.

1. John Herschel, *Preliminary Discourse on the Study of Natural Philosophy* (London, 1831), pp. 104 ff.; W. F. Cannon, "John Herschel and the Idea of Science," *Journ. Hist. Ideas, 22* (1961), 222.

2. *Don Juan,* Canto XV, stanza XVIII.

3. Nathan Reingold, "Alexander Dallas Bache," *Technology and Culture, 11* (1970), 163–177.

4. Charles Minguet, *Alexandre de Humboldt, historien et géographe de l'Amerique espagnole, 1799–1804* (Paris, 1969), pp. 80–81. My other biographical information is also taken from this book. See also the splendid personal biography by Douglas Botting, *Humboldt and the Cosmos* (New York, 1973).

5. Alexander de Humboldt and Aimé Bonpland, *Personal Narrative of Travels to the Equinoctial Regions of the New Continent During the Years 1799–1804,* tr. Helen Williams, 7 vols.-in-9 (London, 1814–29), I, 34; III, 513; esp. IV, 201. All of my citations are to this translation of the *Relation historique du voyage aux regions equinoxiales du Nouveau Continent,* 3 vols. (Paris, 1814, 1819, 1825), since the translation was a major vehicle for carrying Humboldt's approach to English-language people.

   The copy I cite is composed of Vol. I (1814); II (1814); III, 2nd ed. (1822); IV, 2nd ed. (1825); V, 2nd ed. (1827); VI, 2nd ed. (1826 [sic]); VII, 2nd ed. (1829). The copy preserved in the Darwin collection at the Cambridge University Library consists of I–II, 3rd ed. (1822); III, 2nd ed. (1822); IV (1819); V (1821); VI (1826); VII (1829).

   It should be noted that the *Relation historique (Personal Narrative)* was only three of the thirty quarto and folio volumes which made up the complete *Voyage aux regions equinoxiales du Nouveau Continent.* Sixteen of the volumes were on plants and the geography of plants. Many of the botanical monographs were by Aimé Bonpland, Humboldt's companion; some were by both of them. The *Relations historique* is written in the first person as by Humboldt alone. Hence it is hereafter cited as Humboldt, *Personal Narrative.*

6. Humboldt, *Personal Narrative,* II, 122–124.

7. Humboldt, *Personal Narrative,* II, 6.

8. Humboldt, *Personal Narrative,* I, 32–39.

9. Humboldt, *Personal Narrative,* III, 6–7, 25, 97.

10. Humboldt, *Personal Narrative,* I, iii.

11. Nathan Reingold, "Cleveland Abbe at Pulkowa: Theory and Practice in the Nineteenth Century Physical Sciences," *Archives internationales d'histoire des sciences, 17* (1964), pp. 141, 143.

12. William Goetzmann, *Army Exploration in the American West* (New Haven and London, 1965), Index; pp. 18–19 and 422–423.

13. Goetzmann, *Army Exploration,* p. 18; see also p. 451. In his later *Exploration and Empire* (New York, 1967), Goetzmann nowhere, as far as I can see, discusses Humboldtean science as such. He takes it as known that it is romantic, cosmic, wide-ranging, non-specialized, and stressing "enormous" collections (pp. 5, 235, 316).

14. Reingold, "Bache," p. 169.

15. Alexander Humboldt, *A Geognostical Essay on the Superposition of Rocks in Both Hemispheres,* tr. from the original French (London, 1823), pp. 67, 80ff.

16. Humboldt, *Personal Narrative,* III, 326.

17. Humboldt, *Personal Narrative,* VI, Part 2, 591–594; *Geognostical Essay,* pp. 70–75.

18. Humboldt, *Personal Narrative,* I, 230.

19. Humboldt, *Personal Narrative,* III, 188.

20. *E.g.,* Humboldt, *Personal Narrative,* II, 81–82; V, Part 1, 470.

21. Humboldt, *Personal Narrative,* II, 83.

22. Humboldt, *Personal Narrative,* VI, Part 2, 392.

23. Humboldt, *Personal Narrative,* VII, 288.

24. *Cf.* Humboldt, *Personal Narrative,* II, 57–58.

25. *Cf.* Humboldt, *Personal Narrative,* III, 327–330, 540 ff.

26. Humboldt, *Personal Narrative,* VI, Part 1, 190–191 and 191n; the quotation, VII, 123–124.

27. Victor F. Lenzen, "An Unpublished Scientific Monograph by C. S. Peirce," *Transactions of the Charles S. Peirce Society, 1* (1969), 16; see also Kelvin's speech on the award of the Copley Medal, in *Memoir and Scientific Correspondence of the late Sir George Gabriel Stokes,* ed. Joseph Larmor (Cambridge, 1907), I, 268 ff.; reprinted from *Proc. Roy. Soc., 54* (389–391).

28. Charles Lyell, *Life Letters and Journals,* ed. K. Lyell (London, 1881), I, 125–128, 140–141, 146.

29. I am indebted to a discussion with M. J. S. Hodge in arriving at this characterization of "British" geology. As far as I can recall, "local history" was his phrase and "archeology" was mine; but it may have been the other way around.

30. Humboldt, *Personal Narrative,* I, 271n.

31. Humboldt, *Personal Narrative,* III, 145–146. If one believes in Vulcanism and Neptunism, one might want to make Humboldt a Vulcanist. I find these terms sufficiently confusing to justify avoiding them. Catastrophism is not much better, but it may convey some idea (anachronistically) of what Humboldt might have looked like to a young British geologist.

32. Alexander Humboldt, *Essai sur la géographie des plantes* (1807) (Facsimile ed., London, 1959), pp. 19–20, 22–24. His essay "On Isothermal Lines," was published serially in English translation in the *Edinburgh Philosophical Journal* from 1820 to 1822.

33. Humboldt, *Géographie des plantes,* pp. ix–x.

34. Humboldt, *Geognostical Essay,* pp. vii, 17–19, 35–36, 465.

35. Humboldt, *Geognostical Essay,* pp. 44, 53, 64; the quotations, p. 46. Since the time of

Newton's *Opticks*, queries often meant that the author was hoping for the answer "yes" in most cases.

36. Charles Darwin, *Autobiography*, ed. Nora Barlow (London, 1958), p. 77.

37. Humboldt, *Personal Narrative*, VI, Part 2, 590, 592.

38. On isothermal lines, see Lyell, *Life*, I, 270.

39. I was fortunate in seeing an early version of Dr. Herbert's study of Charles Darwin's geological work in South America.

   A good later example of a Humboldtian is Francis Galton, so well described in Ruth Cowan, "Francis Galton's Statistical Ideas," *Isis, 63* (1972), 509–528. Galton was early associated with Kew Observatory, and Kew was a hot-bed of Humboldtianism. (Edward Sabine, another Humboldtian, was associated with it.) I would draw attention to Galton's *transfer of techniques* from one field to another (from meteorology to human beings—Cowan, p. 514) as an especially Humboldtian characteristic.

40. Darwin, *Autobiography*, pp. 67–68.

41. Copy in the Darwin Collection of the Cambridge University Library.

42. Humboldt, *Personal Narrative*, I, xi; VI, Part 2, 650–651; Nora Barlow, "Robert Fitzroy and Charles Darwin," *Comhill Magazine*, n.s. *72* (1932), 505.

43. Humboldt, *Personal Narrative*, II, 15, 11, 15. *Cf.* my article, "Darwin's Vision in *On the Origin of Species*," in *The Art of Victorian Prose*, ed. G. Levine and W. Madden (New York, 1968), pp. 169–170.

44. Humboldt, *Personal Narrative*, II, 55, 55n; III, 493. *Cf.* Cannon, "Darwin's Vision," p. 163, which must now be modified: Darwin's rhetoric is not merely that of the "British colonial empire" but of colonial empires generally.

45. Darwin to Henslow, May 18, 1832, in *Darwin and Henslow*, ed. Nora Barlow (Berkeley and Los Angeles, 1967), p. 55. Compare to Humboldt *Personal Narrative*, III, 355; V, Part 1, 1–2 (these are my favorites; Darwin may have had other passages in mind).

46. *Cf.* as one example, Humboldt, *Personal Narrative*, VII, 263 and passim. Darwin's celebrated passage on slavery is in *The Voyage of the Beagle* (Everyman's, London, 1961), pp. 480–481 (August 19, 1836); see also pp. 18 and 22 (April 8 and April 12, 1832), and Darwin, *Autobiography*, p. 74.

47. Humboldt, *Personal Narrative*, I, V, XXX.

48. Humboldt, *Personal Narrative*, V, Part 1, 180.

49. Humboldt, *Personal Narrative*, III, 491.

50. The letter is given in full in W. F. Cannon, "The Impact of Uniformitarianism: Two Letters from John Herschel to Charles Lyell, 1836–1837," *"Proc. Am. Philos. Soc., 105* (1961), 301–314; see esp. 301–302, 311–312. Darwin's activities are given in Peter Vorzimmer, "Darwin and Malthus," *Journ. Hist. Ideas, 30* (1969), 528–529. The month of publication of the *Ninth Bridgewater Treatise* is given in "List of Mr. Babbage's Printed Papers," in his *Passages From the Life of a Philosopher*, reprinted in *Charles Babbage and his Calculating Engines*, ed. Philip and Emily Morrison (New York, 1961), pp. 372–377. See also W. F. Cannon, "The Problem of Miracles in the 1830's," *Victorian Studies, 4* (1960), 23. My detailed consideration of Darwin's notebooks for 1837–39 is being prepared for separate publication. Lyell knew of the contents of Whewell's *History* by May 24; see Lyell, *Life*, II, 12. His long letter to Whewell of March 7, citing Herschel's letter to him, is on pp. 2–7.

51. Humboldt, *Personal Narrative*, V, Part 1, 88ff.; Part 2, 585.

52. Humboldt, *Personal Narrative*, V, Part 2, 591; Darwin's marginal note on p. 590 is dis-

cussed by Paul H. Barrett and Alain F. Corcos, "A Letter from Alexander Humboldt to Charles Darwin," *Journ. Hist. Med., 27* (1972), 160; *Darwin's Notebooks on Transmutation of Species, Part II. Second Notebook (February to July 1838)*, ed. Gavin de Beer (London, 1960) (Bulletin of the British Museum [Natural History], Historical Series, II, No. 3), pp. 51, 161 excised, 63 [the excised pages are in Part VI, 1967].

53. Quoted in "Account of the great Congress of Philosophers at Berlin on the 18th September 1828," *Edinburgh Journal of Science, 10* (1829), 231–232. The author of this article was Charles Babbage; see the "List of Mr. Babbage's Printed Papers," note 50 above. The number of members is not consistent throughout the article; presumably more came than had pre-registered.

54. Humboldt, in "Account of the Congress," pp. 228–231.

55. John Herschel, *Preliminary Discourse on the Study of Natural Philosphy* (London, 1831), p. 204.

56. Herschel, *Preliminary Discourse,* p. 307.

57. Herschel, *Preliminary Discourse,* pp. 144, 159–163.

58. Humboldt, *Personal Narrative,* II, 48.

59. Humboldt, *Personal Narrative,* II, 52–58.

60. Hans Straub, "Daniel Bernoulli," *Dictionary of Scientific Biography.* One also needs Cook, and perhaps Linnaeus. See J. C. Beaglehole, *The Life of Captain James Cook* (Stanford, 1974).

61. Humboldt, *Personal Narrative,* II, 24–25; VI, Part 2, 523; Florian Cajori, *The Early Mathematical Sciences in North and South America* (Boston, 1928), pp. 58–60, citing Humboldt's *Political Essays on the Kingdom of New Spain,* tr. John Black, I, p. xxxiv. Chappe's quadrant was described both as of 2½′ and 3′ radius, according to Cajori.

62. W. E. Knowles Middleton, *The History of the Barometer* (Baltimore, 1964), pp. 140, 196, 207, 210; Humboldt, *Personal Narrative,* VI, Part 1, 258; Part 2, 660–661, 773. I use the equivalent: 1 toise approximately equals 6′5″.

63. Humboldt, *Personal Narrative,* II, 118–120; V, Part 2, 551. One of the important pre-Humboldtians has been thoroughly investigated: see S. Stewart Gillmor, *Coulomb and the Evolution of Physics and Engineering in Eighteenth-Century France* (Princeton, N.J., 1971). Before Gillmor's book was published, I could only say that, if my arguments about Humboldtian science are correct, some such person as Coulomb should have existed in the 1770's and 1780's. Now I can say he did exist. We need equally full and detailed studies of Borda, Prony, Volta, and Saussure; then it could be seen what a rich field Humboldt had, to choose from, in putting together his over-all approach.

And, I hope, it would be seen that this rich late-18th-century science was *not* physics. Gillmore uses the word in his title, but I can hope that this can be interpreted merely as a convenient translation of *physique;* and French *physique* was by no means what we now call physics.

64. Nevil Maskelyne, *Astronomical Observations Made at the Royal Observatory* (London, 1776), p. i.

65. Maskelyne, *Astronomical Observations,* pp. ix–x. The basic reading for discussions of instruments and accuracy at Greenwich is Arthur Beer and Peter Beer, eds., "The Origins, Achievement and Influence of the Royal Observatory, Greenwich," *Vistas in Astronomy, 20* (1976), 1–272.

66. Andrew Ellicott, "Observations of the eclipse of the sun, June 16, 1806, made at Lancaster," *Trans. Am. Philos. Soc., 6* (1804–*sic*), 260; see also the earlier papers in the same volume, "Astronomical Observation made at Lancaster…as a Test of the Accuracy with which the Longitude may be found by Lunar Observation," pp. 61–69; and "Continuation of Astronomical Observations, made at Lancaster, Pennsylvania," pp. 113–119.

67. I recommend beginning with Thomas Cope and H. W. Robinson, "Charles Mason, Jeremiah Dixon and the Royal Society," *Notes and Records of the Royal Society of London, 9* (1951), 55–78. Then in *Proc. Am. Philos. Soc.* there are, by Cope: "Collecting Source Material about Charles Mason and Jeremiah Dixon," *92* (1948), 111–114; "Degrees Along the West Line, the Parallel between Maryland and Pennsylvania," *93* (1949), 127–133; "A Clock Sent Thither by the Royal Society" with (by H. Alan Lloyd) "Description of a Clock by John Shelton...", *94* (1950), 260–271; "Some Contacts of Benjamin Franklin with Mason and Dixon and their Work," *95* (1951), 232–238; "The Jersey Quadrant Used in Pennsylvania," *97* (1953), 565–570; "Some Local Scholars Who Counselled the Proprietors of Pennsylvania...," *99* (1955), 268–276. By Cope and Robinson: "The Astronomical Manuscripts which Charles Mason gave to Provost the Reverend John Ewing during October 1786," *96* (1952), 417–423; "When the Maryland-Pennsylvania Boundary Survey Changed from a Political and Legal Struggle into a Scientific and Technological Struggle," *98* (1954), 432–441. By Robinson: "A Note on Charles Mason's Ancestry and His Family," *93* (1949), 134–136; "Jeremiah Dixon (1733–1779)", *94* (1950), 272–274.

   In *Proc. Penn. Acad. of Sci., 22* (1958), 151, Cope lists his earlier writings (10, including this article); note esp. "Zenith Sectors and Discoveries Made with Them" in *18* (1944). There are probably others that I am unaware of.

68. My judgment is not exactly in conflict with Brooke Hindle's *David Rittenhouse* (Princeton, N.J., 1964). Hindle does not rate Rittenhouse's scientific work very highly, but Hindle had no framework in the history of science by which to evaluate Rittenhouse's individual works and see what they added up to. Silvio Bedini, *Thinkers and Tinkers* (New York, 1975), p. 323, has now also praised Rittenhouse's large zenith sector but limits himself to "in that period," i.e., the end of the 18th century.

69. Andrew Ellicott, "A letter to Robert Patterson," p. 39, and "A Method of Calculating the Eccentric Anomaly of the Planets," pp. 67–69, in *Trans. Am. Philos. Soc., 4* (1799); and references in note 66, above.

70. Henry Brougham, "Transactions of the American Philosophical Society," *Edinburgh Review, 2* (1803), 348–355.

71. Brougham, *"Transactions,"* pp. 348, 352.

72. Andrew Ellicott, "Astronomical, and Thermometrical Observations, made at the Confluence of the Mississippi, and Ohio Rivers," and "Astronomical and Thermometrical Observations, made on the Boundary between the United States and His Catholic Majesty," *Trans. Am. Philos. Soc., 5* (1802), 162–202, and 203–311. On the astronomical circle: 275; on the sectors, esp. 221, 249, 259; on the sextant, 202, 205; on "impertinent" Indians, 199, 264.

73. Brougham, "Transactions," pp. 348, 355.

74. Humboldt, *Personal Narrative,* I, 50n; Herman Friis, "Baron Alexander von Humboldt's Visit to Washington, D.C., June 1 through June 13, 1804," *Records of the Columbia Historical Society,* 1960–62, pp. 22–35.

75. *Cf.* Goetzmann, *Army Exploration,* p. 14, on West Point; Cajori, *Early Mathematical Sciences,* p. 85–86, on instruments. My narrative expands on Cajori on one point: a Troughton circle *was* tried out in the U.S. before 1816; but it was not successful.

76. Catharine Van Cortlandt Mathews, *Andrew Ellicott* (New York, 1908), pp. 242–243, 243–244.

77. *Cf.* W. F. Cannon, in *Isis, 55* (1964), 447; and the rejoinder by Peter Vorzimmer, in *Isis, 59* (1968), 223.

Chapter Four

The Invention of Physics

Some historians like to look for continuity in human events; a smaller number like to look for discontinuity. Since neither characteristic exists in pure form in history any more than it does in other branches of biology, neither group can prove that its metaphysical preference should be the approved one. Is the child continuous from its parents? Insofar as they are or determine the major part of its environment, yes; to find out what your young woman's dominant character traits will be, observe her mother, is the advice of folk wisdom to the overly enthusiastic young lover. Neuroses are certainly transmissible from parents to child. But the child's genetic inheritance is far more complex than a survey of the parent's somatic appearances would indicate; and we know, since Darwin, that genuine novelty (that is, discontinuity) can develop in what is actually a rather short time, given the geological periods we use as a standard of comparison.

At any given time, however, in any given field we may be able to say that one of the two preferences has been dominant for too long. In the history of science, it is the instinct for continuity which has led to several recent weaknesses. I mention below, in Chapter Eight, the fallacy of studying scientific instruments by going from one successful device to the next more successful device, as though the latter were indeed the offspring of the former. This is not, in many cases, how such devices came into being, but it makes a nice-looking pre-Darwinian sequence of progressive development. The same fallacy is almost as strong in considering scientific ideas. Given general ideas of development in the 18th century; given Lamarck, given Darwin, many historians cannot resist the impulse to arrange them continuously, and then assert that Lamarck must have led to Darwin in some way, even though they cannot specify what way. (Just for the record, on the basis of reading some of Darwin's early notebooks I will assert that he arrived at his evolutionary ideas by a different path, and that Lamarck was a theorist who needed to be disposed of—W. S. Macleay, a non-evolutionist, was another one—in the process of arriving at a valid theory.)

A conceptual scheme, then, which is nowadays used in the history of science primarily for the purpose of asserting continuity is, to me, automati-

111

cally suspect. However, that is not enough of a motive for me to sit down and write an essay attacking every such scheme. A conceptual scheme in history, like a theory in physics, is a mental device for organizing facts, directing research, and then re-organizing facts so as to include the new ones invented on the basis of the research. (A fact, of course, is not a "thing" the historian "discovers," but is a statement which he makes up to account for the data he discovers.) As also happens in physics, in this process recalcitrant data sometimes turns up. At first it may be neglected, or explained away, or included as being "in the right direction" (in physics, the phrase is more grandiloquent: it is "of the right order of magnitude"). Sooner or later there may be enough evidence running counter to the conceptual scheme that the existence of "anomalies" are admitted. And finally, along comes someone for whom anomalies are his principal collections of data, not stray bits to be pushed to one side of the area of attention. Such a person has a motive for scrapping the old conceptual scheme, since it does nothing but disorganize his collections. He usually has trouble in having his new scheme accepted, however, not because it is "wrong" but because no one else much cares about the facts which it is erected to arrange. To take his scheme seriously, other people would have to shift their area of attention, from a place where they are enjoying themselves to the place that he is calling attention to—a rather crabbed, uncivilized place full of strange facts with uncouth, non-law-abiding manners.

This latter is my position. The most common conceptual scheme for the history of the physical sciences since 1600 interferes with my research. It reduces interesting, and apparently important, facts to the status of anomalies. It makes it difficult for me to recognize the patterns that may actually be suggested by considering the evidence itself. I repudiated it, mildly, in a talk in 1962, printed in 1967, and presented in revised form as my Chapter Nine below. I said, "A promising 17th-century candidate [for the role of leading scientific subject], after developing sporadically in the 18th century, blossomed in France following the Revolution. In Ampère, Fresnel, Carnot, Fourier, and the like, physics finally found itself. It soon won eager converts in Germany and England..."[1] But at that time I still, as you can see, tried to maintain some continuity with the 17th century even while feeling in my heart that something really new had happened in the early 19th century. By analyzing my prejudices, of which a desire not to be a revolutionary is one, I realized that I was still suppressing enough of my new approach, in the interests of continuity, to distort my vision not only of the 19th century but of the 17th century as well. Hence since 1970 I have been expressing the discontinuous interpretation given in this chapter.

The common conceptual scheme which scientists, and most historians, believe in goes something like this: physics properly so-called, as opposed to Aristotle's physics, comes in two parts, classical and modern. Classical physics, although it had deep roots in earlier centuries, came together in the 17th century, especially in the developing sequence Galileo-Descartes-Newton. It thereafter reigned until 1895, expanding its coverage greatly in the 19th century but remaining more or less Newtonian. Of course there were problems, such as the wave theory of light, which was non-Newtonian, but on the whole the nature of physics was essentially unchanged. James Clerk Maxwell was one of its last towering figures.

Modern physics began in the 1890's with such things as X-rays and electrons, and almost immediately succeeded with anti-Newtonian theories such as Planck's quantum and Einstein's relativity. It is physics since 1900, then, which is fundamentally non-Newtonian; and we are still in the era of modern physics.

It is not surprising (according to this conceptual scheme) that the word "physics" was not in use in English until the middle of the 19th century to describe this continuous development from the 17th century on, for two reasons. First, it was the name either for Aristotle's approach or, later, for Descartes's approach. Aristotle had to be repudiated altogether for true physics to constitute itself; and Descartes, while being important in the particular developments which Newton used, nevertheless had to be repudiated as a system-builder. Second, people were not so specialized in those days. "Natural philosophy" was a good enough general term at first. Then, as throughout the 18th century various specialties split off, gradually people felt the need for a name for that which remained. By the 19th century, it was safe to use the old word "physics" again; and so it was gradually introduced for that remaining collection of subjects in which resided the essential parts of Newtonianism, plus additions made since Newton's time which somehow belonged there also—electricity, for example. Physics, in short, always has been the fundamental science from Newton's time to our own.

By turning to people who have really studied verbal usage (according to this conceptual scheme), such as the compilers of the *Oxford English Dictionary* of 1902–09, we find the scheme confirmed. According to the *OED*, the word "physics" in Aristotle referred to the whole of nature, organic and inorganic; John Locke even included spirits, such as angels. In the course of the 18th century, it became limited to inorganic nature, and then, by excluding chemistry, it acquired its present meaning. It is now "restricted to the science, or group of sciences, treating of the properties of matter and energy, or of the action of the different forms of energy on matter in general (excluding Chemistry... and Biology)." Thus we find in *Entick's New Spelling Dictionary,* the third edition of 1807 (first edition

1804), which was purchased by my great-great-great grandfather in 1811 for "9/6 Paper Money," physics defined as: "all natural philosophy, divinity." This does not mean that in New Haven, Connecticut, where the book was published, or in the mountains between North Carolina and Tennessee, where it was used, people were intellectually naive. It means that they were still studying authors such as Locke. By accepting the continuous development scheme, that is, we can "place" people in the provinces according to which earlier stage of the main development their current beliefs are correlated with.

To replace this current conceptual scheme, I propose something like this instead: something new and anti-Aristotelian did indeed take shape in the 17th century; but what Galileo was for was not what Descartes was for was not what Newton was for. Newton's *Principia* is, among other things, the great refutation of "physics," that is, of the approach to science of Descartes and his admirers. It is also a rejection of Galileo's approach (or at least an ignoring of it, since it is not clear how thoroughly Newton was acquainted with Galileo's writings[2]). In this three-way contest, Newton won, and thereafter dominated natural philosophy in physical astronomy, dynamics, and optics. Newton's *Opticks* is not an alternative to Huygens' wave theory of light. Both the *Opticks* and Huygens' *Traité de la Lumière* present theories involving combined particle-and-wave mechanisms—combined in different ways, of course. But Newton's work goes so far beyond Huygens' as to reduce it to merely one of the sources Newton used.

Newton also dominated the way of the investigator, not only in general method but also in many suggestive details; the *Principia* is chock-full of all sorts of things that the practising scientist could use, especially in the sections of the book which are usually ignored in general presentations of it for philosophically-minded students. Note that I emphasize the *Principia*, not the *Opticks*. It is not always emphasized that, in England at any rate, in the first half of the 19th century the *Principia* could function as a working handbook for the practising scientist; and finding someone who uses it like that is one way to spot a Newtonian. In the 18th century, so I learn from various recent studies, persons interested in, say, electricity or atomic theory, could consider themselves Newtonians because they were approaching the subject in Newton's way; and that way was not limited merely to finding inverse-square laws.

One of Newton's subjects, or rather one subject which had Newton's seal of approval if done the proper way, was hydrodynamics. The fondness of early 19th-century scientists for waves is so pronounced that it needs some explanation; and Newton's *Principia* is the obvious one. Not only do we have, in the early 19th century, hydrodynamics itself; but also real water waves, wave components as analyzed out of tidal data, acoustics, and, of course, the wave theory of light. As scientific investigations, Thomas Young's writings on optics (I can't speak from first-hand knowledge of Fresnel's) are overwhelmingly an extension, not a refutation, of Newton's optical work. Read Newton's *Opticks,* and then Young's papers: you are continuously in the same mental and experimental realm, with the introduction of one or two additional mechanisms and the extension of observations. Read Huygens' *Traité* and then Young's papers: there is very little continuity. Young himself was apparently not sure whether he would get better publicity by emphasizing his debt to Newton or by trying to magnify his much smaller anti-Newtonian component. So he did both; and got the worst possible publicity.[3] It is true that a number of self-styled "Newtonians" opposed the wave theory of light; but then a number of Newtonians were its strong advocates: John Herschel and George Airy in England, for example.

The real problem with Fresnel's wave theory, as Young saw at once, was that transverse waves were *not* hydrodynamic waves; they must be waves in an elastic solid. This was one of the problems which the new physicists inherited from the Newtonians; and they eventually solved it, as we know, by abandoning the idea that the waves were real waves in anything. I don't know that Newton would have objected to this; but I don't see that this solution was actually derived from the *Principia* or the *Opticks,* so I will call it non-Newtonian from the historian's standpoint, whatever its metaphysical relation to Newton's waves in the *Opticks* may be.

Note that it was not the transverseness of the component which caused problems. Newton knew well that another coordinate was necessary; that was why he had particles as well as waves in his theory. What caused problems was precisely the determination to do the job with waves alone. Since the system had to have a rigid (or quasi-rigid) body in it somewhere—that is, the entity can be called a particle or an ether or anything you like, but it had to supply an equivalent set of coordinates—the early 19th century wave theory of light merely removed these characteristics from the light rays themselves, only to have them pop up as characteristics of the surrounding universe. That anyone really believed in this is evidence, to me, of the great delight which waves produced in some people, regardless of how odd this made the rest of the universe.

In short, the wave theory of light was not a real break with Newtonianism before about 1840 or 1850. Thereafter it began to supply physicists with reasons for breaking with Newtonianism.

Physics itself was invented by the French around the years 1810–30. With regrets that Malus died so young, I would name Ampère, Carnot, Fourier, and Fresnel as among the first physicists; there were others, but I know too little about French scientists to try to pick out all of them. One of the causes, obviously, was the new orientation of French science caused by the Revolution. During the "mobilization of savants," physical scientists had clustered around the mathematician and engineer Gaspard Monge, who more or less ran the affair. The founding of the École Polytechnique, not Monge's idea alone but Monge's institution as it came into being, furnished a center where Monge's circle (which included Laplace, Berthollet, and other former associates of Lavoisier, as well as new men) could regroup and also reach the new generation of students with a highly mathematical, and yet highly down-to-earth brand of science—Monge himself was a man who could think with his hands. From 1795 to 1815 or 1820: just about enough time for a new orientation to reach not one or two but a whole cluster of younger men, and to produce its results. However, this, as I said, was undoubtedly only one of the influences which led to the invention of physics as a new subject, not just to a new group or "community" of scientists.

My actual dating of the invention is not based on high-level theorists but on textbooks. A reading of R. J. Haüy's *Traité elementaire de physique* of 1803 convinces me that, whatever Haüy may have had in mind, it was not a conception of physics, or even an approach thereto. In contrast, I think J.-B. Biot in his *Traité de physique experimentale et mathématique* of 1816 was beginning to grasp at something like our concept of physics, and I would expect it to become somewhat clearer in his later versions and editions. Of course working with French sources gives us the problem of the continued use of the word *physique*. In the middle 18th century, it did not mean "physics;" if you doubt this, you may browse in the *Journal de Physique* from 1773. What you may well conclude is that this is the potpourri of subjects which produced Humboldtian science, not physics. A *physicien* was not a Newtonian analyst like Laplace or Lagrange. He was a somewhat grubby lower order of being, like Borda or Coulomb.

This leads us to the question of the relation between Humboldtian science and physics. They must have been closely related, or how can we place Arago? Biot gave the credit for establishing exactitude in *physique* to Borda and Coulomb;[4] but, exactitude, I shall maintain, was not a defining characteristic of physics, as it was of Humboldtian science. On the whole, Humboldtians were Newtonians, not physicists. Was Biot a physicist? Possibly not; and that makes his near approach in his textbooks all the more interesting.

So much for my ideas about France. At least nothing in Crosland's *Society of Arcueil* makes them impossible.[5] Physics was transferred to Britain by such people as J. D. Forbes, William Thomson (Kelvin), and George Stokes. After reading George Davie's fascinating book *The Democratic Intel-*

lect, which I hope is the beginning of a whole host of studies of the Scottish intellectual scene *after* the days of Hume and Adam Smith, I am inclined to believe that physics was best suited to the Scots intellect and only gradually was domesticated among Englishmen.[6] If Thomson was a physicist, his friend Stokes was Newtonian and Humboldtian as well. At any rate, physics became domesticated, with Lagrange as the hero. Of leading names around 1830, we can say that Michael Faraday and John Herschel were not physicists. Was David Brewster? An interesting question. The new approach developed very rapidly, in some circles, so that of James Clerk Maxwell in the 1850's we can say that we have a fully-formed professional physicist already. That is, he was a physicist who had been trained by physicists. He could philosophize confidently about science, as Davie emphasizes that he did, because he was quite sure that he knew what science, at its best, was.

One interesting feature of physics was that it seemed to drive the mathematicians away. Natural philosophy had been a rich grouping of people and interests in England; the physics grouping was sparser. As I mentioned in Chapter Three, it was mathematically trained scientists who were on the move in a number of fields; but the mathematicians themselves began around mid-century that process of segregating themselves, and repudiating demands that their work be useful, which still characterizes some of them today. The third quarter of the century saw heroes of the independence of mathematics, such as Sylvester and Cayley, arise. Of course in England this may have been due primarily to the transmission of an impulse from the Continent; it needs explaining there too. At any rate, the community of interests among scientists was split again; and physics was a lonelier subject than natural philosophy had been.

Note that I call it a subject, not a discipline. The sociological implications of the latter term seem to me too easily assumed without sufficient evidence, in the case of physics in the early and mid-19th century.

Physics by the 1870's—that is, in its first 50 years—had a number of notable things to its credit. There was the conservation of energy, which in England at any rate had to struggle hard against the tendency to view it as the correlation of forces, the conservation of forces, or of motion, or of matter in motion—as anything except what it was, one single weightless fluid to replace the multiplicity favored in the 18th century. The conservation of energy is not the kind of doctrine which was as a matter of fact discovered by a particular person at a particular time; each discoverer discovered something different that was conserved, and they were all eventually worked together into one doctrine. Well, more or less one doctrine. There was statistical mechanics. There was Maxwell's electromagnetic theory, which swallowed up Ampère's electrodynamics (one of

117

the striking early products of physics), Faraday's brilliant ideas, and all of optics in one great system of Lagrangian-Hamiltonian dynamics. There was, in England, the building of physics laboratories at Cambridge and Oxford. These were not built before, not because these universities were backward in science—Cambridge in particular was forward in science—but because there were not physicists in existence to press for them. Humboldtians wanted observatories, geophysical establishments, and establishments for exact measurements of standards, but not physics labs as we know them. Once there were physicists in existence, and sources of money were located, the labs were forthcoming within a reasonable time, as universities go. The money part is especially worth stressing in any estimate of the performance of universities. Cambridge, for example, was poor as a university throughout much of the century. Having re-established the observatory on a modern footing in the 1820's, it later voted for extensive new science facilities and discovered that it did not have enough money left to pay for them. The goodwill toward science was there; but science is an expensive friend.

Then, too, the presence and personality of particular individuals will determine the timing for particular sciences. The Cambridge Observatory came into being when and as it did because George Peacock was there. The next desideratum was natural history facilities, not because Cambridge was backward in science, but because John Henslow in botany and Adam Sedgwick in geology were highly capable persons. Oxford got a new chemistry laboratory in the 1850's, as part of its new museum, not because it was ahead of Cambridge in the physical sciences generally, but partly because the professor of chemistry, Charles Daubeny, was an active and well-respected university figure. Of course, outside pressures can sometimes produce the introduction of a new subject, and new facilities and staff for it; but for this to happen the subject must first exist. It is only when the pressures are internally generated that facilities are produced for a subject which does not exist (that is, a subject which is only the peculiar way in which a professor wants to spend his time).[7]

Once all of this new science was in being, it then became necessary to adjust the older Newtonian theories to agree with physics. This was done by several people, of whom Albert Einstein is perhaps the most famous, for mechanics, to make it agree with electromagnetic theory. It was done by several people, of whom my favorite is Max Planck, for—what shall we call it? I would say optics. This period of readjusting the older theories was about complete by 1930, and that is where my even superficial knowledge and interest stops.

Now I want to give some examples of the queer verbal behavior of Englishmen in the early 19th century. I put up with it for some time until

118

I decided that I needed an explanation simply in order to understand my sources correctly. For example, the award of the Royal Medals of the Royal Society was subject to a variety of delays, confusions, and controversies from the time that Robert Peel proposed them to the King in 1825[8] I had hoped, however, that by 1840 everything was normal, and that the stated triennial cycle of subjects was being observed: one medal each year for Astronomy, Physics, or Mathematics; and the other each year for Physiology or Natural History, Geology or Mineralogy, or Chemistry. But in that year the Royal Medal "in the department of Physics" was awarded to John Herschel for a paper entitled, "On the Chemical Action of the Rays of the Solar Spectrum on Preparations of Silver and other Substances, both metallic and non-metallic, and on some Photographic Processes."[9] It was not the word "chemical" in the title of the paper that bothered me. Nor was it a belief that photography "really" is a part of chemistry and not of physics. What bothered me was that I knew what Herschel had been doing, and what he had been doing was messing around in his chemistry lab, which he loved to do for relaxation; and he seldom carried an experiment in optics to the point that he would have to stop being involved in chemistry and its products (solutions, crystals), unless he had some other definite use for it (astronomy, writing his encyclopedia article on optics, etc.).

Herschel made a number of contributions to what we now call physics, but he was not a physicist because he did not carry them through extensively for their own sake. They were by-products of some other interest, and he stopped when that other interest was satisfied. It could be that he just wasn't interested in physics. Or it could be that in 1840 he didn't know that such a subject existed. I prefer the latter explanation. It was a Humboldtian characteristic to believe that scientific specialization was a weakness of the past which should be overcome, not fostered.[10]

If, then, what Herschel was doing was not what a physicist does, and if the Royal Medal for "physics" was given to him anyway, it could be that the Royal Society's Council was trying to honor their favorite scientist even if it meant cheating a little bit; their past record was not so impeccable as to rule out this possibility. But then they could have given him the other Royal Medal that year, and assigned it to chemistry; it was available, since it was geology's turn and there was nothing much in geology to award it for. As it turned out, they assigned it to physiology instead. I prefer the alternative, that what Herschel was doing was what they called "physics," even though it was not really what we call "physics."

Rather than presenting a formal definition of the difference, I will give a detailed example. Herschel in 1845 investigated a phenomenon which David Brewster had gone into briefly some years earlier. Herschel noted it first in a liquid, and then went into it especially in fluorspar. He called it "the epipolic dispersion of light." As such, it was not revolutionary. But young George Stokes saw the papers and wrote to Herschel in 1852: "Your 'epipolic' dispersion has given me the clue to a most extensive field of research, which has occupied me during the last year when sunlight permitted. I have now nearly finished a paper on the subject, which I hope soon to present to the Royal Society." The paper was, of course, Stokes's famous work on what he called "fluorescence"—taking the name, obviously, from the prominence of fluorspar in Herschel's work. But the significance of Stokes's work was not in carrying Herschel's work further and in giving it its current name. The title of his paper was, "On the change of refrangibility of light"; and this implied a change of periodicity in the light itself.[11] Such a change went counter to all good Newtonian analogy in mechanics. It was a guiding principle of the Newtonian approach—emphasized, indeed, by Herschel himself—that one can identify a component in a mixed output as being caused by a particular component in a mixed input (the sun and the moon affecting the tides, for example) precisely by the fact that, regardless of how magnitudes of forces may become distorted and how much time lag there may be between cause and result, the result always has the same periodicity as the cause.

Physics was obviously closely related to what went before it. Herschel was one of the more adventurous men of the older way of physical science; Stokes was one of the more Newtonian men of the newer way. The continuity between Herschel and his "epipolic dispersion" in fluorspar, and Stokes and his "change of refrangibility" in fluorspar, and also the discontinuity, gives as finely-drawn a set of distinctions between the two ways as I have been able to locate. Other men and other subjects are much more obviously different: Faraday and Maxwell on electromagnetism, for example. And something like the conservation of energy is simply weird, from the older point of view. In the ordinary Newtonian sense it is actually false; and it seemed perverse to go looking for a new concept just so that you will have something that is conserved.

To see more examples of queer verbal behaviour, let us turn to the *Encyclopaedia Britannica.* The article "Physics" in the third edition continued into the seventh, that is, into the 1830's. It was by John Robison. It involved no mathematical expressions or treatment of the subject. Next to the marginal title, "The term *physics* defined as it is generally understood in Britain," the text said, "Physics, then, is with us the study of the material system, including both natural history and [natural] philosophy....We more generally use the term *naturalist* and *natural knowledge.* The term natural philosophy, in its common acceptation, is of less extent." In more detail: general physics is divided into natural or mechanical philosophy,

120

chemistry, and physiology. Mechanical philosophy includes only the sciences of motion; thus astronomy and hydrodynamics are part of it, but optics became a part only in the 17th century, after Roemer's demonstration of the finite velocity of light showed that something is moved in optics; light is not merely a "quality."

This is certainly a good Newtonian, or perhaps Cartesian, or perhaps Scots way of looking at it. It would fit the *Oxford English Dictionary's* definition: by sluffing off "physiology" and chemistry, we are left with physics equals natural philosophy equals mechanical philosophy. But I don't think that physics was simply mechanical philosophy—although a physicist like Maxwell certainly loved dynamics. Let us see what progress has been made by the 1850's.

In the eighth edition of the *Britannica*, the "Physics" article had become one very short paragraph. Physics is the science of matter, as opposed to the mental sciences and mathematics. It involves both mechanical philosophy (such as mathematics, hydrostatics, and hydraulics) and chemical philosophy (which includes optics, electricity, and magnetism). There is also Concrete Physics, such as natural history, geology, zoology, botany, physical astronomy, descriptive astronomy, etc.

This is a very odd characterization. It is not surprising that the article includes a quotation from Auguste Comte. Even though the whole does not make a great deal of sense, however, there are precedents for various parts of it. Let us consider Thomas Young's *Course of Lectures on Natural Philosophy and the Mechanical Arts* of 1807, a massive two-volume work, including in the second volume a 415-page catalogue of earlier works, some 16,000 entries in all. Of course Young's lectures at the Royal Institution were not a great success, and neither was this work; but it gives us the opinion of an important physical scientist. Young divides Natural Philosophy into three parts: Mechanics; Hydrodynamics (which includes Hydraulics, Acoustics, and Optics); and Physics. Physics "relates to the particular history of the phenomena of nature, and of the affections of bodies actually existing in the universe;" thus it includes observational astronomy, geography, the properties of particles; heat, electricity, and magnetism; meteorology and natural history; and chemistry. That is, it should include chemistry, but that subject is so extended now that it rates a separate professor all its own.[12] Thus it would seem that Young uses physics for everything that is not deducible from fundamental laws expressed mathematically, everything in which our knowledge depends largely on experience.[13] If so, then merely getting rid of chemistry and natural history will not make the residue into what the mathematical Cantabridgians like William Thomson, Stokes, and Clerk Maxwell were going to do.

However, Young also uses "physics" later on in another sense, in his title for Part III: "Of the Subdivision of Physics into Distinct Branches." Here it means primarily dealing with inanimate matter rather than organization and life. There were five groupings, each including several distinct branches. They were: statics, dynamics, pneumatics, hydrostatics, sterostatics, and crystallography; sound, and light; astronomy and celestial mechanics, geology; mineralogy, and chemistry; heat, electricity and magnetism, and a bit of zoology and botany. If we subtract the branches that in 1807 were becoming reasonably well formed—that is, one could be an astronomer, a geologist (just barely), a mineralogist, a chemist, a zoologist, and a botanist—we are left with a hodge-podge, partly mathematical and partly not, which clearly would have had no unity in Young's mind.

The chemist Thomas Thomson arrived at a smaller collection of leftovers from his teaching duties as professor of chemistry at the University of Glasgow. His *System of Chemistry* had six editions from 1802 through 1820; but in 1830 he began to split it into four sections: Heat and Electricity, Mineralogy and Geology, Inorganic Chemistry, and Organic Chemistry. In his preface to *Heat and Electricity* he said that he had tried to teach these subjects but it would be better if a new chair were instituted in each of our Universities, and that it should be the province of him who fills it to explain the principles of *heat, light, electricity,* and *magnetism.*" This he thought was done in France, where "this important branch of science is distinguished by the name of *physique.*"[14] He does call it one branch of science, I note, perhaps because it dealt with imponderable forms of matter; Young had separated light from this trio and associated it with sound, because for him it was true motion of real bodies.

Did physics turn out to be the science of imponderables that no one else wanted? No, but this suggests an idea. Perhaps it was the science that attempted to bring the imponderables under the sway of the laws and procedures of mechanics. But that, I take it, was not a terribly new idea. There must have been something more.

Now let us look at what a group of men decided to do, or rather could not quite decide about. It was a feature of the new British Association for the Advancement of Science that it was divided into subject committees. At the first meeting in 1831, they were: Mathematical and Physical Science, Chemistry, Mineralogy, Geology and Geography, Zoology and Botany, Mechanical Arts. At the second meeting, however, the subject committees were set up as:

I.   Pure Mathematics.—Mechanics, Hydrostatics, Hydraulics.—Plane and Physical Astronomy.—Meteorology, [Terrestrial] Magnetism, Philosophy of Heat, Light, and Sound.

II. Chemistry, Mineralogy, Electricity, Magnetism

III.   Geology and Geography

IV.   Zoology, Botany, Physiology, Anatomy.

The reason for putting electricity and magnetism in with chemistry is quite apparent. It made Committee II a suitable one for such men as Michael Faraday and J. F. Daniell, both of whom were appointed to it. It was not so easy, in England, to separate electricity from chemistry. Otherwise, we are arriving at a recognizable collection of subjects for Committee I, which the president referred to as "Mathematics and General Physics." Unfortunately, the Recommendations of the Committees were listed as coming from the "Committee for Mathematics, & c.," Committee for Chemistry, & c.," and so on. So if physics existed at all in England in 1832, it was reducible to an "& c." whenever it seemed convenient.

At the third meeting, in the *Report* Committee I is still called either "Mathematics and General Physics" or "Mathematical and Physical Science" and Committee II still has the electrochemists. However in the Transactions of the Sections, there is no separate listing for chemistry. The "Mathematics and Physics" section includes papers by Faraday on electro-chemistry, Turner on atomic weights, and Johnston on the analysis of iron. When it is convenient, that is, the word "physics" can be stretched to include chemistry, just as it can be shrunk to "& c."[15]

However, in another publication of the third meeting, Committee I is listed as "Mathematical and Physico-Mathematical Sciences (Astronomy, Mechanics, Hydrostatics, Hydraulics, Light, Heat, Sound, Meteorology, and Mechanical Arts)." Committee II is "Chemistry, Electricity, Galvanism, Magnetism, Mineralogy, and Chemical Arts and Manufactures."[16] This way of putting it certainly suggests that physics could not be properly constituted until the mathematico-scientists took electricity and magnetism away from the chemists. In England, after 1833, the name that jumps into my mind is, of course, William Thomson. But, just thinking of mathematics and currents of electricity, the name that occurs to one is Ampère. So I am back with the men I have favored all along.

Let me conclude my consideration of verbal behaviour with the usage in a source as authoritative in 1831 as was the British Association—perhaps more so. I refer to John Herschel's *Preliminary Discourse on the Study of Natural Philosophy*. For Herschel, "physics" is really a short form of "physical sciences," and that, in turn, is just about the same thing as "natural philosophy." Thus he can speak of "abstract science as a preparation for the study of physics" on one page, and on the next say that "there is yet another recommendation of such [abstract] sciences for the study of natural philosophy." Or—and this is the most common behaviour I have found among scientists around 1830—he can start a paragraph by saying "the physical sciences" and then shorten his references in the paragraph to "physics," even when what he is discussing is exemplified by the discovery of iodine by a "scientific chemist." In the index, "Physics, axioms of" refers to a page on which the word "physics" does not occur; but "physical science" does. His Part III, then, is "Of the Subdivision of Physics Into Distinct Branches, and Their Mutual Relations." He groups them following Young, even to the use of the word "stereostatics," which is, for solids, what pneumatics and hydrostatics are for airs and liquids.[17]

Herschel has a criterion for having separate branches of science at all. "In pursuing the analysis of any phenomenon, the moment we find ourselves stopped by one of which we perceive no analysis, and which, therefore, we are forced to refer (at least provisionally) to the class of ultimate facts, and to regard as elementary, the study of that phenomenon and of its laws becomes a separate branch of science." But of course we hope to be able to trace parallels and analogies between separate branches, and finally see their mutual dependence on a more general phenomenon, as has been done, for example, for electricity and magnetism by Oersted. And we are now seeing the closeness of agreement between sound and light, which will ultimately coincide as the vibratory motion of an elastic medium.[18]

This approach, presumably, justifies Herschel's use of Young's grouping of branches under the more general categories of the Phenomena of Force and the Constitution of Natural Bodies; The Communication of Motion Through Bodies (Sound and Light); Cosmical Phenomena (Astronomy and Geology); the Examination of the Material Constituents of the World; and the Imponderable Forms of Matter. Herschel's idea, it seems to me, is admirable, even if the grouping is borrowed; if it did not work quite as he expected for sound and light, it did work very well for radiant heat and light, and then for light and electromagnetism. The categories for grouping branches, which he took verbatim from Young, were not nearly so important as the idea that there should be a general unity of "physics" to be found at a more basic level than that of the separate branches with which individuals identified themselves—chemists, or astronomers—and that that unity would be one of finding the underlying phenomenon or set of formal laws of which the branches were derivatives. Whereas chemists and astronomers had long been identifiable because of what they did, overtly and obviously, the physicists did not separate out on the basis of an exclusive laboratory technique but on some basis similar to what Herschel was talking about. That is why it is not so easy to identify them by inspection; they did not monopolize the handling of heat, or of electricity, or of gases, at least at first.

One of Herschel's other topics may have been a source of some of the confusion, in the latter part of the century, as to why it was good to measure things very accurately and to expect that this would lead to important discoveries. For Herschel, this was only one of nine methods he sketched, but he thought it was very important, and this is his reasoning:

> Complicated phenomena, in which several causes concurring, opposing, or quite independent of each other, operate at once, so as to produce a compound effect, may be simplified by subducting the effect of all the known causes, as well as the nature of the case permits, either by deductive reasoning or by appeal to experience, and thus leaving, as it were, a *residual phenomenon* to be explained. It is by this process, in fact, that science, in its present advanced state, is chiefly promoted. Most of the phenomena which nature presents are very complicated; and when the effects of all known causes are estimated with exactness, and subducted, the residual facts are constantly appearing in the form of phenomena altogether new, and leading to the most important conclusions.[19]

Put that way, we can see that the scientist not only may use this method primarily, but must use it secondarily at all times. He does it every time he notes the temperature and pressure at which his experiment takes place, the average noise level, and so forth.

John Herschel, then, writing in 1830, certainly had the idea that there ought to be a more general science in the physical sciences, not one that was merely the residue left over after other specialties had firmly established themselves; and the habit he and a number of contemporaries had of using "physics" as an abbreviation for "physical sciences" made the word available for someone to apply to this new scientific venture. But Herschel was not himself a physicist, and did not in 1830 have any very clear idea as to how to constitute such a science—although he knew that Oersted and Ampère had been doing the kind of thing he wanted more of. He also was sure, addressing the British Association in 1845, of what would *not* work: "The time seems to be approaching when a merely mechanical view of nature will become impossible—when the notion of accounting for all the phaenomena of nature and even of mere physics, by simple attractions and repulsions fixedly and unchangeably inherent in material centres (granting any conceivable system of Boscovichian alternations) will be deemed untenable."[20] If that is your view of Newtonianism (it is not mine), then Herschel was against it.

Physics, I am suggesting, did not come into being in the way the *Oxford English Dictionary* suggests, by mechanics, heat, light, sound, electricity, and magnetism finding themselves abandoned by the other sciences and hence clustering together so as to make up a respectable-sized subject. It involved a drive for a deeper, more fundamental unity than the "branches" of science had; and it was not a unity on any simple Newtonian lines that it expected to find.

First, the word "lines" may be taken literally. Physics made use of forces which worked in odd directions—at right angles, for example, as in electromagnetism. Faraday could bend lines of action into all sorts of queer shapes. Of course light rays are bent somewhat by refraction; of course the wave theory of light had transverse vibrations. Physics made the situation much worse.

Second, the physicists did not, by and large, pursue known paths of research with more and more accuracy. By and large, it was the Humboldtians who were interested in accuracy, and the physicists only came to it later in the century, possibly to show they were as respectable as the chemists, astronomers, and geodesists. It is nicely symbolic, if only that, that when the physicist Maxwell carried out one part of the Humboldtian pursuit of accurate standards in physical sciences (the determination of the "British Association" ohm), he got the wrong answer.

Third, workers in the various branches which were to be part of physics often did not pursue subjects very vigorously, in the first half of the century. Spectroscopy could, it might seem, have taken off directly from Fraunhofer's work around 1815. There may be various reasons why it did not, but one of them was that men such as Brewster and Herschel dabbled with it and then dropped it. It was too difficult, did not give up its secrets easily, and was not along their main lines of interest. The shortage of labor in the early and mid-century is not always noted in accounts of physical investigations in that period; to say that the "prevailing opinion" on a subject was such-and-such might easily mean that four men were for it, two were against it, two had no strong convictions, and no one else knew enough about it to have any serious opinions at all. This is, to me, evidence that physics did not exist, because physicists—men who devoted their careers to investigations in physics—did not exist, or were scarce. It was not, I think, the growth of workers in optics and other branches which produced the overall effect of the growth of physics. It was, perhaps, the reverse: the growth of the concept of physics produced workers who would devote themselves to optics because it was part of physics. The paradigm "physics" preceded the community of physics.

Was physics a product of Romanticism? I think not; but we should face the problem squarely in connection with the conservation of energy. Clearly Joule, and Faraday, and Oersted, and others believed that forces *ought* to be inter-convertible. Leaving *Naturphilosophie* to other more capable historians, I will note that it was Coleridge in 1796 who believed in energy: "All-conscious Presence of the Universe!/Nature's vast ever-acting Energy!" This energy, being of God, is obviously not going to be frittered away and lost. And Coleridge, we know, was "the chemical, botanical, geological, astronomical, mathematical, metaphysical, meteorological, anatomical, physiological, galvanistical, musical, pictorial, bibliographical, critical philosopher who had run through the whole circle of the sciences, and understood them all." At least that is how Thomas Love Peacock described him (as Mr. Panscope) in *Headlong Hall*, writtin in 1815. Coleridge himself said that, in order to write an epic poem, he would spend, "Ten years to collect materials and warm my mind with universal science. I would be a tolerable mathematician. I would thoroughly understand Mechanics; Hydrostatics; Optics, and Astronomy; Botany; Metallurgy; Fossilism; Chemistry; Geology; Anatomy; Medicine; then the minds of men, in all Travels, Voyages and Histories."[21] Coleridge would not in 1796 study, and Mr. Panscope did not in 1815 know, I may note, any thing which we could label as physics.

One trouble with speculating this way is that other things than God's energy were thought of as being conserved. This is, after all, the mercantilist notion of wealth: what France loses, England gains. Another problem is: why should God's energy, or Nature's energy, suddenly need conserving? It might seem more Romantic to have it ever being expended, and ever supplied from an inexhaustible source.

Basically, we have to guard against two fallacies. First, Romanticism is not an entity which could "cause" anything. Second, the conservation of energy was not one doctrine which had one source, and there was not necessarily a progressive sequence from the qualitative "correlation of forces" to the quantitative inter-convertibility of forces, to the conservation of a something called energy. In England, for example, one might try to sketch a sequence: the escape from heat as "caloric," a substance, to heat as a mode of motion—hence, with the correlation of forces, all forces as types of motion. But from here to the conservation of energy was a leap so far outside of Newtonianism that some people could not make it.

This problem is neatly illustrated by the contrast between the original 1846 edition of William Grove's *On the Correlation of Physical Forces* and the fourth edition of 1862, which is the one most students encounter, as it was included in an American compilation by Youmans on the correlation and conservation of forces. In the first edition, an essay of fifty pages, Grove is an elegant and consistent believer in force as always being a mode of motion, so much so that he doubted that a static situation existed at all, even in the two arms of a balance; and of course he attacked the concept of "latent" heat.[22] But by 1862 he wanted to show that he was a forerunner of the new doctrine, which with its concept of potential energy depended precisely on that concept of latency, or non-motion which can produce motion, which he wished to deny in 1846. The result is an essay four times as long as the original, completely inconsistent, and loaded with commonplaces borrowed from other authors. Grove was not a fumbling "forerunner" of the true doctrine; he was perfectly clear in 1846—about another doctrine.

127

This concept of potential energy was added to save the doctrine of conservation because kinetic energy, representing actual motion, is not conserved. Where have we encountered something like this, not the actual but a potential to produce the actual? Physics is not Aristotelian, but with this concept it took a giant step away from Newtonianism toward Aristotle. And this, I think, is one of its major characteristics in other fields as well.

Energy, too, did not remain so closely linked with motion. Englishmen kept on thinking of it that way as long as possible, but it took on more of the properties not of a mode of motion but of a thing in itself—more specifically of an (almost) weightless fluid—indeed by the 20th century it seemed very much like that formless primal matter upon which form is imposed, in Greek thought. The forms were, preferably, waves. Not waves of anything, of course. It should not surprise us, then, that it became possible to reduce matter to peculiar concentrated forms of energy—or, rather, to believe for some time that this was possible.

The doctrines of energy, then, were too diverse, or perhaps we should say too many kinds of people contributed to them, for us to blame them on Romanticism, or on *Naturphilosophie*, or on any one philosophic impulse. What we can say is that physicists were willing to understand them in non-Newtonian (and indeed non-Cartesian) ways; and it should not surprise us if some of these ways remind us that the 19th century was a time of a great revival of interest in Greek studies generally, and more particularly in Plato and Aristotle.

If, however, it is questionable to attribute physics to any one positive philosophic impulse, it is much easier to see one negative position which characterized it. A person thinking like David Hume would not be thinking in a way to make it easy to believe in energy and its conservation. Read John Mill's *Three Essays on Religion* against the background of English discussions in the 1860's of kinetic theory, atoms and molecules, ethers, energy and the necessitarian philosophy of the human mind it seemed to justify; and the quaint old-fashioned flavor of British "empiricism" becomes apparent. Hume may not have believed in necessary connection, and W. K. Clifford (a mathematician, not a physicist) may have insisted that science can never demonstrate absolute unchanging molecules from its merely approximate experiments, but John Tyndall believed in the one, and James Clerk Maxwell asserted the existence of the other. For that matter, J. B. Mozley's *Eight Lectures on Miracles* (Oxford, 1865) do not demonstrate any irrationality or special pleading on the part of this traditional Oxford theologian. Rather, they show the bankruptcy of the tradition of British "empiricism" when faced with the conquests of science. Physics has found the chains which hold the world together, even though Hume said that this was impossible.

I will not go into a second great accomplishment of physics, James Clerk Maxwell's electromagnetic theory, except to point out that its significance was not primarily for electromagnetic theory. Its significance was in showing that electromagnetism, looked at from Faraday's point of view, could be treated as part of William Rowan Hamilton's system of Lagrangian dynamics. It is not too surprising that the same system turned out to contain statements about optics also, since optics was one of the things Hamilton had been most interested in. But this new kind of dynamics was not consistent with Newtonian mechanics; so Newton finally had to be corrected, not merely tolerated.

Not all of physics was quite so overtly a challenge to the past, although such a characteristic does not tell us how ultimately upsetting a set of doctrines may be. A third great accomplishment of physics was statistical mechanics, and the statistical interpretation of the Second Law of Thermodynamics.

Now it probably only creates confusion to speak of Newtonian science, or of the Truth-Complex I sketch in Chapter Nine, as "mechanistic." When E. J. Dijksterhuis in his *Mechanization of the World Picture* says that, "although the rigorous mechanicism of the seventeenth century had reached its climax in Huygens, it is Newton above all who has to be regarded as the founder of the mechanistic world-picture of classical physics," I simply become puzzled.[23] It then becomes necessary for him to explain at length that this world-picture did involve random atomism, did not picture the universe as a machine; that true believers in mechanistic philosophy did not like Newton's theory of gravitation, and so on.[24] It seems simpler to deny that "classical physics" had a mechanistic world-picture. So that is what I do. A machine is always teleological, and usually necessitarian as well. Men in the 18th century liked teleology, but they did not put it into the explanation to begin with. Their procedure had the happy result that it was still necessary to explain order, after the science was done; hence they could postulate God to explain it. As Hume pointed out, the "laws" of nature are merely regularities which show that there is order; they are not causal connections which act to bring it about. In science taken by itself, facts are truer than laws are truer than theories of mechanism. This is really what British "empiricism," and Baconian induction, are about. It did not, as I observed above, satisfy the new scientists of the 19th century; they wanted some of God's attributes in the scientific system itself. One solution was Kant's solution: to find the necessary absolute qualities in men's minds, in the mental categories they used to put science together. But this did not satisfy most 19th-century scientists I have studied. They wanted God's attributes to be out there, in the world as studied by observation and experiment and apart from themselves. They were moving God's attributes from theology into science; but they wanted them to be real attributes of the natural world, not products of their own study. This was a problem they eventually had to face up to squarely after 1900. They solved it by admitting only that they *interacted* with nature to produce the observational results; but this interaction was limited and could itself be quantified. A certain amount of uncertainty was what they would admit to.

Of course uncertainty had been a prominent feature of one of the sciences since the 18th century: observational astronomy. That it is usually treated under the heading of "instrumental error" should not lead us to think that it was all that different from the problems involved in observing an electron in the 20th century. *Metaphysically* it may have been; but that depends on your metaphysics; and I see no reason to accept 20th-century physicists as especially good judges of the novelty of their metaphysics. What I observe is that many of the scientific techniques used to cope with the problems were the same; in particular, one form or another of the somewhat varied ideas which go under the name of "statistics" or "probability."

This kind of approach, then, became important to astronomy around 1800, first negatively as a way of dealing with error—not casual error, but the error implicit in all observational science. Then positively, as a way of arriving at an answer, as in Carl Gauss's *Theoria Motus Corporum Coelestium* of 1809. Overall, then, it seems likely that probabilistic thinking was picked up by the physicists from, or at least because of, its use in astronomy. I believe this supports Professor Mary Hesse's position, or at least is not incompatible with it.

Once they had a bit of statistical mechanics, one of the early things the physicists did with it was to counteract the conservation of energy (the First Law of Thermodynamics) with the notion that the *availability* of energy is not conserved (the Second Law of Thermodynamics). This notion, based on the simple observation that steam engines consume fuel, might have been difficult to make plausible when extrapolated to the entire universe, had not a statistical explanation showed that it was really only a basic idea of the Newtonian Truth-Complex reappearing. The most natural state of the constituents of the universe is randomness, and it will statistically prevail, except where locally countered by organizing centers. But these centers are themselves subject to the forces of randomness and so—without God—the universe will eventually run down. Any average 18th-century natural theologian could have told the physicists that; but they preferred to reintroduce the idea as though it were a new discovery of science. So the anti-Newtonianism of the First Law was to some extent countered by the pro-Newtonianism of the Second Law.

This kind of thing happens often enough that it deserves a set of general statements of its own. In the system of European thought since the early Middle Ages (which I find it convenient to call "Christian," although I don't insist on it) there is always an area of the universe which is random, arbitrary, uncontrolled by any categories such as reason, science, law, justice, or mercy. But it is not always the same area. The tactic of the anti-theist, then, is to position it so that it seems to be "fundamental" or "basic" and then claim that God is not all-powerful.

The pattern of debate is standard. "Knowledge" discovers such an area. The anti-theists claim it is an area of chaos, free from purpose (that is, God). Science moves in and establishes laws. The theists then claim that these laws show rationality, design, purpose. They are correct: the laws for the new area show as much rationality, design, and purpose as any of the laws for older areas ever did. The area is therefore conquered for God.

But then two things happen. First, the new laws become familiar and begin to be felt as necessary, as "mechanical," hence needing no external power to maintain them. (Anti-theists can never live with the psychology of Hume's explanations for very long). Second, a new area of randomness is discovered "more basic" than what is now an old area. Thus if the theists are beginning to accept Darwinian evolution as God's way with organisms and therefore not really random, the anti-theists can begin to emphasize the random causes of genetic change. Thus the argument can continue on its way.

The cycle can of course go the other way. "Knowledge" discovers, let us say, microscopic organisms. Here is a great new area for showing design. Then the anti-theists show that materialist laws can be substituted for design; and behind these laws is an area of randomness; and so forth.

Thus it is always possible to believe that chance, or Fortune, or Nature, or materialism are independent of God; and it is always possible to believe that God really rules them or through them. All that is necessary to make either position plausible is care to make sure that you are not caught in territory that is being over-run by the opposition.

In the same way the irrational in human psychology can be claimed for God (Coleridge, and many existentialists) or used to explain away God (Darwinians and Freudians).

Such debates are not useless, because they sometimes stimulate the production of good science, both directly and in response to the challenge from the other side. Statistical mechanics was perhaps a direct invasion of physics by Newtonian Christianity. Darwin was stimulated by the Romantic Christians into producing a counter-system which would explain all of man's faculties on a materialistic basis; elucidating one biological theory such as natural selection was not enough for him. And so forth. The endless debate is one which keeps Western thought moving.

There are two other such abstract characteristics of this thought-system that I have been able to identify. One is that all versions of the system have a large discontinuity in them somewhere. By definition, a miracle is that which goes across a discontinuity. Systems therefore do not differ as to whether or not they have a miracle. They all do. They differ as to where the discontinuity is. Overt Christians often have their miracles out in the open; others often shove theirs as far away from the center of attention as possible, near the edge of the universe somewhere. A scientist will believe anything he has to about the rest of the universe, so long as it enables him to cultivate the part he is interested in in the way he deems suitable.

(My notion is that this characteristic is related to the fact that arithmetic and geometry are incompatible; but that is only a speculation.)

The other such characteristic is that, by going a bit further, one can always prove that an apparently symmetrical situation is actually asymmetrical but that this does no good, because people will always assume that the next situation they encounter is symmetrical. Thus for every particle there must be an anti-particle; so if our current machines produce an asymmetrical set, let us build more powerful machines to restore symmetry. Or, having investigated Being, let us with the existentialists insist that Non-Being must be around somehow.

The reason why physical science, at any rate, can always be pushed to give an asymmetrical result would seem to be because of the Platonism in it (including Aristotle's as one version of Platonism). Plato's philosophy goes from zero up. I don't know why people want a symmetrical system in spite of Plato, but the impulse is certainly strong. Not merely has it produced a belief in negative numbers (from which anti-particles derive, by a simple addition of Romantic emotion: the rushing together of two opposites produces not a quiet zero, but a spectacular annihilation); but also the continued use of the Laws of Sufficient Reason, of Uniform Distribution, of Equipartition. These are the most often disproved laws in the history of the physical sciences; and their repeated failures have made exactly no difference in the tendency of scientists to resort to them at the first new opportunity.

The invention of physics, then, involved the absorption and re-deployment of a good many intellectual impulses of the early and mid-19th century, so that it is not safe to generalize about the importance of any one of them until you see how its characteristic new doctrines settled down in the new complex. To me, it is its reversion to some concepts of an Aristotelian nature which most obviously distinguishes it both from Newtonianism and from the Humboldtian way of being somewhat Romantic. Its own whiff of Romanticism, energy, has long since been converted into Aristotelian form and matter—probability waves in an ocean of primordial stuff. (Note that perfectly good alternatives to this version exist; physicists have deliberately chosen to favor this one merely because they liked it.)

You will note that I have indicated some sociological reasons for saying that a new group emerged and began to build characteristic institutions for itself (physics "labs"). But I have said nothing about the arrival of physics as a profession. My chapters on professionalization indicate why. Of course I think that William Thomson and George Stokes were professional physicists, and that James Clerk Maxwell was even more thoroughly one also. But there are several studies of the professionalization of physics going on now—and what period they will choose and what major scientists they will rule out of consideration will depend largely on whether they use one or another of the positions I sketch in my professionalization chapters. It is predictable that there will be arguments and controversies while those who claim that physics became a profession when and only when it produced technicians for industry and government attempt to convince the rest of us that Ampère and Clerk Maxwell were amateurs. From my point of view, physics is one area that was largely professional almost from the beginning. Not entirely—but more so than most sciences were at their beginning. Since it had a strong mathematical orientation from the beginning, was put together as a general subject using already developed topics such as acoustics, optics, etc., and since its period of emergence was the one in which other sciences were becoming professional, this is not surprising. We should expect that each of the sciences would have its own history within the larger framework of the time, and not that a pattern taken from the emergence of chemistry or astronomy would also neatly fit the emergence of geology or physics. Ideal Type thinking in this, as in other areas of history, is by now rather too simplistic to be satisfactory. There were professional physicists, in my opinion, before there was a sociological group or "profession" of physicists. The sociology clearly existed by 1900; when and under what impulses it came into being are still to be determined.[25]

Please note that when I say that Michael Faraday and John Herschel were not physicists, I do not mean that they were involved in physics but were not "professional." I mean that they were professional scientists; but their science was not physics.

A final warning. In my experience, almost all erroneous views of what went on in the 19th century are related to particular ideas as to what went on in the 18th century; and for the history of science and the history of ideas in the 18th century you can trust almost no one. The

133

amount of "hard" history of science for that period is so lacking that one simply leaps, as I did above, from Newton in optics to Young in optics. Who else observed what, how well and how thoroughly, and with what results? It is hard to tell. It is no reproach to my friends who are trying to do something with the 18th century to tell them that their labors have not yet reached the point at which a 19th-century historian can confidently go ahead from the stable platform they have erected. And the same is emphatically true of the history of ideas related to the history of science. Trust no one, except possibly Marjorie Hope Nicholson and me; and don't trust me very much.[26]

Notes to Chapter Four

The first version of this chapter was as a talk to the colloquium of the Department of Physics, University of Maryland, February 17, 1970; repeated, with some changes, for the philosophy and history of science section at Leeds University in May 1971. I believe I also included the topic in a talk to the historians of science at the University of Pennsylvania in December 1970.

D. S. L. Cardwell's *From Watt to Clausius* (Ithaca, N. Y., 1971), the first significant publication I know of on the subject, was not available when I first wrote down my ideas, and I did not read it until I had finished writing this chapter. Cardwell believes that physics was a new thing invented in France in the early 19th century. He traces the interesting connections between improvements in the steam engine and attempts to understand them theoretically. Thus his concentration in physics is on heat, and eventually on the first and second laws of thermodynamics.

Insofar as modern physics *is* "a science concerned with energy and its manifestations and transformations" (p. 165), this is deservedly the central line of approach; but I question the complete validity of the characterization. There was a school of thought at the end of the 19th century which tried to establish this definition; but it did not fully succeed. Thus I think Cardwell has the most suggestive treatment of that approach to physics; and his detailed chapters may make my briefly stated opinions seem rather lightweight. But even on the subject of the first law of thermodynamics there is more to be explained, I think, than the papers of Joule, Clausius, Kelvin, and such men. By adding Cannon to Cardwell, the reader may get, not yet a full treatment of the appearance of physics, but a wider treatment than either author gives individually. Neither of us puts the creation of a physics profession or community as primary; these are derivative from the invention of the thing itself.

To me, Cardwell's most fascinating comments throughout the book are those concerning the reliance of heat theorists on concepts taken from water technology, which reached its own peak in the late 18th century. Others may prefer his proposal that heat furnished the basis of a whole new cosmology for some of its theorists (Chapter 4). These two topics alone would make his book recommended reading for anyone who has been exposed to the subjects of heat and energy only in the traditional history of physics framework—or for anyone else, for that matter.

1. W. F. Cannon, "P. S. If I find out what Truth is, I'll drop you a line," *Smithsonian Journal of History, 2* (1967), 12.

2. *Cf.* D. T. Whiteside, "Before the *Principia:* the Maturing of Newton's Thoughts on Dynamical Astronomy, 1664–1684," *Journal for the History of Astronomy, 1* (1970), 7.

3.   G. N. Cantor, "Thomas Young's Lectures at the Royal Institution," *Notes and Records of the Royal Society, 25* (1970), 100; see also p. 88. (I don't mean to imply that Cantor would approve of my judgment regarding Newton and Young.)

4.   C. Stewart Gillmor, "Borda, Jean-Charles," in *Dictionary of Scientific Biography,* II, 300. See in general Robert Fox, "The Rise and Fall of Laplacian Physics," *Historical Studies in the Physical Sciences, 4* (1974), 89–136, only noting that "physics" is not the French word used by the participants.

5.   Maurice Crosland, *The Society of Arcueil: A View of French Science at the Time of Napoleon I* (Cambridge, Mass., 1967). See also John Herivel, *Joseph Fourier* (Oxford, 1975), pp. 149–241.

6.   George Davie, *The Democratic Intellect* (Edinburgh, 1961); see esp. pp. 191–197. Of course I don't agree with his characterization of William Whewell as an English empiricist (p. 185); Whewell was a Lancashire Idealist. There is an interesting preliminary attempt to trace the Scottish influence: Richard Olson, *Scottish Philosophy and British Physics, 1750–1880* (Princeton, N. J., 1975). Because of the complexity of the invention of physics, it is an unusually difficult problem.

7.   There are several "transmission" and "reception" studies now, two of the more interesting of which are R. G. A. Dolby, "The Transmission of Science," *History of Science, 15* (1977), 1–43; and Geoffrey Cantor, "The Reception of the Wave Theory of Light in Britain," *Historical Studies in the Physical Sciences 6* (1975), 109–132.

8.   *Sir Robert Peel from his Private Papers,* ed. Charles S. Parker (London, 1891), I, 387. See also Roy MacLeod, "Of Medals and Men," *Notes and Records of the Royal Society of London, 26* (1971), 81–105.

9.   "Adjudication of the Medals," *Phil. Trans. 130* (1840), Part II, facing title page; see also p. vii.

10. John Herschel, *Preliminary Discourse on the Study of Natural Philosophy* (London, 1831), pp. 93–94, 259.

11. John Herschel, "On a case of superficial colour presented by a homogenous liquid internally colourless," and, "On the Epipolic dispersion of light," *Phil. Trans., 135* (1845), 143–146 and 147–154; George Stokes, "On the change of refrangibility of light," *Phil. Trans., 142* (1852), 463–563, and *143* (1853), 385–396; Stokes to Herschel, 6 April 1852 (John Herschel Papers, Royal Society Library). The letter has a note on it, probably by Herschel's son, "he ought to have said so," and it is true that Stokes did not give Herschel enough credit in his publications.

     A few more details, set in a somewhat different and suggestive context, are to be found in Michael Sutton, "Sir John Herschel and the Development of Spectroscopy in Britain," *British Journal for the History of Science, 7* (1974), 56–57.

     The Royal Society's referee report on Stokes' paper was by Herschel, who praised it to the skies (Herschel Papers, Stark Library, University of Texas, Box 37).

12. Thomas Young, *A Course of Lectures on Natural Philosophy and the Mechanical Arts* (London, 1807), I, pp. 9, 14.

13. Young, *Lectures,* I, 606.

14. Quoted in J. R. Partington, "Thomas Thomson, 1773–1852," *Annals of Science, 6* (1948–50), 121–122. See also J. B. Morrell, "Thomas Thomson." *Brit. Jour. Hist. Sci., 4* (1969), 245–265, for Thomson's teaching at Glasgow.

15. *Brit. Assoc. Report, 1 and 2* (1831–32), 46–47, 97–98, 112–113, 115–123; *3* (1833), v–vii, xxxix–xl, 469.

16. *Lithographed Signatures of the Members of the British Association...with a Report of the Proceedings* (Cambridge, 1833), p. 64. See also William Whewell, "Report on the Recent Progress and Present Condition of the Mathematical Theories of Electricity, Magnetism, and Heat," *Brit. Assoc. Report, 5* (1835), 1–2. Whewell does not recognize the existence of one major new science but of several lesser ones—none equal to good old physical astronomy.

17. Herschel, *Preliminary Discourse,* pp. 18–19, 50, 228.

18. Herschel, *Preliminary Discourse,* pp. 93–94.

19. Herschel, *Preliminary Discourse,* p. 156. Some of the later results of Herschel's doctrine can be traced on into the 20th century: see Lawrence Badash, *The Completeness of Nineteenth-Century Science, 63* (1972), 48–58.

20. John Herschel, "Address of the President," *Brit. Assoc. Report, 15* (1845), xli. See also Thomas Carlyle, *Sartor Resartus,* (London, 1975). pp. 12, 64, for another feeling that the times needed a new "Science of Things in General." This was published in 1833–34.

21. *The Destiny of Nations:* see Joseph Warren Beach, *The Concept of Nature in Nineteenth-Century English Poetry* (New York, 1936), pp. 47–50; Thomas Peacock, *Headlong Hall,* ch. 3; B. Ifor Evans, *Literature and Science* (London, 1954), p. 59. My presentation does not rely on recent accounts such as Yehuda Elkana, *The Discovery of the Conservation of Energy* (London, 1975).

22. W. R. Grove, *On the Correlation of Physical Forces: Being the Substance of a Course of Lectures Delivered in the London Institution in the Year 1843* (London, 1846), pp. 35, 15ff.

23. Edward Jan Dijksterhuis, *The Mechanization of the World Picture* (Oxford, 1961), p. 490.

24. Dijksterhuis, *Mechanization,* pp. 479, 495–496.

25. See Paul Forman, John L. Heilbron, and Spencer Weart, "Physics circa 1900," *Historical Studies in the Physical Sciences, 5* (1975), 1–185; and the most interesting unpublished Harvard Ph.D. thesis by Kenneth Nier, *The Emergence of Physics in Nineteenth Century Britain as a Socially Organized Category of Knowledge: Preliminary Studies.* On geology see Roy Porter, *The Making of Geology* (Cambridge, 1977).

26. The most directly relevant new contribution to this subject is Thomas Kuhn, "Mathematical vs. Experimental Traditions in the Development of Physical Science," *Journal of Interdisciplinary History, 7* (1976), 1–31; but one should not ignore the general suggestiveness of Gerald Holton, *Thematic Origins of Scientific Thought: Kepler to Einstein* (Cambridge, Mass., 1973). My notion has been tested and approved by Pierluigi Pizzamiglio, *L'Insegnamento della Fisica tra il Settecento e l'Ottocento* (Bologna, 1975). If my approach is correct, however, it is the content of the subject "physics" which takes priority, and so we should pay careful attention to such studies as Daniel M. Siegel, "Completeness as a Goal in Maxwell's Electromagnetic Theory," *Isis, 66* (1975), 361–368; and Stephen G. Brush, "Irreversibility and Indeterminisms: Fourier to Heisenberg," *Journal of the History of Ideas, 37* (1976), 603–630; as well as the latter's comprehensive *The Kind of Motion We Call Heat* (2 vols., New York, 1976).

Chapter Five

Professionalization

In the history of science, as in other branches of history, it is possible, and often useless, to argue as to what is the basic level of explanation. Should we be concerned with overarching intellectual constructs or specific mental tools, with the functioning of intellectual groups or with experimental techniques and actual field work done by individuals? In truth, all of these partial methods are necessary. Debates over which is more necessary are sometimes expressions of the determination of one set of scholars to keep a journal available for their researches, and the determination of a new set to justify publishing what *they* like to do.

Enterprising scholars often have a collection of rejection letters somewhat more numerous than the entries in their list of publications. (I don't; eccentric or not, I have been shamefully successful in placing articles.) It is amusing to note how many of those shown to me by friends amount to statements that the author should be ashamed of himself for thinking that what he has chosen to do is really scholarship at all. The more established the author becomes, the shriller are the rejections. Having been a scholarly referee for a number of years, I must confess that I have composed a number of such evaluations myself (of which I am now ashamed). Such judgments (if other referees are like me) are largely the expression of the territorial imperative by a group of challenged possessors. A primary function of the referee system is to make sure that any given vehicle of publication remains as limited a way of expressing truth as it always has been, so that its readers may feel secure. If they are also a bit bored, well, security *is* a bit boring.

(What about weeding out simple incompetence? Actually, an intelligent assistant editor can do much of that. More articles are deflected by the editorial staff, before they reach the stage of being considered by scholarly referees, than an editor usually likes to admit to an open meeting of his scholarly society. At least this was true of the journal I edited, and I hear that it is true of others.)

There is, however, a way we can criticize new fads in scholarly interest without falling into this pattern, and it is a way suitable to historians. The history of science seems fated to go through all of the stages of development that other branches of history have gone through, and probably in the same order. We can learn from historiography not so much what to avoid as what we shall have to put up with, hoping only that we can speed the transition from one stage to the next by just a little. Or rather— since a stage once reached is never abandoned, but is always full of prac- titioners, we can only hope to increase the population of the higher levels by two or three. In history as in quantum physics, the lower levels are the most attractive, and are usually filled up first.

I don't pretend to any exhaustive study of past historical writing, but for my purposes a simple scheme is adequate. (This is true of all of my analytical schemes. I erect one sufficient for the material I have at hand, and do not invest it with categorical finality or exhaustiveness. If it is use- ful for assimilating further material, well and good. Then it is a theory. If not, it remains a useful classification for the original material.)

If we let Stage 1 be *Sheer over-simplification* and thus dispose of heroes, nationalisms, schoolbooks, and history written to prove a world-view, then we come to:

Stage 2. *Rationalistic and religious orthodoxy.* All stages are nicely exem- plified in writing about Isaac Newton. Stage 2 is represented by the inter- pretation of Newton as really being a rational mechanist, who cast teleology, purpose, goal-seeking, out of science. Therefore he was a means whereby the Good triumphed. This rationalistic orthodoxy is equivalent to that of the Pilgrim historian William Bradford's *Of Plimouth Plantation,* where, however, the Good is pro-teleology rather than anti-teleology. The notion that science must expel religious ideas from its own pure set of concepts is a current popular form of the orthodoxy.

It is easy to see how Stage 2 thought can be softened down by the addition of a time process and thus give us the ever-popular:

Stage 3. *The Whig interpretation of history:* science has been becoming more and more like what we today know to be better. If one is squeamish about saying that it has become truer (which would involve an outside standard of truth), one may say it has become more "objective" (whatever that means). Or more comprehensive. Or more fruitful. Or something. Perhaps "more true to itself" is a good general expression: science should and will develop this way if properly aided and if not opposed so much as to stunt its growth or deform it. That is, the Whig interpretation is nicely Aristotelian. Science, like liberty, is like an oak tree.

Since an obvious characteristic of science today is not that it is shel- tering but that it is dangerous, Whig history does not much appeal to the philosophical-minded, but its popular appeal is great. It justifies what is happening now, and that is what young people are always in search of.

I won't try to derive the next stage from Stage 3, but I will locate it in history as being prominent in Western thought in the first half of the 19th century. Not only history but also the material sciences experienced:

Stage 4. *The search for Ideal Types.* This gave us Richard Owen's archetypes in biology, some versions of Catastrophism in geology, and lots of Type hypotheses in chemistry. If I use the word "patterns" as equivalent to Types, you can doubtless think of various historical interpreters who fall into this stage. In the history of science I will draw attention to two: William Whewell, who in his *History of the Inductive Sciences* of 1837 could maintain that he was in conformity with much of the advanced scientific thinking of his time; and Thomas Kuhn, who in his *Structure of Scientific Revolutions* of 1962 could maintain that he was in conformity with much of Whewell.

Thomas's approach is, I think, over-used by sociologists and under-used by historians. It is true that in his system both a "paradigm" and a "revolution" may be entities of any magnitude and any significance. He himself is reported as saying (*Science*, 8 July 1977, p. 145) that "it is not going to apply in any detailed way to any particular episode." In my terminology, that means that it is not a theory at all. It is rather a new emotional assertion as to where the soul of the scientific process is; it is a Romantic manifesto. As such, the proper use of it is to re-orient one's attitude to the history of science, and then seek for actual theories (hypotheses—derivations—comparisons to data) compatible with this attitude.

Simply as a matter of intellectual history, the approach has affinities to Ideal Type thinking, a type which has become popular with sociologists but was in the 1840's characteristic of a number of British natural scientists. I consider some of these—William Whewell, for example—to be best described as Romantics. Whewell, however, was enough of a Lancashireman to try to rescue the objectivity of science, even though this called for (in his estimation) strong post-Kantian spectacles. The impact of Tom's writing is to destroy objectivity altogether. This is very Romantic, but it is not my position. I am analagous more to the English response to the Scots Common Sence School; so that persons like John Herschel, not any of the Romantics, Idealist, or Ideal Type thinkers are my *sympatico* characters in the 1840's. I had expressed my own position in publications by 1964, when I first read the *Structure;* and in general have not yielded to it much.

In general; but in detail it is different. A Romantic often expands a characteristic of a limited part of the world, say his own soul, into a pattern for the universe. A good Scots-Irish-English intellect reduces the scheme to one of the small areas from which it originated, and finds that there it works. The "paradigm" approach works well for a limited set of reformers in a given context. One should not look for "the" paradigm for all of geology in a century, for example. But Charles Lyell's *Principles of*

Geology yielded a paradigm for a number of other aspiring scientists: not too many, but some. That is, they could do work of the kind the *Principles* suggested, and then feel that they had done what a scientist should do to justify his social existence.

Alternatively, a follower of Lyell who found he could *not* practice geology as Lyell indicated it should be practiced, might drop out of geology altogether; and one young scholar, Barry Gale, is investigating an important case of this kind.

I have also used the term above, in Chapter Four, to describe the new subject "physics"—but this must be taken as a tentative suggestion.

What is involved, then, is the personal feeling of the social validity of spending one's efforts in a particular way, and this is most important in the early stages of the activity, when it may be the determining factor as to whether enough work is done to advance the area significantly.

As for "revolutions," I have proposed my own in this book; I see them as big things, not as processes at every level. No one, I think, is really very much like the others. I recommend the magisterial article of I. Bernard Cohen, "The Eighteenth-Century Origins of the Concept of Scientific Revolutions" in the *Journal of the History of Ideas, 37* (1976), 257–288, as required reading for everyone before he or she believes in *any* revolutions in the history of science, Tom's or mine or even the traditional ones.

A characteristic of Stage 4 schemes is that a newer Ideal Type is not obviously better than an older Ideal Type. This is fine in paleontology: why should one want the Devonian to be better than the Silurian? It is fine in political history: why should one want to admire Macmillan as prime minister more than Melbourne? But in the history of science, it remains true that in some sense Darwin's evolution in the 19th century was "better" than evolutionary schemes in the 18th; and one is in danger of falling back into Whiggism with such phrases as "more complex," "greater explanatory power," or etc. He is a sturdy metaphysician indeed who will consistently maintain that science in all these centuries has never really advanced at all.

Stage 5 is occupied by persons who have gone beyond Whewell and Kuhn, but there are not so many of these that they can be readily analyzed as a group. One technique, however, may be expected. In Stage 4, the archetype or paradigm or pattern is in some sense felt to be implicit in the events, perhaps logically prior to them (if it is God's pattern for history) but in practice discoverable or verifiable by inspection. Its exact status is a fertile field for debates between, e.g., Kantians and Hegelians and Schellingites.

On Stage 5, one may expect a metaphor to be dragged in from outside the field, ears and all. That it is a metaphor, and even a crude one, is not disguised. After all, "fields" and "boundary conditions," or "sinks" and "sources," are not very sophisticated borrowings from agriculture and land drainage, but they served 19th-century physicists well.

My example, by the way, of a Stage 5 thinker in science is Charles Darwin, all of whose metaphors failed to be complete expressions of his theories. I am not sure that I have encountered a Darwin among historians lately.

In this hierarchy of mine, most studies of the professionalization of science in the 19th century fit neatly into Stage 3, Whiggism. Science once was not professionalized; it now is; the process whereby it became so is a path to the present; and that is good. Here we encounter a verbal deceit which many professionalization scholars use. They insist that they are not praising the process, they are not insisting that professionals are better than amateurs. *But they are saying just that anyway.* Their disavowals are contradicted in their texts. They show throughout their treatments that professionals know more, know better, work harder, practice more, and in general are the mechanism whereby science today is at a larger, higher, more influential, and more sophisticated level. Refined leisurely gentlemanly science was wonderful (they will admit) but it could not keep up with the hard-driving professionals. This is refined Whiggism; perhaps nostalgic Whiggism; but Whiggism no less.

I first became aware of professionalization as a problem in the history of science in 1965, when, at the History of Science Society's meeting in San Francisco, there was a joint session with the Society for the History of Technology on the "Professionalization of American Science and Technology." Such studies, of course, did not originate in the history of science; the fad began in other fields, and I will discuss one of the better treatments in sociology, Geoffrey Millerson's *The Qualifying Associations* of 1964, further on. One advantage of the approach, indeed, is that it can be applied to anything. One can study the process without saying very much about the science being professionalized, or saying very much new about the scientific work of the men who are becoming professionals. One can study the organization and politics of a society or association without considering very intensively the subject of which it is a society or association. To be sure, it is then rather difficult to show that the process, or the society, had any effect in producing or stimulating science.

For whatever reason, the movement grew. Rather than to single out individuals, I have constructed a composite definition of what a professional scientist is considered to be in recent studies of the professionalization of science in the 19th century. It is an Ideal Type, if you wish—but not a straw man, since it is derived from actually published

141

articles and actually delivered lectures. It represents, shall we say, the set of ideas which an unwary entrant into the field might pick up and use as the basis for his new study.

What, then, is a professional scientist in current professionalization studies?

1. *He is one trained in an institution for training specialists.*
   Implication A: he is a specialist.
   Implication B: he has learned to use an esoteric technique (mathematical, logical, laboratory, linguistic, etc.) which makes his results more "reliable" or "precise" or "scientific" than those of other people.

2. *He earns his living largely by selling his professional competence.*
   Implication: he works at his competence more regularly than do other people.

3. *He belongs to a group which stands for exclusiveness and ethics.*
   Implication: the exclusiveness is necessary to enforce the ethics. (The ethics is one of scientific conduct, not of general conduct: he does not lie about his experimental results, but he might pad his expense account.)

What is wrong with this set of ideas?

First, it is used quite regularly to classify some of the best scientists as not professionals, usually with the implication that therefore their scientific findings need not be considered as very definitive after all. In particular, it is used to classify English scientists in the first half of the 19th century in that fashion, because the English did not have nationwide, routinized, institutionalized methods for training and employing specialists.

To take some specific examples I have encountered:

If you learned conchology by talking with members of the Sowerby family and by studying with Deshayes in Paris (that is, by getting together with some of the best men in the field), the professionalizationist will not admit that you had a specialized competence in conchology, because you did not take a registered course in the subject. (Charles Lyell)

If you learned biology by talking to the professor of botany and listening to his lectures but not taking his course, and then by going on a five-year field trip, the results of which were good enough to be published with the aid of a special government grant, you are not a professional biologist, since you have no science degree. (Charles Darwin)

If you have enough money to spend your time exclusively on research, and have turned down almost every relevant major position in Britain, you are not a professional compared to the young observatory assistant who took a job as government astronomer at Capetown as his first real scientific opportunity. (John Herschel)

If you did go through normal training procedures but impress a historian as having been successfully proficient in a wide variety of researches, you will be cited as the last member of the great amateur era. (Thomas Huxley) (The amateur era, by the way, usually ends in whatever decade the historian either begins or ends his study.)

These, then, are some of the men who have been declaréd to be "not professional scientists." I can give a few more examples of men who ought to be excluded, if we hold to the definition given above. George Boole read Lagrange's *Mécanique analytique* at age 18 on his own, and hence did not have to go to an institution to have it explained to him. Arthur Cayley went into law for fourteen years, during which time he produced 200 mathematical papers, invented matrices, expanded the subjects of invariance and of n-dimensional space, and developed a new form of projective geometry. James Clerk Maxwell quit teaching in 1865 and spent six years on his estate not earning a living; during this "amateur" period he composed most of his *Treatise on Electricity and Magnetism*. If we consider the geological co-workers Adam Sedgwick and Roderick Murchison, the former (although he taught himself) at least earned money as professor of geology at Cambridge. Murchison not only was self-taught but also lived on his wife's money, so he was less professional than Sedgwick.

In short—remembering always that "professional" does imply "better," regardless of what the author says—the application of these criteria produces nonsensical results. There are men who are known to their contemporaries as masters of their craft, whose opinions on an experiment or a bone fragment or adjusting a telescope are worth four times as much as the average scientist's. Their opinions in a society or association have much more weight than those of an average member, not so much because they are correct on a particular point as because their standards of evidence or of proper scientific debate are the more reliable ones available. When is a speculation a theory? When the suggester is important.

These men are professionals, whoever else may be also, and they are recognizable at any given time. We may therefore start with a list for early 19th-century Britain: say, Richard Owen, John Herschel, Robert Brown, Adam Sedgwick, Joseph Hooker, George Stokes—and make sure that our definition does include them at any rate.

A second objection to the scheme given above is that it classifies men as professional scientists *because* they spent most of their time doing something other than science. A common thing they did other than science was teaching. Now teaching may be enjoyable, and it may contribute to a scientific culture, but it certainly takes time away from research. It is a well-known assertion about American scientists in the second half of the century that some of them would turn down a good job at, say, the University of Kentucky because its research facilities were small and its teaching load substantial. They were willing to stay in relative poverty near a research center. (The same phenomenon has been observable in England, at Cambridge.) Here is how the American scientist Cleveland Abbe put it in a letter to his mother in 1865:

> I do not think it has ever been my highest desire to be a Professor in a College which most of my friends seem to have thought the practical end of my study. When I have thought of such a position it has only been as a means of livelihood by no means entirely accordant with my tastes. The unvarying experience of teachers is that their work is too exhausting to leave them energy and time for investigations. They are rarely able to keep up with the progress made by others.[1]

Another common thing people with scientific training did instead of science was *industrial chemistry*. The science of application is just as much science as is the science of abstract inquiry (at least by the standards of the Enlightenment); but when one finds university-trained chemists hired at Manchester in the 1840's and 1850's, one may ask whether they were supposed to advance profits rather than applied science. Young graduates may have gone to Manchester hoping to be the new Bertholets; the question in my mind is how many new and fruitful methods had they devised by the 1860's—and preferably published?

It is not my intention to tackle once again the unremunerative task of distinguishing between science and technology. I want to distinguish between innovators, whether they be called scientists or engineers, and those who are well-trained enough to administer an already-established process, or at most to introduce a process invented elsewhere into their own establishments. For the time being I will not endorse a special set of terms. "Technician" is now felt to be down-grading, and "practitioner" brings in a comparison to, e.g., lawyers, the validity of which has not been established. It is, however, the confusion of these with the first group that has led to much of the confusion about "professionalization," both in the 19th century and in studies made today.

In mid-19th-century England a propaganda line was developed by one group of educational reformers which remained essentially unchanged at least through the First World War. Whether it was the need for scholarship in the universities, technicians for the chemical industries, the founding of new colleges, or secondary education on a large scale, the same praise of the same characteristics of the German system came forth.[2] The Germans, these reformers said, were professionals; the English were amateurs. The stress should be upon education, that is, going through a rigorous directed process certified by a diploma or degree at the end. The old ways were lazy ways. Fellows of English colleges had leisure to waste their time; true professionals worked hard. In science, the ideal was experimental or laboratory science, and the exemplar of laboratory science was chemistry. In practice, the emphasis of the reformers was in producing a considerable number of scientifically adequate routinists for industry.

Here is where the ideas that a professional must have a specific technical education, that a professional works at a job, that a professional routinist is to be classified together with a professional innovator, and that the old English way did not produce professionals, got all mixed up together. This was, of course, delightful for the routinists, and they set about devising ways to secure their professional status.

Let me stop to emphasize that the ranking "professions" in England were, in the 18th century, the military (that is, the officers), the Church, the law, and medicine. One was a gentleman in order to be an officer. But the Church, law, and medicine conferred social status on men involved in them, and that is why the word "professional" means something good— but not too good, of course. There was no such thing as a professional shopkeeper. The professions offered *routes through the middle class* from the one towards the other. Holding a Church position was compatible with being a gentleman, but it is true that many clergymen were of fairly low social status in the 18th century. That is because many of them were of low social origin. That they had any status at all was due to the fact that all clergymen had some status. The largest group of students at Oxford and Cambridge were nobodies who, through a university education and a cleric's position, hoped to become somebodies; the process converted them from persons of no status into persons of low status. Sometimes they did better: in the early 19th century William Whewell, son of a master carpenter, became Master of Trinity College, Cambridge, and then outranked his wife Lady Affleck at a royal marriage ceremony. In general the social level of Cambridge and Oxford rose throughout the 19th century as compared to the 18th, as they attracted more and more students from the substantial middle class.

Science was a proper pursuit for a gentleman in the 18th century (and therefore also for a clergyman). A socially-aspiring London physician might well want to add F.R.S. to his name, without of course attending meetings of the Royal Society or writing scientific papers. The question in the early 19th century was whether a man could devote himself *exclusively* to science and retain status. Here the efforts of Joseph Banks at the Royal Society and Humphry Davy at the Royal Institution paid off. Banks's social efforts were sneered at by many of the younger men after 1800, but (perhaps because) they were the beneficiaries. The uprising of scientists at the Royal Society in 1830 was in part a proclamation that science could now stand on its own, that it had the prestige of a profession.

Thus as late as 1815 John Herschel, although he was the son of a famous scientist and had been a brilliant student at Cambridge, nevertheless rejected the Church and failed as a law student before settling down as a scientist. Richard Sheepshanks around 1825 qualified both in the law and as a clergyman before settling down as a scientist. Charles Darwin around 1830, son of a successful physician, did not like his exposure to a medical education. His father's middle-class objection that going into science would encourage Charles to continue wasting his time (that is, that it was a gentlemanly avocation) was overridden by a man of higher status, his uncle Wedgwood, who could see that science, although not part of the professional *duties* of a clergyman, was socially compatible with being one.

In the early 19th century, then, science was becoming more of a profession in England: that is, a person who spent most of his energy on this activity might initiate, maintain, or improve his middle-class status; might devote his life to this activity without needing the justification of being a doctor or lawyer or clergyman in addition; and might feel that his success was determined by the reputation he gained among his peers, not by monetary returns.

If one wants a decade to specify (it was certainly not that specific), I would suggest the 1830's, in part because of the usage of one word. This is the decade in which I have noted that the way the word "science" was being used began to annoy some people. Young Arthur Stanley attended the 1838 meeting of the British Association for the Advancement of Science. His father, the Bishop of Norwich, was much interested in such things, but Arthur was not. He wrote of meeting Robertson, editor of the *Westminster Review:* "My heart warmed the moment he entered the room, at his beginning to criticise the appropriation of the name 'science' by physical philosophers, wherein I cordially joined, and managed to sit by him at dinner."[3]

Perhaps this appropriation of the word "science" was made easier by the fact that a number of leading thinkers were Romantics of the Coleridgean school who did not want to save the word for use in morals, theology, and so forth, because they thought these subjects were *higher* things than science; Evangelicals might feel the same way. And Romantics might have no motive for saving the word for use in political economy, since that would give prestige to Ricardo and Benthamites. And there was no well-developed body of historical scholarship in England which clearly had to be considered on a par with physical philosophy. For whatever complex of reasons, at any rate "science" was tending in the 1830's to mean "natural philosophy" or "physical science" if used without modifiers.

In the first half of the 19th century, science in Britain produced its own characteristic organizations in considerable number, and these can be put crudely into three categories:

1. the one-subject scientific society, usually metropolitan;
2. the Literary and Philosophical society, usually provincial;
3. The British Association for the Advancement of Science, which I shall discuss at some length in the next two chapters.

It is not the novelty of these organizations but their proliferation which is most striking. The one-subject society had its notable models in the Linnean Society (1788) and then the Geological Society (1807). The Lit & Phil had its prototype in the Manchester Lit & Phil (1781). The Linnean might be considered not as a "forerunner" but as a parallel to Joseph Banks's equivocal position. The Lit & Phils can be studied as interesting combinations of all sorts of things, often perhaps being places where one or a few professionals found arenas for action before, during, or after going nationwide. Note that I do not put statistical societies in any of these categories. They are certainly characteristic of the 1830's and 1840's but not, perhaps, characteristic of science.

Let me now return to Geoffrey Millerson's sociological study of professionalization in Britain today, *The Qualifying Associations*. His title is chosen to help untangle the confusion caused by grouping together all groups which lay claim to the word "professional." After rating fourteen elements included in various previous definitions of the word by two or more authors, and nine additional ones which only a single author has used, he distinguishes among four different kinds of professional organizations. "Prestige" and "Study" associations are exemplified by the Royal Society and by single-subject societies such as the Astronomical and Zoological Societies. Essentially all "scientific societies" are in these two categories. On the other hand, "Qualifying" associations actually examine and certify individuals wishing to practice; and "Occupational" associations work for the interests of the persons so qualified.[4] These examined and qualified people are all ones who sell their services. Some, as doctors and lawyers, may be independent. Most of them are employees. One does not go to the Institute of Bankers (which has a membership of over 48,000 and examines 30,000 candidates a year[5]) to look for the head of a banking firm.

The apparent exceptions to my generalization that "scientific societies" are not "qualifying associations" are the Institutes of Chemistry (1877) and of Physics (1918). But Millerson shows the reason for these apparent anomalies when he says, of the latter: "By the beginning of the twentieth century, and as a result of the First World War, the physicist emerged as a distinct professional scientist."[6] By this he surely does not mean that Kelvin, Stokes, Clerk Maxwell, Rayleigh, and J. J. Thomson were amateurs. He must mean that physicists were beginning to be employees hired by government and industry in considerable number, probably to work on applied science subjects.

Millerson's different kinds of groups help to show the confusion in the use of the word "professional." When I say that Michael Faraday was acting as a professional scientist when he decided to refuse consulting jobs, I am not discussing the same attitudes or practices which led to the Society of Cosmetic Chemists (1947). Nor, I imagine, to those which led to the British Medical Association (1832 as the "Provincial Medical & Surgical Association," renamed 1856). With the bankers, again, we can show the verbal confusion. Congruous to the explanation I gave above of the professional scientist in the early 19th century, a professional banker would be the head of a bank or manager of a local branch, whose interest was not merely in becoming rich but also in making himself and his establishment into trusted pillars of the community. According to an alternative definition, all of his employees who had passed the examination of the Institute of Bankers would be the professionals; he, if he had not taken the examination, would not.

This is not a matter of mere verbal quibbling. In the United States, for example, most university teachers are legally incapable of teaching in most secondary schools, because they do not have the teacher's certificate which "professional" teachers are required to have by most public Boards of Education.

The push for such "qualifying" associations in Britain, especially ones with formal examinations for admission, was a phenomenon of the second half of the 19th century.[7] Scientific societies, Millerson's "study" associations, proliferated in the first half of the century. Different kinds of organizations, different relations to employment, different time periods: clearly two quite different processes have been mixed up in definitions of "professionalization" in the 19th century.

This mixing up was done at the time by the two groups of people I have identified (and perhaps by others). One was the collection of Germanophile reformers within and just outside of the establishment, who thought that scholarship needed rigorous training and continuous discipline to make it more modern. Thus "professional" training to them meant a hierarchy of specialized course work, and a degree. It is these people who, probably in all honesty, initiated the fiction that Charles Darwin was not a professional scientist. How could he be, if he did not attend a German-style university with laboratory courses, and if he did not earn his living as a scientist? Actually, of course, this approach to "professional" training simply made it more compatible with the instincts of a new kind of student, the hardworking ones lower on the class scale—Gladstone's people, let us say, instead of Melbourne's supporters. The class bias is very clear in many sources one reads.

The other group successfully manipulating the meaning of the word "professional" was the collection of vocational organizations striving to gain some of the prestige of the higher-ranking groups. Their mechanism was exclusiveness (always necessary[8]), gained by the relatively clear-cut device of an examination—examinations having gained prestige, perhaps, as the "university" way especially from the Cambridge tripos since about 1800. Scientific societies such as the Geological and the Astronomical were of course exclusive, but their mechanisms for exclusion were too fuzzy for a group of hard-working achievers to be satisfied with copying.

Thus a rivalry developed between attending courses, with a certificate or "degree" awarded for successful completion of them, and merely passing an examination; this rivalry continues to the present.

The confusion between the two processes called "professionalization" also continues to the present in Millerson's habitual usage, even though his analysis is a beginning of clarification (especially with respect to ethics *vis-a-vis* the public, which, he notes, has been of little concern to qualifying associations, in spite of what they say.[9]). A typical example of such usage occurs when he discusses the rise of qualifying examinations after 1850. He calls it "an astonishing and bold step forward."[10] Why not a step sideways? Or simply "an astonishingly different mechanism for that class of society?" Why not? Because professionalization authors are Whigs.

A more serious example occurs in Millerson's own definition of a professional. As explained at length in six paragraphs, the definition includes the phrase "knowledge and experience must be obtained." On the next page, in a one-sentence summary, this provision has been transformed into "after advanced training and education."[11] Thus begins the slipping towards requiring some sort of formalized course-taking; the next author will begin to look for an advanced training institute, and the next for a diploma proving the education. When one encounters this process in one of the best authors simply by turning the page, it is no wonder that lesser ones so often wind up with requirement 1 of the unsatisfactory definition I gave above: trained in an institution for training specialists.

It is not the function of the historian to decree who really are "professionals" in our own society and who are not. He can, however, object to definitions framed to fit every group which now claims the word, being applied to the 19th century so as to hide the fact that there were two distinct processes at work. Millerson's observation that "any collection of people forming a Qualifying Association acquires 'professional status'" is enlightening, if not at all satisfactory.[12] Any such group is *claiming* status of some kind, and, if the association is successful, it shows that they were claiming it correctly. But use of the blanket eulogistic word "professional" does not mean that they all have the same status; and a group does not have to have a qualifying association to be "professional." Scientists who

149

are members of, let us say, both the Zoological Society and the Royal Society (neither of which are "qualifying" in Millerson's use of the term) need not fear losing ground socially to members of the Association of Dispensing Opticians.

These, then, are some of the things wrong with current professionalization studies as applied to the history of science.

My First Way Out is to accept the usual procedure of external definitions, but to divide up the original unsatisfactory definitions.

Let us separate out the two characteristics of specialized training in an esoteric technique as testified to by a piece of paper, plus earning a living largely by selling the competence so learned. Let us say that it is not especially related to science or technology but is a general process which appealed to the new middle classes generally. Let us call it the process of *certification*. One earns a certificate, and shows it in order to get a job.

This definition enables us to note the similarities among public accountants, persons with teaching certificates, industrial chemists certified by a university degree, clerics who have actually been seriously examined by their bishop as well as graduating from Oxford, and Civil Servants certified by a public examination tailored to the studies required for an Oxford degree in classics, with a suitable adjustment for Cambridge mathematics students. I mention the Civil Service especially, as it is an example of how far Victorians could go in thinking "generalists" could be produced by a specialized education.

The definition also enables us to note that the process is not one to produce a greater number of experts, but to mass-produce a greater number of routinely competent office-holders.

This process of certification is one process advocated much earler but used especially in the second half of the 19th century and the first half of the 20th century by some groups wanting to raise their status. Other groups have used other processes. This process was not the characteristic one for high status before the 19th century, and it was not characteristic of all groups in the 19th century. Nor is it today.

We are therefore encouraged to isolate the "real" characteristics of a professional. I will arrive at my version of these by uncovering the qualities which certification was supposed to ensure. A professional scientist was characterized by the facts that he was involved in an endeavour of recognized social status and that he recognized its own standards of merit (he did not appeal to a general "public"; it was a "self-reviewing circle"), but he was distinguished from the gentleman, the amateur, the patron, the dilettante, the teacher, the interested cleric, and the retired naval officer, by his long-term attention to science (not necessarily to *a* science) as his major activity, by his technical expertise, and by the number and quality of his accomplishments.

To give examples:

Rat-catching was not an endeavour of recognized social status.

Robert Chambers's *Vestiges of the Natural History of Creation* was not written for fellow-scientists.

Darwin's independent income enabled him to devote all of his available energy to science.

When Charles Lyell first thought of writing on geology in the 1820's it was because he wanted to write something that would be read by the public, preferably of a reforming nature. He was therefore no more of a professional than was Rowland Hill at the same time, who wanted to reform *something;* and since he couldn't reform the paupers, or colonize Australia, or revolutionize newspaper printing, he reformed the postal system.[13]

While writing his *Principles of Geology,* Lyell became so deeply involved in the subject that he abandoned the law and devoted his life to geology; he became a professional. He tried teaching at King's College, London, but abandoned it, in part because it took up too much energy that could be devoted to science itself.

When Darwin came back from his *Beagle* voyage, he took his fossil bones to Richard Owen. Earlier, geologists had often taken theirs to William Clift. Such referrals give us evidence as to who the professionals were.

John Herschel had too many good papers on too many subjects for anyone to think of him as a gentlemanly dabbler in this and that; rather, he was considered a professional in *all* of his subjects, or, rather, he was considered the model of a professional *scientist,* not astronomer or chemist or whatever.

Social advancement of the subject is taking place when the kind of persons I have described as professional scientists are becoming dominant. They need not be in the numerical majority in an organization to be dominant. When debates among these men are almost the only important debates, then the entity in question has become professionalized.

Note that I completely avoid "specialization" as a distinctive trait. The various disciplines or "specialties" change so much over time that a man in the early 19th century might think he was a specialist when a modern author would consider him rather general. For example: a concern with geodesy, magnetism, meteorology, geology, and botany might indicate, not a generalist, but a specialist in Humboldtian science—especially if all the subjects were pursued according to the same scheme. And vice versa: a man might think he was a generalist if he worked with acoustics *and* crystal structure *and* solar radiation *and* fluorspar *and* spectral lines *and* atomic ideas *and* electromagnetism *and* Lagrangian analysis. We would simply say he was a physicist.

Given the technical expertise, his contemporaries didn't much care. The

151

distinction they noted was that between the scientific observer, who worked according to a general scheme which he understood, and the undisciplined observer, who merely noted everything that occurred to him. There were various proposals to make these latter useful, by providing them with particular instructions, or prepared charts to fill in, or whatever; these schemes only emphasized the distinction. Young Darwin collecting and classifying beetles at Cambridge was the latter; it was on his field trip with Adam Sedgwick that he first realized that science depends for its existence on the former.[14]

Anyone can see what I mean by reading the classic book written for these undisciplined observers, the Admiralty's *Manual of Scientific Enquiry, Prepared for the Use of Officers in Her Majesty's Navy; and Travellers in General*, edited by John Herschel and published in 1849. And Martin Rudwick in his study of the founding of the Geological Society of London notes the desire of its founders to encourage observations from men scattered about the country, men not necessarily having much formal scientific knowledge.[15] A. D. Orange points out a somewhat similar idea expressed by William Vernon Harcourt in 1831 as one justification of the British Association for the Advancement of Science, one I will return to in Chapter Seven: the members of the provincial Philosophical Societies, he said, were "'a multitude of hands'" which only needed to be shown "'in what manner they may employ their industry to the best advantage.'"[16] So the impulse to make some use of all these "cultivators of science" (Vernon Harcourt's phrase) was present throughout the first half of the century; and I have not mentioned all such schemes. What all the schemes had in common was an agreement on the distinction between the few men who are really scientists and the many who may be useful to science. Whether they were very useful, whether any of the schemes really contributed very much, is an open question.

Jacob Gruber, in his article, "Who Was the *Beagle's* Naturalist?" makes a distinction which may be similar to the one I am making, and you may prefer his approach.[17] Gruber is looking from 1830 and observing the decay of the Linnaean impulse in natural history into a justification for collecting; the newer naturalists (of whom young Darwin was one) wanted to relate their facts to some more general scheme.

The scientific observer, then, had the technical expertise to evaluate the significance of an observation or experiment in the light of more general theory. That technical expertise has not (contrary to the general impression) become more difficult to master since 1800. It has not forced scientists to become more and more specialized. Laplace's *Mécanique céleste* was just as obscure then as any treatise on theoretical physics is today. (For that matter, the mathematical portions of Newton's *Principia* were considered rather difficult in its day.) Geology, to my way of thinking, was more difficult then, because of the localization of terminology; a description of rocks on the Isle of Wight might be very hard to correlate with one's own knowledge of strata in Saxony.

If specialization has taken place since 1800, it does not have any one simple cause. The demand for it varies from culture to culture as well as over time. Thus the German historian Bunsen wrote in 1816 that, when he came to England, he would hide the fact that he knew any non-European languages except Greek and Latin. "You know that the English demand of every man to make one thing his profession, and one who pretends to many at a time is considered empty and not to be trusted."[18] Personally I would doubt that very many important scientists have been very specialized in their interests; this may not show in their monographs but it often does in their lectures.

Another problem separate from professionalization is "elitism." I use Nathan Reingold's sensible definition, that elitists want standards determining who is a scientist which will also indicate who is a *good* scientist and therefore should receive various appointments and control various organizations. Therefore I do not use the term for a mixed organization in which the professionals control the amateurs: I use it for a professional group in which some of the scientists dominate the rest. Reingold's article, incidentally, contains a clear warning to historians against confusing professionalization, specialization, elitism, prejudice for "pure" science over applied science, and preference for national over local organizations.[19] A given scientist's influence on any one or a number of these separate issues does not enable us to predict his position on the remaining ones.

I will end my First Way Out with a letter by John Herschel which shows elitism in practice, and its difference from professionalization. In 1831 the government was looking for a replacement for the ailing Fearon Fallows as astronomer at the Cape of Good Hope. Naturally they wanted Herschel's advice. So Francis Beaufort, hydrographer to the Navy, asked him for it, and here is a summary of his response:

> You need a man who both knows enough about astronomy as a science to set up and guide his own operations, and also has a practical knowledge of instruments and observing. Your choice is very limited, because the really satisfactory men would not take the post: George Airy (the best), Dr. Robinson, or Richard Sheepshanks. W. R. Hamilton's "rare talents are, I fear sadly buried in Dublin;" it would be a loss to send him still further away; also I think he has turned away from practical astronomy.
>
> At the other extreme, the multitude of "men who buy fine instruments and even use them to tolerable purpose" almost never learn the principles of astronomy as a science.
>
> A group worth considering is that of officers of the armed services who

have done surveying, who know practical work and some theoretical principles, and who have enough ability to learn more. There are several of these. Lt. Drummond of the Irish Survey would make a superb observer, as would Captain W. H. Smyth (but he has his own observatory).

A final group is the assistants at Greenwich Observatory. None of these men really have the advanced science needed. Richardson has industry and is a thorough computer; "moreover Astronomy is really the idol of his sould which he loves devotedly;" but he does not know high theory. Henderson has great abilities as a computer; I don't know if he has any of the other requisites of an observer. I don't know Maclear well enough to judge "whether the promise which he exhibits as a computer and an observer is such as to establish any ground for taking him into consideration." Of the other assistants, I know still less.[20]

Herschel's letter certainly has a familiar sound to anyone who has advised on fitting the right man to the right professional job today. In his hierarchy, the top professionals are known by name. Their employment situation may have something to do with their availability, but not their ranking. Airy was in charge of the Cambridge Observatory, and T. R. Robinson was in charge of Armagh Observatory; Herschel does not bother to mention these facts. Richard Sheepshanks, who was qualified both as a lawyer and a clergyman, never worked at a job of any kind.

The lower professionals have some but not all of the expertise desired. Herschel knows only some of them. Not all of them are employed in science; W. H. Smyth was a scientific-minded naval officer who gave up sea duty in 1824 and devoted himself to astronomy. My own experience in writing letters of general recommendation leads me to believe that the main advantage of being employed was that it gave Herschel a category to think about when trying to remember names of suitable persons. "Now who do I know at Cambridge who might do? who at the Science Museum? is there someone in that bunch around Leeds?" is the way I do it. Of course major publication, or major visibility at the Astronomical Society, might be better, but the young professionals might not have achieved those distinctions.

The amateurs are dismissed completely. They are mere observers. A professional, even in an observer's position, must have some theoretical expertise.

(Incidentally, Maclear got the job; and Herschel in South Africa had the ticklish problem of how to tell him when he was doing something wrong without pulling professional rank on him. Herschel, there on his own project, had no official connection with the South African Observatory; but Maclear was continually involving him in its affairs. Herschel seems to have succeeded; at least Maclear remained a life-long friend, and perhaps never suspected what Herschel really thought about the way he handled his various projects.)

My Second Way Out is, to me, more interesting than the first. If the first is a needed refinement of external sociological criteria, the Second Way takes us inside science itself (or history, or any other scholarly subject). I suggest:

one is acting as a professional when one concentrates on problems that are posed within the field—without any necessary motivation from the outside.

Here are some examples:

1. A few years ago, students of British history, in order to get an advanced degree, studied the rise of the gentry in the 16th and 17th centuries. This was not because relevant modern issues are mirrored in the rise of the gentry, but because that is what their examiners thought was one of the burning issues in scholarly debate. This was professional training.

American students today sometimes demand "black" or "women's" or "gay" history. This demand is closely related to their own socio-political situation and/or their personal ideals. It is therefore not a professional demand.

2. Charles Lyell in 1825 wanted to make a name for himself by writing something liberal, so he wrote an article on geology. He was not acting as a professional geologist.

Charles Lyell in 1835 wanted to establish his theory of climate change, which had been attacked by other geologists, even though it made "liberal" geology harder to establish. But it was a nice theory on a problem which puzzled geologists, and it was his own. He was acting as a professional geologist.

3. In the 1840's, George Airy was much interested in governors for speed regulation. This was because they were needed to run large equatorial telescopes smoothly. Airy's interest was that of a professional astronomer, but not that of a professional physicist.

Around 1868, James Clerk Maxwell was interested in the theory of governors. His interest had nothing much to do with their practical use. Governors were by then a valid intellectual problem for a professional physicist.[21]

4. In 1856, Clerk Maxwell wrote to George Stokes, "I have been trying to simplify the theory of compound Optical Instruments....I thus render the theory of instruments independent of the particular optical apparatus used in the instrument...."[22] He has no use in mind for the solution; he wishes to solve the problem because it is there. He is acting as a professional theoretical physicist.

With Maxwell in mind, I would like to introduce a phrase. The "fully-formed professional" is one who is fully equipped both with the expertise and the attitudes to devote himself to solving problems in the given professional field, with no need for monetary or ideological motivation from outside the field. Going to a university is one way to gain the expertise; but it is perhaps more important as a way to gain these attitudes. Generally speaking, one needs to feel that there is a group or community which approves of devoting oneself to solving these problems. Any research-oriented group with prestige will do.

However, it is important to note that there does not have to exist a true sociological group or community for this purpose, nor does the individual have to belong to it if there is such. He will usually seek for one, and move towards it if he can. Normally—and here I must disagree with Thomas Kuhn—there is no "well-defined community of the scientist's professional compeers."[23] There are a number of sheltered arenas for professional expression, of differing and varying prestige, membership, and interconnection with one another. The situation is notoriously ill defined. In particular, normally it is not centered around an actual community; that is, the scientist does not live, work, play, go to church, and intermarry largely with his professional compeers. He may do these things with his *university* compeers—which include historians, classicists, theologians, even a poet or two—but he meets many of his fellow scientists only once or twice a year. (Students may think that their university department *is* the center of the professional world; but the faculty know better.) It is these fellow professionals scattered all over, and not the members of his actual living community, who validate his claim to be a professional scientist.

Scientists are always trying to form real communities (one of Bacon's ideals) but such groups have been either quite partial—as at Woods Hole, Massachusetts, in the summertime—or have not lasted for very long, if only because a university or other organization is not willing or able to hire all of the good men in one subject, and it would be hard to get them all into one structure of prestige and authority anyway.

As this discussion shows, one reason I like Way Out Two is that it raises further problems. Why is a professional problem "there?" If there is no actual community, how and by whom are problems made into major professional problems? This process, and these men, are as important as the processes and men who solve the problems.

This suggests something to do with men earlier than the fully-formed professionals in each field. Instead of treating obviously good men as "precursors" or "forerunners" who were fumbling around in the right area but got the wrong answers, we can treat them as men who were shaping the field by bringing to prominence the "right" questions. For men earlier than Darwin, for example, we need not ask whether they had a more or less vague idea of evolution which may or may not entitle them to partial credit. (Historians who treat Buffon or Lamarck this way always strike me as self-assured examiners passing out marks to rather confused students who nevertheless have tried hard). We may ask instead, how much did each one contribute towards making the origin of species a central problem which a professional naturalist might well devote himself to solving?[24] It may be that a scientist opposed to evolutionary ideas as they existed in his time nevertheless helped push the species problem to center stage.

Way Out Two teaches us to look earlier, and more widely, for external influences on scientific theories. The process of making a problem, or a concept, into a professional one takes time. To explain why Maxwell felt free to study governors, we need not look to the influence of steam-engines in the 1860's on physics. We do need to explain why, in the earlier 19th century, astronomers wanted more precisely driven equatorial telescopes. Of course, if A. R. Hall finds no significant relation between practical gunnery and the science of ballistics in the 17th century, this does not tell us about the origin of scientists' interests in ballistics a century or two earlier. It is possible that scientists were originally motivated to see what they could do about real gunnery. As the problems became internalized, the professionals lost the need to have any motivation from practical applications of their theories.

Way Out Two suggests that we trace the stages of importance of a problem. Thus from William McGucken's study we can learn that spectroscopy limped throughout much of the 19th century in part because only a few scientists took it up, and they did so only for short periods. The problems were too difficult to be solved by small investments of energy; and they were not of sufficiently major importance to motivate large investments of energy. The transition to a situation such that several professionals worked at spectroscopy in a major way is a topic of considerable importance. The heroes of this transition may not be heroes because of new theories or solutions in the convention of the ideas-developing-in-a-row intellectualist approach to the history of science.[25] Here I think a "paradigm" approach is called for.

To return to social considerations, what is it that permits the degree of professional isolation which exists at any given time and place? When and how do current concerns break into the sheltered arenas? And with what results? Alternatively, who or what keeps them out? Who expended how much effort to keep open an arena for the consideration of the naturalistic origin of species in an increasingly Evangelical society?

Note that this approach can include applied science. One may work

157

away at perfecting bridge-building even though society is not clamoring for new and better bridge-building methods in this decade. Or better, rail systems in a decade when jet planes were all the rage.

My approach does, however, rule out civic improvement or social welfare motivations as a proof of increasing professionalization. The professional is content with a general feeling that his profession is inherently in the public interest. The direct motivation for individual researches is not a personal desire to cure this or that ill of the community—unless, of course, the curing of such ills has been internalized by his group as being what a professional parasitologist, say, does. An army officer who chose his wars according to the amount of social justice they promoted would not be considered, in that respect, strictly professional. His nation always fights just wars: that should be enough justification for the fully-formed professional.

Finally, I cannot resist the temptation to use Way Out Two to illustrate my conception of a scientific "revolution." A revolution occurs when a new set of interests (or concepts) is brought into or forced into a professional field from outside, so that the new particular subjects become fit subjects for ordinary professionals to make a reputation by dealing with. And the revolutionaries are the people who succeed in bringing these new interests into the field, not the ones who "solve" the new problems. That the innovators do not perform their task neatly, that their specific theories are often crude, ill formed, or "amateurish" has often been noted. In the 1830's, for example, everyone saw that Alexander Humboldt's papers on almost any subject were unsatisfactory, old-fashioned, devoid of convincing theory, etc., etc., etc. But the people who noted this were involved in doing better work on the subjects which Humboldt had insisted it was important to work on. They were the ordinary professionals; Humboldt was (relatively speaking) the revolutionary professional.

The history of science has largely proceeded from professional solution to professional solution, and has never quite known what to do with these uncouth giants who impress their contemporaries enormously but whose published work never seems quite to justify their contemporary reputation. Their abilities were engaged in forcing new subjects into professional science; the fully-formed professionals could then produce an array of "revolutionary" solutions.

This approach, then, suggests that great names in the history of science have often been the Napoleons of their fields, not the true revolutionaries. They were the winners in the fight to control the new. Such a hero, like Napoleon Bonaparte himself, need not represent all of the revolution; he is usually a resultant of only some of the new trends, mixed with some of the old. It is always questionable how much such a person *diverts* the course of events from the path already being followed—if one can define such a path, which I think is possible. Thus I prefer to speak of the astronomical revolution, not the Copernican revolution; the revolution preceded Copernicus. It is not a "Darwinian" revolution but the revolution in natural history of the first half of the 19th century that we should speak of. Darwin blended the revolutionary and the traditional in his synthesis. And it was the revolution in physical science of the earlier part of the nineteenth century which furnished the directions for the later, fully-formed professionals such as Maxwell, Planck, and Einstein.

A revolution may produce a new discipline, or several of them; it need not merely change the old discipline. But what it does and how it does it is, in my approach, individual in each case. It depends on the culture of which it is part. Not characteristics general to all science, but characteristics peculiar to early 19th-century England shaped the revolution and its impact on young Charles Darwin. One can speak of the nature of English scientific culture in the 1830's, when Darwin began to speculate. This can be described, analyzed, theorized about. Scientific revolutions divorced from time, place, subject, size and importance,[26] seem to me to be forever fated to be Ideal Types, tantalizingly flexible, never capable of being made precise and usable. They are not suitably definite and describable abstractions to be the subjects of a proper theory, with deductions and all. Of course this depends on what one accepts as a "theory;" I like systems of hypotheses and deductions.

Another approach is for historians of science to abandon the analogy of "revolution" altogether. It has served in several fields; perhaps it is worn out by now. By a "revolution," at any rate, I mean something like a new level of activity, an upraising of the plain, a high plateau. These are concepts which, we know, can be the subjects of a theory, since geologists have constructed such theories. The word "structure," like the word "framework," should be taken casually in my writings. Both words imply a definiteness of connection, hint at a stability or rigidity, which is not very characteristic of the entities which the historian investigates, at least as far as I have investigated them. I prefer "frame," meaning "frame of reference," a set of arbitrary parameters to make comparisons possible, or meaning a "loom" on which one may weave many patterns. The frame, in either case, is ours; it is not in the events themselves.

Way Out Two can be seen to suggest approaches to the problems of how and over what period of time an external factor becomes internalized, becomes accepted as part of the internal dialectic of a discipline. This seems to me more interesting than making stout affirmations or denials that external or internal factors are or are not dominant. External factors of any kind can lead to scientific theory. The way they do it is by being forced into the professional arena by a revolutionary (or a series of reformers) who has the status to do it (or the appeal to the younger generation to do it).

Note, by the way, that in Stage 4 (Ideal Type) thinking, even a major "revolution" is not necessarily a very great thing. It may be a revolution in the old non-political sense, a half-turn of the Wheel of Fortune. Phlogiston is up and kinetic theory is down; a century later, kinetic theory is up and subtle fluids are down. This is not my view of what happened. If new subjects to concentrate on are introduced, and new institutions to support the men doing the concentrating become relatively permanent, then a real change has taken place. It makes no difference that scientific theories in some conceptual sense oscillate back and forth; all those meteorologists, say, with all their maps and instruments and data and ideas are nevertheless an addition to the total body of science. The landscape has changed; we have a new topography for science.

My Third Way Out may seem more limited than the first two ways, but it applies directly to a kind of study which has been and is now being made. The Third Way is to study individual societies and associations before generalizing too much about their roles, and to study them not as though they were rational structures with determined goals and procedures for reaching those goals, but as though they were informal personal groups. Goal-directed activity is a favorite with some sociologists; but to me it is a rationalistic myth imposed by Aristotle.

Who can say that there is a single factor which causes several men to get together regularly, even if they always play poker when they do? One needs to study the men's individual characters and backgrounds, the parts of society they come from and are in, the status of poker-playing as a male pursuit in the community, whether each one is present at every session, which ones are recent additions to the group, what other pursuits they have, what other pursuits are available to such as they, what level of poker is played for what stakes and how the stakes relate to their incomes, where the sessions are held, and so forth. If social conditions are sufficiently adverse, poker sessions will not take place at all. If they do, it is not clear that they have a single purpose. Some men may be there primarily for the beer, some for the company, some for the conversation, some for the absence of women. Few will be there solely and simply for the poker; that is, almost any one of them will drop out if enough of the conditions are changed. But the group might recognizably continue even if its "subject" changed to bridge, or bowling.

The existence of any one such group is not fully explained, and its history is not deducible, from considerations of the role of poker-playing in society. On the contrary, the latter is a generalization justified by a knowledge of the details of the individual groups.

Note that I am not merely using a pleasant analogy. After writing the last three paragraphs, I re-read Martin Rudwick's account of the founding of the Geological Society of London, and this is what I discovered:

The first meetings were of fourteen men as subscribers to finance publication of a book on mineralogy. "'These meetings were held at tea time once a week at the hospitable dwelling of Dr. Babington, and when the object of them had ceased they were found too agreeable to be discontinued.'"

Then the meetings seem to have been held at William Phillips's house. Some of the original members dropped out, and a few non-subscribers joined, including Humphry Davy.

Davy proposed that the meetings become dinner meetings. "'The chills of Novr mornings are very unfavourable to ardor in the pursuit of science, and I conceive we should all think better and talk better after experiencing the effects of Roast Beef and wine than in preparing for tea, coffee and Buttered Buns.'"

Although most of the members were originally interested in mineralogy, the dominance of a few men turned it into a geological group.

From a small dinner club, it was quickly changed into a "society" with all the usual paraphernalia. It actively recruited a large membership. The changes disgusted Davy, who wanted a small conversational dinner group of friends, not bad dinners in a crowd and boring formal papers.[27]

Our friendly poker game has turned into a formal bridge tournament.

I am not opposed to generalizations about such organizations. On the contrary, I will generalize readily, thus: after perusing a confessedly incomplete list of Victorian publications which Lionel Madden prepared, I will say that the formation of Lit & Phil societies was characteristic of the early 19th century in Britain, and that the formation of Archeology and Naturalists' Clubs was characteristic of the late 19th century. In Shropshire alone, I discover from the Geological Society's pamphlets, we have in the 1880's the Caradoc Field Club, Shrewsbury; the Ludlow Natural History Society; the Oswestry and Welshpool Naturalists' Field Club; the Severn Valley Naturalists' Field Club, Bridgnorth; and the Shropshire Archaeological and Natural History Society, Shrewsbury. There may well have been others.

But before generalizing further, about professionalization or anything else, we need to consider a significant number of societies in their specific time, locality, and personnel; and study their actual operations over time. There is no reason to believe that the ideas of the proposers, the statements made at the first meeting, and what was happening five or ten years later are at all the same. Thus the Cambridge Philosophical Society was founded principally by a mineralogist, a geologist, and a botanist (Clarke, Sedgwick, and Henslow), but it soon became a center for mathematical physicists; and a stratigraphical geologist like Sedgwick could not follow most of the papers presented at the meetings.[28]

We must also remember to sort out the various characteristics noted above which interest students of scientific groups but which are distinct from professionalization. These are, to repeat: elitism *vs.* equality within the organization; specialization *vs.* a general approach in the subject-matter covered by the organization; national or metropolitan (these are not always identical) *vs.* local or provincial tendencies; and an emphasis on "pure" science *vs.* an emphasis on applied science. Thus among metropolitan British societies, the Astronomical Society of London in its early decades was elitist in its organization, being controlled by Francis Baily and his four followers, Richard Sheepshanks, John Herschel, George Airy, and Augustus De Morgan. Was it also elitist in the work it encouraged? Did it encourage certain specialties in astronomy far more than others? Was it truly national, and not merely metropolitan? These are questions worth answering.

The Geographical Society, founded in 1830 with the help of several prominent Astronomical Society members, notably W. H. Smyth but also including Francis Baily, was soon equally under the firm control of John Barrow; but it spread its net so wide in the quest for support that some felt it was not professional enough. Thus young Charles Darwin joined it in 1838, was elected to the Council two years later (he had to act as formal auditor of accounts on one occasion),[29] and refereed several papers. But he came to feel that its affairs involved too much publicity, were not conducted with proper scientific decorum.[30] The professional, we recall, shuns appeals to general public opinion.

Thus one must be careful in making general statements that one-subject metropolitan scientific societies are direct evidence of professionalization. Or that they are obvious agencies of increasing specialization: are persons interested in geology led to become interested only in one part of geology because they attend meetings with lots of other geologists? Scientific organizations are sometimes proposed as ways of *bringing together* specialists so that they can encounter the field as a whole. The existence of the society shows that some degree of specialization has already taken place; but the society itself may work to counteract further specialization.

A third alternative is that a society may be active in spreading one technique or approach throughout its subject; this is specialization of a kind, but it may not be felt as such. It may be felt as simply "modernizing" the subject generally. More accurate measurements in astronomy might seem to be a goal for improvement generally; but, since it led George Airy to refuse to enter the race for more powerful and far-ranging telescopes, we can see that it constituted an emphasis against the farther exploration of the nebulae. Likewise John Herschel wanted to measure more double stars and nebulae more accurately than William Herschel had done; he did not yearn for a more powerful telescope than his father's, but showed how the Herschelian 20-foot reflector could be used for accurate measurements as satisfactorily as his friend Struve's pet refracting telescopes. The emphasis on accuracy therefore—since good men and resources are always finite—like any other emphasis necessarily worked for one kind of astronomy and against another kind.

Since simple contrasted terminology has been so important in past discussions, at the risk of confusing matters still farther I shall introduce a new opposition. The verbal opposite of *professionalism,* I shall say, is not amateurism (this has caused too many problems in the past), but *parochialism.* I use *parochialism* in the widest possible number of meanings: identification with the aggrandizement of this group, this town, this sect, this technique, even this philosophy of scientific methodology. The professional should always try not to be parochial in his orientation: that sounds about right.

I will conclude with a final generalization. There is one law which can be called the First Law of Society Dynamics. That is, *societies breed societies.* Beyond a certain point, there does not have to be much of a justification for starting a society. Given the Linnean, the Geological, the Astronomical, even the Asiatic, why *shouldn't* we geographers have a society? is the way the reasoning runs.[31]

Then how do we explain the waning of the society-founding impulse? Is this because, at a certain point, all feasible subjects have been societized? I think not. Remember that a society needs only eight members to flourish. My father belonged to one such, and a very good time they had at the old Princeton Inn once a year. The subject was defined *so that* there were not more than eight or ten people with that specialty. I think, therefore, we should rather look to change in the environment in which societies live, a change so that conditions are unfavorable for the survival of an increasing population of this type—but perhaps favorable for a population of a related but distinct type. That is why I have called attention to the vogue for Naturalists' Clubs as indicating something important happening in the environment in which British science lived in the latter 19th century. Simply dismissing them as "amateur" will not tell us what that something important was.

Notes to Chapter Five

1. Quoted in Nathan Reingold, "Cleveland Abbe at Pulkowa," *Archives internationales d'histoire des sciences, 17* (1964), 139.

    The important article for historians, on the subject of professionalization, is Reingold's "Definitions and Speculations: The Professionalization of Science in America in the Nineteenth Century," in Alexandra Oleson and Sanborn Brown, eds., *The Pursuit of Knowledge in the Early American Republic* (Baltimore, 1976), 33–69. But see also his article cited in note 19.

It is important to remember that the concept "profession" changed its content drastically between the mid-17th century and the end of the 18th century in England. Any static definition taken from modern sociology will not only be anachronistic but will be so in different directions at different periods. One cannot pretend that up until such-and-such a period one state of things prevailed and then there was a movement more and more leading up to us. Thus there are two substantial weaknesses in W. J. Reader, *Professional Men* (London, 1966), which is nevertheless very much worth studying. One is shown by the sub-title: "The *Rise* of the Professional Classes in Nineteenth-Century England" (my italics). A very normal Whiggish assumption. But I will assert on the basis of evidence not yet published, that there was a *decline* in the importance of the professional classes from 1800 for a number of decades. The other is a problem endemic among historians: to use census figures too readily. The census records occupation, not profession, and the conversion of one into the other is not without theoretical problems. Thus in Reader's treatment there are no scientists; that is, such a category simply does not exist.

The sociological literature on the professionalization of science is substantial, and I cannot do it justice in a brief account here. The historian of science may learn as much by studies of a different vocation in its historical period, such as Brian Heeney, *A Different Kind of Gentleman: Parish Clergy as Professional Men in Early and Mid-Victorian England* (Hamden, Conn., 1976); and Harry W. Paul, "Scholarship and Ideology: The Chair of the General History of Science at the Collège de France, 1892–1913," *Isis, 67* (1976), 376–97.

2. See George Haines IV, *Essays on German Influence upon English Education and Science, 1850–1919* (New Haven, Conn., 1969), *passim.* I note that we have the sociologist Joseph Ben-David *attacking* the German system, in his "The Universities and the Growth of Science in Germany and the United States," *Minerva, 7* (1968–69), 1–35.

3. Rowland E. Prothero. *The Life and Correspondence of Arthur Penrhyn Stanley* (New York, 1894), I, 203.

4. Geoffrey Millerson, *The Qualifying Associations: A Study in Professionalization* (London, 1964), pp. 5, 33–41.

5. Millerson, pp. 117–118 (1960 figures).

6. Millerson, p. 70.

7. Millerson, pp. 121–122, 124–125.

8. Millerson, p. 10.

9. Millerson, p. 148 and chapter 6 generally.

10. Millerson, p. 125. We learn from the *Times Higher Education Supplement,* 1 Jan. 1974, p. 13, that Ivar Berg in his *Education and Jobs: The Great Training Robbery* (London, 1970) found (in the reviewer's words) that "the higher the level of certification, the worse the actual job-performance." A step forward?

11. Millerson, pp. 9–10.

12. Millerson, p. 8.

13. Howard Robinson, *The British Post Office: A History* (Princeton, N. J., 1948), p. 260.

14. Charles Darwin, *Autobiography,* ed. Nora Barlow (London, 1958), pp. 68–70. Access to Darwin as professional is available through Thaddeus J. Trenn, "Charles Darwin, Fossil Cirripedes, and Robert Fitch," *Proc. Am. Philos. Soc., 118* (1974), 471–491.

15. M. J. S. Rudwick, "The Foundation of the Geological Society of London: Its Scheme for Co-operative Research and Its Struggle for Independence," *Brit. Journ. Hist. Science, 1* (1963), 333–338, 354.

16. A. D. Orange, "The British Association for the Advancement of Science: the Provincial Background," *Science Studies, 1* (1971), 327.

17. *Brit. Journ. Hist. Science, 4* (1969), 277–278.

18. Frances Bunsen, *Memoir of Baron Bunsen* (London, 1868) I, 95.

19. Nathan Reingold, "National Aspirations and Local Purposes," *Trans. Kansas Acad. of Science, 71* (1968), 235–246.

20. Herschel to Beaufort, 19 September 1831 (Carbon Book II, John Herschel Papers, Royal Society).

21. *Cf.* Otto Mayr, "Victorian Physicists and Speed Regulation," *Notes & Records of the Royal Society, 26* (1971), 205–228, and his "Maxwell and the Origins of Cybernetics," *Isis, 62* (1971), 425–444.

22. *Memoirs and Scientific Correspondence of the Late Sir George Gabriel Stokes,* ed. Joseph Larmor (Cambridge, 1907), II, 2.

23. Thomas Kuhn, *The Structure of Scientific Revolutions* (Chicago, 1962), pp. 166–167.

24. I am not sure whether to call Darwin a professional biologist or a professional naturalist. This is not because I am hesitant about whether Darwin was really a professional. It is because "biology" was a new Lamarckian term, and I am not sure that *biology* was professional. Around 1840 it gives me the feeling that it is something that somebody not quite sound, like Geoffrey St. Hilaire, might do—*probably* all right, but tending toward the idealogue. (Darwin was also a professional geologist, of course.)

25. My generalizations are based on the narrative in William McGucken, *Nineteenth-Century Spectroscopy* (Baltimore and London, 1969).

26. Thomas Kuhn, "Postscript" dated 1969, *The Structure of Scientific Revolutions,* 2nd ed. (Chicago, 1970), p. 181: "A revolution...need not be a large change, nor need it seem revolutionary to those outside a single community, consisting perhaps of fewer than twenty-five people."

27. Rudwick, "Geological Society," pp. 327–329, 336, 341.

28. J. W. Clark and T. M. Hughes, *The Life and Letters of the Reverend Adam Sedgwick* (Cambridge, 1890), I, 208.

29. Charles Darwin to the Geological Society, 31 January 1842, in Letter Books of the Geological Society, VII, #28 (Geological Society Library).

30. Hugh Robert Mill, *The Record of the Royal Geographical Society, 1830–1930* (London, 1930), pp. 48–49, 80.

31. On the origins of the Asiatic Society, see Garland Cannon, "Sir William Jones, Sir Joseph Banks, and the Royal Society," *Notes and Records of the Royal Society of London, 29* (1975), 205–230. Such studies are necessary; but so too are detailed sociological studies such as Michael Hunter, "The Social Basis and Changing Fortunes of an Early Scientific Institution: an Analysis of the Membership of the Royal Society, 1660–1685," *Notes and Records, 31* (1976), 9–114.

Chapter Six

Professionalization and the B.A.A.S.

I began the last chapter by saying that one should not dismiss explanations at one level merely because there are also explanations available at another level. I went on, with Nathan Reingold's help, to say that professionalization, specialization, elitism, national organization, and dreams of "pure" science are separable variables which occur in different combinations in different persons. I said that subjectively a professional yearns for approval from a self-reviewing circle rather than from the general public. And I said that the study of the process whereby external factors become internalized is a good way to approach important change in science.

It will be helpful if you keep these assertions in mind while I go on to consider arguments about the founding of the British Association for the Advancement of Science in 1831. These arguments are an interesting example of professionalization in historical studies. That is, the British Association is not currently a burning issue in our society, and there are no pro- and anti-Association groups which use the nature of its founding as a sign of fidelity to larger issues. Nor is the subject eagerly studied for clues as to how to start a new organization. It is, rather, a nice subject for some historians to use in displaying their notions of expertise, and in forcing their interests into the attention-range of historians of science generally.

That they argue among themselves should not be permitted to obscure the fact that they are all contributing to sociologizing the history of science. History students can then study the positions, pro and con. For students, such debates are usually arranged so that there are an equal number of good things on each side, regardless of evidence and abilities. The students may, after several such exposures, decide that such debates are futile; but that means that they have missed the point of the training. The point is not to decide the truth about the subject, or even to distinguish good historical techniques from poor ones. The point is rather to make it clear that this *kind* of topic is what professional historians of science may properly concern themselves with.

Undergraduates who are not planning to become professional historians are also exposed to this procedure. They often come to the general conclusion that historians never decide anything, since there are always professionals on both sides of every topic they encounter. They are wrong: the historians decide what topics they will encounter.

I am aware, therefore, that simply by discussing my subject I am aiding in this indoctrination; but my interest in displaying my own expertise outweighs any scruples I might otherwise have. In order to show off my own abilities to best advantage, it is necessary that the historians I discuss be good historians, so let that be postulated for all of them at the start.

My story begins in 1951, not because there were no previous accounts of the founding of the British Association, but because professionals don't like to recognize scholarly interpretations that are more that twenty years old unless they led directly to a flourishing crop of articles. It is simply embarrassing for a professional to have to admit that his new view is the same as that of, say, Thomas Huxley. This is one reason why historians of science are so bothered by "precursors" in science itself. If I admit that 18th-century ideas on evolution really are the basis for Darwin, I might then have to admit that in history also, any stray interpretations by non-professional predecessors—Huxley, or Alfred North Whitehead, or Andrew Macfarlane—should receive the credit which I want for myself. We professionals have some justification for our snobbishness. Many of these predecessors did not do what I said in the last chapter a major reformer does, that is, force the subject into the professional arena. One uses the information in old accounts; one may be largely validating their suggestions; but it takes a live person to risk part of her professional career in the attempt to make the interpretation a part of current professional concern. The penalty for failure is severe: to sit neglected in the rear building of a second-rate college or institute while promotions, status, and groups of eager and admiring young seekers-after-professional-status flock elsewhere. It is nicer to sit at the top of a new multi-million dollar building in a world-famous institution and let the seekers come to you. This is what I do.

Thus O. J. R. Howarth's *The British Association for the Advancement of Science,* dating from 1922 and therefore now a half-century old, has been of value to, I think, all of the scholars I discuss. But none of us have liked his thesis that:

> The fundamental idea in the minds of the leading founders of the Association was undoubtedly to create a body which should form a mouthpiece of appeal from science to the world at large.

Scientists who objected to the Association did so "partly on account of some vague sense that science might degrade itself by making such an appeal," and partly because making science comprehensible to a general audience was too much hard work.[1]

That this public relations view was that of Charles Lyell at any rate, Howarth shows by quoting Lyell's letter to young Charles Darwin, recently back from the *Beagle* voyage, in 1838. Lyell wrote that "in this country no importance is attached to any body of men who do not make occasional demonstrations of their strength in public meetings.... As Dan O'Connell well knows, nothing is to be got in the way of homage or influence, or even a fair share of power, without agitation."[2]

The Lyell-Howarth interpretation, I would say, is that the Association was a scientists' general publicity and pressure group, the publicity being directed at, and the pressure exerted against, general public opinion rather than specifically the government or other men of science. The task of the Association was to make science into one of the elements of public opinion which politicians and other decision-makers throughout society (say, school boards) would naturally take into consideration thereafter. The task was not to professionalize science internally, but to have professional scientists convince the public of the value of science, in part simply by teaching them the values of science. Or, as the first president of the Association said at the third meeting, "The great object of the Association was to incorporate a love of knowledge with the very frame of the country."[3]

That such a role is a most necessary one in paving the way for specific favorable policy decisions, however vague it may seem on the surface, is a common speculation. This has been a standard "liberal" way of proceeding, from anti-slavery propagandists to current anti-pollution groups. That the British Association did serve this end to a certain extent is also, I think, admitted; references to the Lyell-Howarth interpretation flit in and around the presentations of the authors I am going to discuss. What the authors question are the two beliefs that this was the *primary* idea in the minds of the founders, and that this was the *primary* role of the Association played in the early and mid-19th century (regardless of what its primary role may have become by the 1920's).

In 1951, then, George Foote's article on "The Place of Science in the British Reform Movement 1830-1850" was published in *Isis,* the journal of the History of Science Society in the United States.

At the most general level, Foote suggested that the "great reforming energy in England in the early nineteenth century" had one of its manifestations in desires to do something about the state of science. He noted that "the period most marked by agitation [concerning science] coincided exactly with that of the period of the first reform bill." At the next level, Foote identified the agitation as that of the "decline of science" controversy: "Science had its reform movement.... What brought it about? It was the cry that science in England was in decline."[4] Foote discussed the "cry" as it was manifested in four review articles, then in a footnote by

John Herschel in his article on "Sound" for the *Encyclopaedia Metropolitana* in 1829. Charles Babbage, with his book of 1830 called *Reflections on the Decline of Science in England and On Some of Its Causes,* was the real agitator, in Foote's opinions; and Babbage was followed by David Brewster.

Then Foote made a controversial leap. He said that the decline of science controversy "directly caused the founding of the British Association and the gradual reform of the Royal Society." There was agitation within the Society to do something to make it more effective. It is not clear whether Foote believed that the agitation was a part of or derived totally from the declinist position, or whether it had an independent component; but he asserted that "out of this agitation concerning the Royal Society a new scientific group developed, the British Association for the Advancement of Science. And the main movers in its organization were Charles Babbage, David Brewster, and other declinist controversialists."[5]

Foote did not, however, give much detail to support these assertions. Indeed, he was quite fair in noting that William Vernon Harcourt, one of the principal founders of the Association, stated that he did not subscribe to the declinist opinions. Harcourt asserted that a new association was needed because there was *more* scientific activity taking place than there had been. Foote agreed with this, at least to the extent that there was more interest in science; the decline of science controversy, he thought, was evidence of interest, not of apathy.[6]

We can, perhaps, erect a schematic for Foote's exposition something like this:

Reform energy in society
Decline of science controversy
Agitation concerning Royal Society
Founding of the British Association   Reform of the Royal Society

(The arrows may be read as "caused" or "led to" or "manifested itself in" or "was logically prior to," depending on your own preferences concerning the metaphysics of historical connection.)

In 1961, L. Pearce Williams's article on "The Royal Society and the Founding of the British Association for the Advancement of Science" was published in *Notes and Records of the Royal Society.* One manifestation of agitation in the Society was the contested election for president in 1830, with John Herschel being the candidate of the "scientists," and His Royal Highness the Duke of Sussex being the candidate of the retiring president, Davies Gilbert, representing the continuation of rule by gentlemen. The "scientists," to my way of thinking, were in effect insisting that by being scientists they were automatically gentlemanly enough, without needing any other marks of status. They lost the election by a close vote, 119 to 111; quite a few members did not vote (there were 680 members in all, many totally inactive).

Williams's intention was to replace the overly simple explanation of the founding of the British Association as deriving from the declinist controversy. He derived it from growing professionalism, and more specifically from the loss of the election of 1830. The article gave a general narrative of the campaign in 1829–30 to reform the Royal Society, with much new and interesting material. Babbage's *Decline of Science* appeared as a major factor *in* this campaign, of which the declinist controversy was only a part. Babbage seemed, in Williams's narrative, to be the central agitator and, as a "supposedly experienced politician," even as the important worker for Herschel's election.[7]

Williams's chief theme was professionalism. The serious scientists in England envied other countries such as France, where science *was* a profession. The Royal Society was a club run by the amateurs. The reformers criticized Thomas Young's and Michael Faraday's Admiralty appointments because the men were not "professionally competent." [If this sounds odd, remember that the terminology is that of Williams. In my terminology, the criticisms were that the men were not "specialists" in the appropriate subjects.] Professionalism is what Babbage's *Decline of Science* was basically about: "The substance of his arguments was that other countries—most notably France and Prussia—had recognized the professional status of science.... the citing of rare geniuses like Davy by his opponents served only to obscure the fact that the foundations of science in England were not broad or strong enough to support the new professionalism that was slowly emerging."[8] It was the professionals, the "working scientists" (in context this means primarily "publishing scientists," apparently) who seized upon Sussex's candidacy to force the issue.[9]

One problem was that Williams presented no clear meaning for the word "professional" except what his readers could synthesize from the contemporary sources he quoted. This is not a bad procedure—better than that of imposing a definition coined by a 20th-century sociologist interested in the cosmetics industry—*except* when the dominant sources are two arch-trouble-makers like James South and Charles Babbage. Thus Michael Faraday did not yearn for the French system, did not believe in the virtue of specialization, did not believe in the decline of science in England, and was not a charter member of the British Association. Yet he clearly was a Royal Society reformer, and a professional. Indeed he was the very professional against whose prestige Babbage's declinist agitation shipwrecked itself, as we shall see later on.

If Williams did not mean that a professional scientist was defined as one whom "the government supported, encouraged, and consulted"[10]— and I think that is not what he meant—then "professional" was a convenient general term used to give a unity to the events he narrated. It was a self-evident explanatory term which needed no explanation—rather like "natural" in the 18th century.

In his final paragraph, Williams used this approach to explain the founding of the British Association, without, however, giving any evidence drawn from that event itself. What follows is literally his complete text, including all documentation:

> On 21 February 1831, David Brewster wrote Babbage that he had just finished reading 'a very interesting account of the different meetings of the German Naturalists.' What did Babbage think of founding a similar Association in Great Britain, Brewster asked? Those members of the Royal Society who had just seen their hopes for reform dashed, eagerly seized upon Brewster's idea and the British Association for the Advancement of Science was born.

(I will interrupt here to note that on 23 February, Brewster wrote to the secretary of the Yorkshire Philosophical Society, stating the general aims of a proposed British Association of Men of Science, proposing York as the meeting-place, and asking if the YPS, the Mayor of York, and influential local persons would promote the meeting, hopefully for the 18th or 25th of July. The letter is a formal proposal; one would expect that preliminary discussions had already taken place.[11] Brewster, apparently, did not feel that he had to wait for a reply from Babbage, with assurances of support from London, before beginning to set up the Association.)

Williams continued:

> All those reforms which it had been hoped to introduce into the constitution of the Royal Society were here embodied. The possibility of the kind of political manoeuvring that had marred the history of the Royal Society was eliminated by keeping the Presidency on a yearly basis and reducing it almost to an honorary post. The exigencies of professional specialization were met by dividing the association into sections where specialists could speak to specialists and profit from their mutual discussions. Membership was open to interested persons, but it was simply assumed that they would work at science.

(I interrupt again. Was this membership what the reformers wanted for the Royal Society? It seems to me that it is what they were opposed to about the Royal Society. At any rate, the statement is not correct about the British Association. As we shall see, no such assumption was made, it was explicitly assumed that many, perhaps the majority, of the members would not work at science.)

Williams concluded:

> In short, the B. A. was the direct reaction of the Royal Society reformers to

their defeat by the amateurs. The British Association was a professional group dedicated to the advance of science and the scientific profession.

On the founding of the British Association, then, Williams diverged from Foote's analysis in three significant ways. (1) It was the loss of the election of 1830, and not the decline of science controversy, which led directly to the British Association. (2) The *details* of the Association derived from the reforms desired for the Royal Society. (3) The Association derived from a desire for professionalism; this being the kind of desire which was expressed in Charles Babbage's *Decline of Science,* as well as in other ways.

Given time to consider the matter, Williams did not modify his position, or produce new evidence, but in his book, *Michael Faraday,* emphasized some points even more strongly. Thus he wrote: "The party of reform was headed by Babbage.... The campaign, however, was lost by Babbage. He had managed Herschel's candidacy.... The result of this failure was the creation of the British Association for the Advancement of Science in 1831. The members of the reform group were charter members and they set out to create an association run for professionals by professionals.... Although he was a charter member of the British Association, Faraday did not play a prominent role in either its organization or its development."[12]

This procedure of converting qualified suggestions, in a professional article, into unqualified assertions of truth in a later book, is so common that it may be said to be our chief vice as historians. This is why professionals read articles; their colleagues don't feel the need to be quite so careful about making general statements without much evidence in books. Williams is by no means a major offender in this respect. My Chapter Four is a splendid example. We all engage in speculations to fill up gaps in narratives, to provide motives, to suggest consequences. The trouble comes when the reader does not notice, or often does not care, whether an assertion is well-documented or only strikingly asserted. More of the history of science depends on assertions that never have been documented and yet have been received as well-established generalizations by other historians, than most historians are willing to admit. As a rule of thumb, I think that one should have *three* pieces of evidence to believe in an event, and *four* or more people to signify a "group," "school," or "trend." (I have violated these rules often in this book, but here I have provided a standard whereby the reader can judge whether to believe what I say or only to become interested in it.)

In 1964, my article on "History in Depth: the Early Victorian Period" was published in *History of Science,* an annual co-edited at Oxford and Cambridge. The article was about the first half of the 19th century generally—it has furnished the basis for Chapter Eight, below—and had only one paragraph about the founding of the British Association. Since Williams did not give any evidence for his assertions about that event, I saw no need to give any for my denials, and simply contradicted him:

> It is not true that the founding of the British Association was intimately bound up with the revolt in the Royal Society. The Association had its own sources of vitality, drawn largely from the 'provinces' (if one is ever justified in using that word to include Edinburgh). . . . What *is* true is that the Association clearly fulfilled a felt want of British scientists, so that even the initial disapproval of one of the chief victims of Royal Society politics, John Herschel, could not diminish its appeal for very long. By its third meeting the Association had attracted most of the 'urban' London-Cambridge group of scientists and could claim thereafter to be fully representative. It never became the dictator of scientific research as Vernon Harcourt had originally proposed, but it did become something more: the arena of free discussion and debate in which the great Victorian scientists felt at home.[13]

Here we have the introduction of *provincialism* into the debate. This has the effect of not arguing about which influence of the same man was more important—Babbage was both a "declinist" and a Royal Society electioneer—but of insisting on the importance of other men altogether, and thus of the influences on them from *their* environments. Note that I skirt around the formal problem of professionalization, by implying that the Association didn't do much, but actually saying that great scientists liked it.

The young professional scholar may like to know what gave me the self-assurance, not to hint at an inadequacy in, but flatly to contradict Williams's assertions. The fact is that I had read unpublished manuscripts about the Royal Society election too. I knew that Babbage was not as central a figure as Williams thought he was, and I knew that my manuscripts *were* by a central figure: the candidate himself, John Herschel. According to these letters, the organizer of the party in the Royal Society to head off the election of the Duke of Sussex was William Fitton, and his chief associates in the early stages were Robert Brown and William Somerville. After Herschel had agreed to run, he asked Francis Beaufort and Charles Babbage to get together with Fitton to map out the rest of the campaign. I will discuss the significance of these choices toward the end of the chapter.

And I had what seemed to be a trump card: the very negative attitude of Herschel toward the proposed new Association. There is nothing like knowing unpublished material which the other fellow hasn't seen to give a scholar self-confidence. This is the primary reason, in my opinion, why young scholars are well advised to dive into archival material as soon as possible. They may not uncover much of importance, but it gives them a dug-in position from which to conduct offensive sorties.

However, note the dangers of manuscript sources. By starting with the assumption that Babbage was important, you are led to read Babbage manuscripts. These confirm that Babbage was important, because Babbage thought so himself. You can then erect a whole explanatory system based on Babbage's motivations. Why he did what he did became reasons why scientists generally did what they all did. Babbage thus becomes a typical figure.

But whatever Babbage was, he was not typical. The whole system derives from a faulty premise, and even ten times as many manuscript sources which are chosen on the basis of that premise will not justify the system. You need better premises, which will motivate you to look at other sources as well, ones which will help you to limit and assess the importance of any one batch of sources. Putting Babbage at the center of things is, I think, the procedural error which has done most to mislead historians about all of these events.

In 1968, Nathan Reingold's article on "Babbage and Moll on the State of Science in Great Britain" was published in the *British Journal for the History of Science*, the organ of the British Society for the History of Science. The article begins: "Charles Babbage's *Reflections on the Decline of Science in England* is very well known to historians of science who are aware of its role in the movement to found the British Association for the Advancement of Science and to reform the Royal Society." Reingold's footnote to this sentence begins with a citation of Williams's article.[14]

It may safely be assumed that most readers understood the sentence to mean that Babbage's book had a role of some significance, not that its role was negligible. And Reingold says later, "The result of the agitation against the Society by Babbage, Brewster, Davy and others was the formation of the British Association for the Advancement of Science ... by a fusion of provincial and professional scientific interests." Most readers would understand "provincial" to be here contrasted to "professional" so that they would take the statement to mean: some professional scientists (who were from the metropolis) and some provincial scientists (who were not professionals).

Reingold's position on our subjects was not that of Williams—he avoided the contested election of 1830 altogether, and, in an article on Babbage used the word "professional" only in the sentence I have quoted—but was similar to that of Foote, leaving a similar uncertainty in my mind as to the exact relation between agitation concerning the Royal Society and the decline of science controversy. My provincials and Williams's professionals were thrown in casually. The Association was still the result of "agitation against" the Royal Society; there was no notion that the impetus for the Association came from a positive growth of vitality in the provinces, or that its nature was determined largely by provincial needs and desires.

Of course, if you read Reingold's article to find out more about Babbage and Moll on the state of science in Great Britain, it is rather interesting. I cite it as showing a certain type of uncertainty present in 1968 as to just what position to accept in the Foote-Williams-Cannon affair.

In 1970, George Basalla, William Coleman, and Robert Kargon published a paperback anthology of excerpts from addresses to the British Association. It was called *Victorian Science*. Its historical introduction, to which my remarks are confined, contained several generalizations on the subjects I have been discussing in this and the last chapter. Professionalization was the chief message; at least that is what Basalla, Coleman and Kargon chose to underline. Here are their positions on some of our subjects:

On the founding of the British Association: "Above all, the British Association was a *professional* organization as compared to the *amateur* Royal Society." When they lost the election of 1830, "the professionals turned to the organization of a new scientific society. This was the B.A.A.S."[15]

Pure Williams.

On the structure of the British Association: Basalla, Coleman and Kargon noted that its membership was open but that, "Its governing board, however, was to be drawn from the ranks of contributing scientists, thus insuring professional control over policy matters."

The actual regulation was that the annual General Committee to exercise executive power was made up of all members present who had had a paper printed by any philosophical society in the country. It did not take much professional science to get one paper of some sort in some Lit & Phil Society's publications somewhere. Howarth called it a "wise though not difficult qualification."[16] If this were what "professional" means, then Bishop Edward Stanley of Norwich would have been a professional scientist several times over.

On professionalization: "It is wrong to conclude that science achieved *full* professional status in nineteenth-century Britain. Yet there were definite signs of progress by the end of the century." For example, the electrical and chemical industries were offering posts to scientists and technologists.[17]

Here we have acceptance of the definition of "professionalization" useful for technicians, not scientists, as explained in the last chapter.

On science at Oxford and Cambridge: "Prior to the 1850's, the Universities of Oxford and Cambridge paid little heed to science." Basalla, Coleman and Kargon admitted only that, at Cambridge, one could study some mathematics, but not science.

The authors do not discuss such matters as the contents of a Cambridge tripos exam paper, or how expensive the re-founding of the Cambridge Observatory was when compared to the fiscal resources of the university. But, I imagine, they were aware of William Thomson and other physicists and astronomers produced by Cambridge. It seems that they did not consider the whole range of mathematical natural philosophy to be real science.

How is such a belief possible? Well, these subjects are not laboratory science. They do not produce technicians for industry—how can they be professional science? Cambridge should have built a chemistry laboratory rather than an observatory in order to be recognized as a place where science is fostered.

As a matter of fact, Basalla, Coleman and Kargon went futher in their condemnation of places which favored mechanics and physical astronomy: "The ancient universities were not only indifferent to science, they were decidedly hostile." No evidence is given for Cambridge; such generalizations, in my experience, are *always* based on Oxford. (Why Oxford should have concentrated heavily on science, when that is what Cambridge was doing, is not clear to me; the student could often choose either one). Basalla, Coleman and Kargon's example of hostility was the incident related by the political leader Lord Salisbury in 1894. He pointed out that in 1832, when the British Association met at Oxford, John Keble wrote to Edward Pusey that the members were a "hodge-podge of philosophers" and that the university had "truckled sadly" to current opinion. Basalla, Coleman and Kargon conclude, "Keble's remarks could be forgiven if they were made by an Oxford professor of the seventeenth century, but not one of the nineteenth century."[18]

It is not clear that a university which permits a new scientific organization with hundreds of members to meet on its premises, and which gives four of them honorary degrees, is showing hostility to science because one humanities professor objects. Keble, we remember, was not only a rather reactionary leader of the nascent Oxford Movement in religion (Pusey, less reactionary, was another) but was also a poetic follower of Wordsworth. *That a professor of poetry thinks his university is giving in too much to these science fellows* is, I imagine, a phenomenon documentable at every university in every century in which universities have existed.

Keble was correct, of course: Oxford *was* "truckling" to the scientists, that is, lending its prestige to them. Oxford was certainly not especially devoted to science in 1832, but the new Grecophile classicists had not yet come to dominate the university. I think Howarth's comment on the incident is sound: Oxford "contained notable elements of disaffection."[19] The visit of the British Association was yet another of those affairs which may have convinced Keble and his friends that they must soon go on the offensive if they were to have any chance to stem the creeping domination of modernity.

On Germanophilia: After Chapter Three, we know enough to recognize, by now, the cluster of ideas which Basalla, Coleman and Kargon used. They read one school of late Victorian reformers. We can expect, then, that they would have nothing but praise for the Germans' "superb educational system that took into account the needs of pure and applied science as well as technology." And praise for those late Victorian scientists "who saw clearly the secret of Germany's industrial hegemony."[20] The historians have decided that the Germanophile reformers were correct.

The reader has perhaps noticed by now that no one of these historians generalizing about the founding of the British Association had done enough new research since Howarth's time specifically on the founding of the British Association to write more than a few paragraphs or pages containing his opinions. Nothing has really been demonstrated at all. Nevertheless, enough positions have been advocated so that we can see more clearly what it is our final two contestants, who have done some needed investigating, are contending for.

In 1971, J. B. Morrell's article on "Individualism and the Structure of British Science in 1830" was published in *Historical Studies in the Physical Sciences,* an annual which at that time came from the University of Pennsylvania. Morrell is primarily concerned to assert that the British were professionalizing in science around 1830, even though they were individualistic and did not erect central state structures to administer patronage and education, as the French and Germans did in different ways. The British Association, then, was "the first national pressure group for professionalizing science." Charles Babbage is, for Morrell, the originator of the wrong interpretation of British science, since he did not realize "the sheer variety of ways in which science was organized" in the British Isles. The British Association, like the British Medical Association, "aimed to improve professional status," among other things. It "strikingly shows the continuing vitality of local private enterprise in its origins and crucially successful first meeting. Its foundation primarily reflected provincial Northern initiative and not the struggle for power which split the Royal Society of London during 1830."[21]

Thus Morrell flatly rejects Williams's explanation of the origins of the Association. He also does not favor a direct origin from the "declinists," but stresses that various people were concerned to improve the status of science and scientists. The Northern provincial-minded men were the founders; the metropolitan scientists and established university scientists held off at first "through resentment of embarrassing allegations about the decline of science and the accompanying state attention to be given to science, which to them constituted explicit criticism of what they regarded as their own adequacy and success."[22] In brief, Morrell is here saying that Babbage's writings *hindered* the founding of the Association rather than helped it.

As a matter of fact, Morrell gives me credit for his position concerning the Association, stating that, "My interpretation is merely an extension of W. F. Cannon, 'History in Depth.'"[23] I appreciate the credit, and I welcome the support, but I am glad that Morrell used the word "extension." Provincialism I will take credit for; but I had not heretofore maintained that the Association was a professionalizing agency to the extent that Morrell seems to propose; nor had I condemned Babbage's declinist efforts as actually hindering the founding of the Association. This last seems to me a delightful proposal which I hope is true, simply so that I can admire the bravado of the man who proposed it.

I *had* called attention to the danger of using Babbage's *Decline* as an interpretation of the state of science, but only for one particular subject. In "History in Depth" I wrote that, "Charles Babbage's *Decline of science* has tended to lead historians away from what is probably the truth, that the British government patronized science perhaps more in the first half of the nineteenth century than in either half-century on either side (except for the development of the museum concept in science as in other fields)." This says nothing leading to Morrell's point about the variety of scientific organization in Great Britain; on the other hand, he does not care to accept my financial challenge to Babbage, which indeed he implicitly denies.[24]

The idea that professionalization had a strong connection with the British Association was thrown into the current historical arena by Williams, and it needs to be examined more closely. But Morrell has at least provided a statement as to what he means by it. He has, however, singled out only one characteristic from the rich array he found available in Geoffrey Millerson's *The Qualifying Associations.* He chooses, as the most important element, "the growth of self-consciousness—the realization of collective identity."[25] If that is the element he likes, I think that is fine; but he has no basis for objecting if I use more complex criteria such as I have given in the last chapter.

In 1971 also, A. D. Orange's article, "The British Association for the Advancement of Science: the Provincial Background" was published in *Science Studies,* a new mixed-topic journal from Redbrick. It is not surprising that Orange, being the youngest of our authors professionally, has done the most research specifically on our subject. He comes out emphatically for the provincial character of the founding of the Association, so I look to be in a fair way to winning that argument, with both Orange and

Morrell as backers. What is more important, perhaps, is that Orange has gone over to the offensive in a major way. He has inquired as to what was going on in the provinces in general, and in Yorkshire more particularly, which could provide enough impetus to found the Association. This gives a new rationale for studies of local societies; they can be tied in to a major professional problem. Orange did not, perhaps, give a complete enough answer in his first short article so that other historians would be dissuaded from pursuing the inquiry.

Orange is sceptical about going too far with the professionalization argument. He writes:

> The British Association was certainly dedicated, from the time of its birth, to the cause of advancing the profession of science, but these variations of attitude on the critical question of membership, with Babbage, the uninhibited declinist, favouring an open society and Whewell, from the opposite camp, warning of the attendant dangers, make it difficult to sustain the contrast which has sometimes been drawn between the 'professional' British Association and the 'amateur' Royal Society.

Orange identifies, in a footnote, the historians of whom he is sceptical: Williams, and Basalla, Coleman and Kargon.[26]

What a lovely array of subjects has thus been developed to tempt historians into further research! Just by itself, Foote's suggestion that the agitation for reform of scientific institutions was a part of the great reforming impulse in society as a whole leads to all sorts of interesting speculations. How are political movements related to professionalization in science? Are both the manifestations of middle class upward striving? Does the general political situation contribute directly to the founding of specific kinds of science institutions? Or does it merely determine which ones will thrive? How are new science institutions and the economic growth of industrial communities related? Surely not in a simple manner. Would answers to any of these questions apply to Scotland and America as well as to England? And so forth.

For my particular new contributions, I want to add a little information to the decline of science controversy; present one scientist's reactions to Harcourt's recruiting letter for the proposed Association; and analyze the proceedings of the first meeting in some detail, to see if these show by whom and for whom the Association was set up. I shall conclude with some thoughts on the various controverted points we have encountered, adding a bit of new information on some of them. The rather odd oscillation between analyses and narrative which results is an example of what often results when a historian is not sure whether she is telling the reader what happened, or arguing with him about it.

Charles Babbage's book of 1830, then, is the most-cited document in historians' discussions of the decline of science controversy, for which indeed it provided the name. David Brewster, editor of the *Edinburgh Journal* (once the *Edinburgh Philosophical Journal* but, at the time we are discussing, the *Edinburgh Journal of Science*), was Babbage's strongest ally. Indeed, Brewster was critical of English science before, during, and after the controversy; so it might be more suitable to say that Babbage was Brewster's ally. Brewster in Scotland needed no special occasion to want to free British scientific organization from what he conceived to be the unjustified dominance of London and the old universities in England.

However, it is common to recognize the controversy as being started by John Herschel, by a footnote to paragraph 320 of his long article on "Sound" for the *Encyclopaedia Metropolitana,* published in 1829. Foote, indeed, cites earlier articles in the major reviews, but I am reluctant to accept these as the beginning of a real controversy, since I don't know who got excited about them and argued back. What is not so commonly recognized is that Herschel may also have been the person who stopped the controversy; and that he opposed the founding meeting of the British Association in 1831, even though he was the defeated candidate in the Royal Society election of 1830.

A footnote in an encyclopaedia article—what an unusual way to start things off! But that is the way Herschel's influence worked around 1830. The note has been reprinted by Foote and by Williams, but it expressed so many attitudes relevant to the professionalization problem that I want to reconsider it, section by section.

The note began with praise for the French *Annales de Chimie,* from which, Herschel said, he had drawn heavily for this article as well as for his earlier classic article on "Light." (That in itself says something about the absence of "physics" as an independent subject.)

Then, in one of those characteristic comments which left his contemporaries wondering whether he was really aiming at their ventures or not, he said: "Unlike the crude and undigested scientific matter which suffices (we are ashamed to say it) for the monthly and quarterly amusement of our own countrymen, whatever is admitted into its pages has at least been taken pains with, and, with few exceptions, has sterling merit." Was this a criticism of Brewster's journal? of the Proceedings of various societies? of the science articles in the general review journals? No one exactly knew, I imagine.

Some of the articles in the *Annales de Chimie,* Herschel noted, were abstracts of Memoirs of the French Institute, or were taken from reports of the committees assigned to review articles for inclusion in the Memoirs. This led Herschel to express the professional's yearning for an ideal professional audience: "What author, indeed, but will write his best when he knows that his work, if it have merit, will immediately be reported on by a committee who will enter into all its meaning, understand it however profound, and not content with *merely* understanding it, pursue the trains of thought to which it leads, place its discoveries and principles in new and unexpected lights, and bring the whole of their knowledge of collateral subjects to bear upon it."

Herschel went on to assure his readers that there was a large professional audience available for them in Europe, which could be reached through journals such as those of Poggendorff and Schweigger covering all of the physical sciences. "This ought to encourage our men of science. They have a larger audience, and a wider sympathy than they are, perhaps, aware of; and however disheartening the general diffusion of smatterings of a number of subjects, and the almost equally general indifference to a profound knowledge in any, among their own countrymen, may be, they may rest assured that not a fact they may discover, nor a good experiment they may make, but is instantly repeated, verified, and commented upon in Germany, and we may add too in Italy."

Note that Herschel was not praising specialization for its own sake. He was praising the expertise needed to appreciate advanced work. And he was for internationalization simply because it produced a larger group of experts. He did not suggest that there was a special German science or German approach to science which was enlightening, only that there were more expert and interested scientists there.

Then came the statements which stimulated some readers more than a plea for better scientific journals, and which account for the emphasis on mathematics and chemistry in the ensuing controversy: "Here, whole branches of continental discovery are unstudied, and indeed almost unknown even by name. It is in vain to conceal the melancholy truth. We are fast dropping behind." He singled out two of his own favorite subjects: "In Mathematics we have long since drawn the rein and given over a hopeless race. In chemistry the case is not much better. Who can tell us anything of the sulfo-salts? Who will explain to us the laws of Isomorphism? Nay, who among us has even verified Thenard's experiments on the oxygenated Acids—Oersted's and Berzelius's on the radicals of the Earth—Balard's and Serrulas's on the combinations of Brome—and a hundred other splendid trains of research in that fascinating science?"

(To me, this seems a rather unconvincing list of examples. Herschel seems to have been calling for a larger number of expert laboratory workers, not for profound theorists or radical new departures, not for more Daltons or more innovations like electro-chemistry. I invite the comments of historians of chemistry on this point.)

Herschel ended with another of his cryptic remarks: "Nor need we stop here. There are, indeed, few sciences which would not furnish matter for similar remark. The causes are at once obvious and deep seated. But this is not the place to discuss them." By "deep seated," Herschel meant that it was not the fault of a few individuals. But what he thought the "obvious" causes were has never yet been fully explained.

It was clear to Babbage, however, that Herschel meant the causes that were obvious to Babbage. He in his book of 1830, and Brewster in his own journal and in his review of Babbage's book in the *Quarterly Review,* thus felt free to attack whatever they had been wanting to attack all along. Moreover, Herschel had concentrated on chemistry. This led Babbage and Brewster into the undiplomatic venture of attacking Michael Faraday; and it led J. F. Daniell, professor of chemistry at King's College, London, to attack Herschel. James South, Herschel's erstwhile observing partner in astronomy, was the worst. In a pamphlet of *Charges against the President and Councils of the Royal Society* (with two editions), he lashed out at everyone in sight, asserting among other things that the Society had wasted thousands of pounds of public money on attempts to manufacture optical glass. This, it so happened, was a project on which Faraday was the chief scientific worker—and Herschel was one of the prime movers. It was becoming clear that Sir James South was *not* a gentleman.[27]

Herschel avoided taking any public part in the controversy. In writing privately to friends, however, he did not conceal his disapproval. To Babbage, to be sure, he wrote nicely in May 1830, saying that the *Decline of Science* book would do good, but would have done more good if it had not been so bitter; he hoped that it would not produce a controversy between the Royal Society and the Astronomical Society. But in writing to William Fitton in October to encourage the choice of Francis Baily as candidate for president of the Royal Society, he said that Babbage's book, "by systematically lowering the motives of all scientific men," had made it necessary that no one put himself in a position where his motives could possibly be questioned.[28] Babbage's demands for money and honors for men of science was, it seems, having a definite effect; but on men such as Herschel it was just the reverse of what Babbage had wanted.

Writing to the Reverend T. J. Hussey, whom he knew through the Astronomical Society, Herschel nine months later spelled out his views on government grants, a subject with which he had had some experience, having been instrumental in getting grants for Babbage for his calculating engine. Herschel said that he did not agree with the recent general outcry against the government for giving little money to science. He thought that lately the government had been most liberal, and had shown a desire to be useful to science. But its grants had often failed to produce anything, because of a lack of a good system of choosing among applicants, and because of too much concentration on astronomy. He was unwilling to use the Astronomical Society to push for grants for astronomers alone. If a public body were to advise the government on grants, it should not say, we are chemists and we want all we can get for chemistry, regardless of what happens to all the other sciences. I can't said Herschel, ask for £500 for you without being prepared to ask for half a million pounds for all the others—which I would be willing to do if I thought the nation would give that much to science.[29]

Herschel's position, in short, was that scientists appeared too centered on their own specialty and too jealous of other scientists; if they would put their own house in order, he seemed to suggest, the government was ready to listen to reasonable proposals. Demands like those of Babbage merely made scientists look greedy. Thus in writing to William Whewell the following month (September 1831) in order to denounce Harcourt's proposal for a new Association, he referred to "that unlucky note which served as a test for Babbage to preach on and thus (by a strange perversity) has served to identify me with all the grovelling and degrading views of the objects & idols of scientific men which have lately made such a hubbub."[30] We today might see the proposed Association as a way to set up a system of screening grant proposals and making sure that all sciences had an equal chance, to do just what Herschel said was needed in order to give the government a good set of requests; but the declinist controversy had so poisoned the atmosphere that Herschel could see it only as an embodiment of the selfish approach he disliked.

Herschel's self-imposed silence on the declinist controversy (after all, Babbage was one of his oldest friends) was finally broken in the autumn of 1831, when he found himself praised in an anonymous article in Brewster's *Journal* at the expense of Michael Faraday, and attacked by J. F. Daniell in the latter's introductory lecture as professor of chemistry at King's College, London. As it happened, Herschel liked and admired Faraday personally, having worked with him on the Optical Glass Committee of the Royal Society in the late 1820's. On November 24 he wrote to Daniell that he had had no idea that his footnote would become so prominent in the controversy, "which I have from its outset regarded with regret, and in which I have abstained from taking any part." A brief set of explanations and formal assurances apparently satisfied both parties.[31]

On November 25, Herschel wrote a formal Letter to the Editor of the *Edinburgh Journal* (with a covering note to Brewster personally) in his stiffest—that is to say, angriest—manner, disavowing the article. He wrote that he regretted "the tone of invidious contrast by which...I find myself exalted into a rank of preeminence above others who are characterized as constituting an inferior class in the scale of scientific estimation, and among whom, by direct implication, Professor Faraday seems intended to be included....I must decline, as foreign to all my feelings, applause thus bestowed whether at the express or implied expense of others." Faraday, he said, was "an estimable man and a sound philosopher."

Brewster refused to print the letter. He revealed that *he* was the author of the anonymous article; he did not want to disown himself in his own journal, he said. Herschel explained in a second letter that he did not despise the Royal Society, as had been suggested; after all, almost half of its members had been his supporters the previous year. He had omitted "F. R. S." after his name on the title-page of his *Preliminary Discourse* only because he would then have had to list twenty-odd other memberships of scientific societies, domestic and foreign, as well, and such a string of letters after his name would have looked pretentious in a popular book. Brewster stoutly defended his own interpretations of Herschel's actions; and Herschel ended the correspondence in January 1832, writing: "You must excuse my entering into any further discussion by letter respecting those topics in your note on which it does not appear probable that a prolonged correspondence will produce any approximation of opinion."[32] This effectively ended Herschel's long-standing connection with Brewster, which went back to 1817, when Brewster had solicited mathematics articles from the young enthusiast to go into Brewster's *Edinburgh Encyclopaedia;* and to 1819, when the young chemistry enthusiast had sent his articles to the newly-founded *Edinburgh Journal.*

Thus by the end of 1831 both the declinists and the anti-declinists knew emphatically where Herschel stood. Such quasi-public letters of Herschel as those to Daniell usually became known in the proper circles. And Babbage's authority was based to a considerable extent on his long close friendship with Herschel. It would not be surprising, therefore, to find not much enthusiasm left among the declinists by the end of 1831.

The Brewster-Herschel correspondence has one final chapter. The following November, Brewster (with his usual lack of tact) solicited Herschel's support in the contest for the chair of natural philosophy at Edinburgh. Since Herschel had been offered the position and declined it, Brewster wrote, his recommendation to the electors would be especially valuable. Brewster's own claim on Herschel, he said, was that in the Royal Society election he had come 400 miles to vote against a Royal Prince, "and I have experienced the consequences of it in more ways than one."

At this, Herschel really exploded. In a draft of a letter which may never have been sent, he blistered Brewster for thinking that being supported in the election obligated him in any way. He could not have avoided the duty of being PRS, but his sacrifice would have been far greater than Brewster's, "had I unfortunately for my own happiness and for the success of my most cherished prospects been placed in the chair of the Royal Society." Herschel was referring to his long-hoped-for trip to the southern hemisphere to complete his (and his father's) survey of the whole nebular heavens.[33] He was quite right in his judgment; the astronomical results of the trip secured his position in science far more than holding any office could have done.

As it turned out, Herschel supported, not Brewster but Brewster's young rival James David Forbes for the Edinburgh chair. I don't know when he decided to do so, but his public testimonial was received too late to be printed in the main sequence of Forbes's pamphlet of testimonials (thus giving it more prominence) and it was dated 17 December 1832— that is, well *after* he had received Brewster's request.[34]

These events suggest certain speculations to me. In spite of the fact that Babbage distributed at least 204 free copies of his *Decline of Science*, presumably where he thought it would do the most good (including 53 to the Continent and 8 to the United States),[35] it is not clear that the book had any important influence in producing any scientific reform. I follow Morrell on this; if other men felt as Herschel did, Babbage's open desire to get government money for science may have made scientific leaders thereafter more embarrassed to approach politicians for grants.[36]

The book may have produced one change, although it is not clear that it was a reform. Perhaps, as Babbage thought, he influenced the government in its decision to give Hanoverian knighthoods to Herschel, Brewster, Ivory, Leslie, and Bell. Herschel, however, was not happy about being so honored, writing to William Somerville in September 1831 that the recent discussions concerning scientific patronage made him not merely reluctant but averse to receiving the distinction. However, he said, this particular distinction was the best possible one for him personally, as it was the one his father had received for his science.[37] William Herschel had come from Hanover, so a Hanoverian order was appropriate for members of the Herschel family. My guess is that Herschel thought it would cause controversy if he turned it down, as his omission of F. R. S. after his name in the *Preliminary Discourse* was doing; it would certainly make him look like a sore loser, since the man honoring him (the King) was the brother of the man who had defeated him in the election; but that he would much rather have received it under some other circumstances.

Did the "decline" controversy influence anyone to do anything he would not otherwise have done, with respect to the founding of the British Association? Babbage, Brewster, and Brewster's ally in Edinburgh, J. F. Johnston, were already interested in an institution similar to that of the German *Naturforscher*.[38] Did their declinist agitation convince anyone but themselves that a new organization was needed? The actual organizer of the first Association meeting, William Vernon Harcourt of the York Philosophical Society, was openly looking for an alternative to the Royal Society (not avowedly a nation-wide organization) as early as February 1830, in a speech to his own society printed in the *Philosophical Magazine*.[39] And Harcourt, as Foote mentioned in his 1951 article, was a non-declinist; he thought that science in the nation had grown beyond the resources of the Royal Society.

It seems to me likely—again I am following Morrell—that the decline controversy had the effect of splitting the supporters of Herschel so that many of them would not support an organization proposed by David Brewster and his *Edinburgh Journal* cronies. One job of the organizers was to overcome this initial disadvantage.

My next speculation is, I think, an expansion rather than merely an endorsement of Morrell's position. I agree with him that, in order to recruit the London-Cambridge scientists, it was necessary to abandon Harcourt's ideas of making the Association into a central director of research, and of making it into a body exerting strong pressure "'to claim for Science what is due to it and to the interests of Society depending upon it.'"[40] We might paraphrase this quotation from Harcourt as expressing a desire to influence policy and extract money by claiming to be the voice of the scientific community. Morrell explains the rejection of these ideas by using his theme of individualism: "Quite simply the very diverse local and personal interests in the Association wanted to preserve opportunities in which their own initiative could be exercised; any profound surrender of autonomy was fundamentally anathema to their own concerns. . . ." In this sentence, "local" refers both to Londoners and to provincials; none wished to be subject to the others. Equally, neither scientists nor amateurs (if "amateurs" is a proper translation of Morrell's term "sciolists") wanted to be dominated by the others. It would seem that the Association had to be weak in order to include all of these interests. This, Morrell suggests, was a limitation, for as time passed it became apparent that such individualistic organizations could not cope with "the problems of scale and expense of professionalized science."[41]

My documents suggest a more positive emphasis. I think that the lesser men (I am going to call them "cultivators" of science, since that is what they called themselves) did want to be dominated by the professional scientists. It was the professionals, I think, who felt that any profound surrender of autonomy could *not* cope with the problems of professionalized science. They did not want a dictator of projects or a spokesman for science because these would hamper not merely their own concerns but also the growth of good science in general. This is still individualism, still within Morrell's general approach, but it makes the Association sound more interesting. The professionals did not oppose a strong form of organization and therefore miss an opportunity for science; they insisted on a weak form of organization and therefore gained an organization which science needed.

Let me begin my discussion by disposing of David Brewster's desires. The fact is that these, at one time or another, involved the whole spectrum of goals we have discussed. Thus his first call, in his *Quarterly Review* article on Babbage's *Decline of Science,* was for a great organization to mobilize all gentlemanly ranks against the politicians:

> An association of our nobility, clergy, gentry, and philosophers, can alone draw the attention of the sovereign and the nation to this blot upon its fame. Our aristocracy will not decline to resume their proud station as the patrons of genius; and our Boyles, and Cavendishes, and Montagus, and Howards, will not renounce their place in the scientific annals of England. The prelates of our national Church will not refuse to promote that knowledge which is the foundation of pure religion, and those noble inquiries which elevate the mind, and prepare it for its immortal destination.[42]

Brewster's second call, in the formal letter I mentioned above to John Phillips, secretary of the Yorkshire Philosophical Society on February 23, 1831, was for "a British Association of Men of Science similar to that which has existed for eight years in Germany and which is now patronized by the most powerful Sovereigns in that part of Europe." He said that "the principal objects of the Society would be to make the cultivators of science acquainted with each other, to stimulate one another to new exertions—to bring the objects of science more before the public eye and to take measures for advancing its interests and accelerating its progress." He stressed that "the Society would possess no funds—make no collection and hold no property." He wanted to know if the first meeting could be held at York with the cooperation of the Philosophical Society, the Mayor, and influential persons in the vicinity.[43]

This is certainly a different set of proposals. As Orange noted and I mentioned above, it sounds like the outcome of prior informal discussions with persons at the Yorkshire Philosophical Society. Direct mobilization of nobility and clergy as equal partners in the association is clearly not contemplated. I note in particular that, without funds, the new society would be hampered in attempts to subsidize scientific men or their projects. It could only ask someone else to subsidize them.

Power passed quickly, and by agreement, to the Yorkshire Philosophical Society;[44] so I will go over Brewster's immediately subsequent proposals. But he never abandoned his desire for official subsidies for men of science; and brought it up in 1850, when he was president of the Association. He praised the Institute of France as "the noblest and most effective institution that was ever organized for the promotion of science," objected to the situation in Britain with its belief that "the intellectual interests of the country should, in great measure, be left to voluntary support and individual zeal," and proposed a Royal Academy with "a class of resident members enabled to devote themselves wholly to science."[45]

This was a good 18th-century ideal (or perhaps a 17th-century one); subsidizing men of science had been one of the functions of the old Royal Academy of Sciences in France. The Revolution and Napoleon were not original in thus "professionalizing" science; their new Institute continued the practice of the *ancien régime,* but eliminated the aristocratic cultivators of science who had also been members of the Academy.

It is not surprising, therefore, that Brewster was directly opposed the following year by the next president of the Association, a man built wholly upon the ideals of the 19th century. Morrell has drawn attention to the speech of George Airy, the Astronomer Royal, but not in its context as a reply to Brewster. "I gratefully acknowledge," said Airy, "the services which Government has rendered to science by acceding to the requests of this and other bodies who have indisputably established claims to their attention.... But I trust that in all cases the initiative will be left to individuals or to independent associations."[46] The last phrase is worth noting: the individualism we are concerned with is not solitary self-help; it is a stress on voluntary associations as opposed to official organizations. It is analogous to (and was in part derived from) the methods of dissenting Protestant churches, as opposed to those of an established church; that is my speculation, at any rate.

These contrasted speeches seem to me to give us a useful dividing-point, coming as they do so neatly at the end of the first twenty years of the Association's existence. I will call these the "early years" of the Association. In these years, it is not possible to say whether Brewster had won or lost; it depends on which of his positions you consider. He did not have his great national higher-class pressure group, on the one hand; he did not have his subsidized scientists, on the other hand; but he did have just about what he proposed to the Yorkshire Philosophical Society in February 1831.

And Airy, the Cambridge-trained professional scientist and government scientist, had just the kind of organization he wanted. Let us consider how it was that it was this one of Brewster's proposals, rather than some of his more ambitious ones, which held the field.

My next step is to consider two stages in the proposals of William Vernon Harcourt, the central figure in the Yorkshire Philosophical Society. In his remarks to the Society from 1830 on, the direction of scientific inquiry meant the direction of the hosts of cultivators of science in the various provincial societies. They were to make observations according to one or another common plan, the plan itself to be initiated and supervised by full-time scientists. He drew special attention to Bacon's plea, in the *New Atlantis,* for a "'whole and entire man'" who was supported so that he could devote all of his labor throughout his life to science, and was supplied with money and equipment for experiments. But he also stressed the value of those "who unite scientific pursuits with the active business of life," and in his 1830 speech assigned to his "whole man" the business of collecting the observations and ideas of others. He seemed to envisage a limited number of useful but not dominant men as full-time arrangers and collectors, lecturers, experimenters and observers. He was, in context, obviously generalizing from the role played by Keepers of Museums, such as the YPS's own John Phillips.[47]

But in the late summer of 1831, Harcourt sent out a number of recruiting letters for the proposed meeting of the new Association; these were in addition to the official announcement of July 12 sent by the Council of the Yorkshire Philosophical Society to 39 other "Scientific Institutions in England, metropolitan and provincial" (13 in London, 26 elsewhere), and several hundred individuals—John Phillips at the first meeting said the 39 were all the scientific institutions they knew of. The official announcement did not explain very much, and it was from Harcourt's personal recruiting letters, presumably, that many of their recipients learned authoritatively and in detail of the aims of the new group. (Brewster had had an announcement in his *Journal,* but Brewster's statements were no longer authoritative.)[48]

Harcourt's statements in his recruiting letters were far stronger than the mere proposal to make cultivators of science useful. He was talking about influencing or even commissioning professionals. The proposed society would "indicate points requiring elucidation, propose problems to be solved, data to be determined, and either charge such of its members as may be able and willing to undertake the task as an honourable commission, or offer a prize for a particular investigation, or defray the expenses of a specific process."[49] In my terminology, the society was to give to problems the status of being truly professional.

The British Association did do something along these lines in the decades to come, but not as much as Harcourt had seemingly envisaged. One obvious problem was that it had no good source of money. Harcourt in 1833 convinced the General Committee that it should devote what was available to subsidizing specific research projects, but this was primarily only the surplus from the sale of membership tickets. As Howarth observed in 1931, "A total sum of £92,000, spread over a term of ninety-six years, is not, truly, an imposing figure." The other elements of society did not see the Association as an obvious agent for administering funds; and the Association did not go out to get money in a big way. John Herschel's attitude (as expressed below) was still prevalent in 1931, when Howarth explained that "to the national well-being science contributes much more than it receives; it has not been its habit to bargain."[50]

Just why the Association took on certain researches and not others would be an interesting way to study its history in what I have called the "early period" (1831–50) and also in what I now designate as the "middle period" (1850–80). After the 1870's, as Howarth observes, there was a relaxation of sorts, especially in the matter of cooperation with the government. He suggests that this may have been due to lack of interest in science on the part of politicians; if we add in the new Civil Servants as well, I think he is probably correct. In the early years, the pattern of researches seems Humboldtian: a star catalogue, tides, meteorology, terrestrial magnetism, fossil zoology, early marine biology, and the reestablishment of Kew Observatory. But such an assertion requires more study before being considered as established. A grant might be small but very significant, as any young enthusiast can testify. Or the researches may have been dominated by a group of men, many of whose members but not all were Humboldtians, so that the pattern is that of the group, and only fuzzily a Humboldtian one.[51]

Clearly, the Association did not become the central judge of which problems throughout science merited more professional attention and which ones did not. What it did become was an arena to which professionals came, carrying their interests and their status with them. If these men (and not necessarily the Association itself, or one of its subject committees) then became interested in a paper or a subject, the paper or subject was thus endowed with professional importance. It was not the Association as an organization, I would guess, so much as an active element at its sessions, which made it a professional center.

Here is an example of what I mean. The geologist Adam Sedgwick is writing to his niece about the Association meeting at Oxford in 1847. Prince Albert and the Grand Duke of Saxe Weimar visited the meeting. "When the Royal visitors entered the Physical and Mathematical Section, the great French mathematician LeVerrier was on his legs discussing the orbit of a comet. When he sat down, Airy, Challis, Herschel, Adams, Sir W. Hamilton, and Sir David Brewster took part in a short discussion. In all Europe it would not have been possible to bring greater astronomical names together. This was not got up but mere accident, for they took their chance, and interrupted no business."[52]

Sedgwick was not quite correct. It was not got up, but it was not mere accident. The dispute over the priority for the discovery of Neptune had surely helped to draw just these men together at this time—LeVerrier and Adams were the heroes, Airy and Challis and Herschel were all involved. It was at the British Association that LeVerrier encountered the British forces; and it was at the Association that Herschel arranged for LeVerrier and Adams to come home with him to Kent after the meeting. The two men liked each other, and became friends.[53]

This is an obvious role which a professional organization plays—for some to which I belong it is the only role of any importance—and it needs to be stressed in any discussion of whether an organization is professional or not. Formal papers are less important than discussions; but the real question is, "Where is the party when the session is over?" The importance of the paper is that it may lead to an invitation to the party.

Returning to the late summer of 1831, we can gain some idea of the professional forces which helped to push the Association away from trying to become a strong centralizing body, by considering John Herschel's reactions to the recruiting letter of August 25 which Harcourt sent him. He replied, on September 5, that he did not believe in the usefulness of a "great central & presiding power" which would "chalk out districts for individual or combined diligence to explore, subdue, & fertilize,—to distribute to every class of mind its appropriate task and assign it limits," and, "as you say 'point out the lines of direction in which the higher researches of science should move.'" A society for all of science is probably no longer useful, Herschel said. A society for a particular subject may be valid because *"the whole body* can form an opinion" of any proposed project; but the utility of such groups is in exact proportion to their moderation, to the absence of attempts to control research. They should only provide facilities, make information available, promote an interchange of scientific opinion. They should not try to influence the government, but should gain such a reputation that the government will naturally turn to them.

Furthermore, Herschel said, the men who run a large organization must sacrifice their own research to do so, and thus lose all chance of "posthumous fame." He himself had already given as much of his life to scientific organizations as he intended to, so he personally had "only the power of such participation in the prospects of the proposed Institution as consists in sincere wishes for its utility & consequent success."[54]

Thus Herschel's primary overt reason for rejection was that a large organization could not be professional enough; that is, could not be a group all of whose members were at the same level of expertise. But in a small group which was a set of inter-communicating peers, the idea of an elite which controlled research was unthinkable.

Herschel was clearly affected by, and was reacting to, his experiences with the Royal Society, those of the 1820's as well as those of the election of 1830. But these experiences did not turn him toward the proposed British Association. They turned him away from all general science societies whatsoever.

That was what he said politely, to the organizer of the Association. Later in September, just a week before the meeting, he wrote to his friend William Whewell at Cambridge and, among other matters, expressed himself more openly on this subject.[55] He said that "a certain Mr. Vernon Harcourt has written to me about the York meeting." They intend a central power to give a systematic direction to research, "and moreover & here seems to me to be the cream of the thing 'to be sufficiently powerful, independent & deliberative to possess influence over the Government of the Country to claim for science what is due to it through the medium of public opinion to lead to a more enlightened & creditable national dealing with men of science & their objects.'" How pretentious, said Herschel; simply because of the last point alone, he would refuse to have anything to do with the project. These things are much better done by individuals; even public statements are better as coming from individuals than *ex cathedra* from an organization.

Whewell's idea of annual reports on the progress of various sciences was excellent, Herschel said; he had had something of the kind in mind, (and I repeat this part of the letter deliberately) "when I wrote that unlucky note which served as a text for Babbage to preach on and thus (by a strange perversity) has served to identify me with all the grovelling & degrading views of the objects & idols of scientific men which have lately made such a hubbub." This is not precisely a statement that he would have nothing to do with the Association because it seemed to want to put Babbage's approach into practice, but it comes rather close to suggesting just that.

Herschel tactfully reminded Whewell that Whewell had alluded to progress reports in his review of Herschel's *Preliminary Discourse on the Study of Natural Philosophy,* which had recently been published. I say "tactfully" because, as a matter of fact, it was Herschel in his book who had drawn attention to the value of such reports. In his final chapter, "Of the causes of the actual rapid advance of the physical sciences compared with their progress at an earlier period," Herschel had written:

> We are not to forget the powerful effect which must in future be produced
> by the spread of elementary works and [also] digests of what is actually

known in each particular branch of science. . . . By giving a connected view of what has been done, and what remains to be accomplished in every branch, these digests and bodies of science, which from time to time appear, have, in fact, a very important weight in determining its future progress.

The importance, Herschel felt, was for the psychology of the individual researcher, who can tell from such a digest whether his work is new or not.[56]

Dissertations on the progress of science were not a new idea in England—those of the *Encyclopaedia Britannica* were often cited—but having them done frequently and for individual departments of science may have been new. Herschel's own treatises—no modern editor would call them "articles"—for the *Encyclopaedia Metropolitana* on "Sound" and "Light" may have helped him toward the idea, as may have his reliance on Thomas Thomson's *Chemistry* in his chemical work. None of these were exactly arranged as progress reports, however, so the British Association Reports really were (it seems to me) something of an innovation in England. At any rate, in 1831 it was Herschel who asserted in his *Preliminary Discourse* what was desirable, and it was Whewell who saw in the proposed Association a way to carry it out.

Returning to Herschel's letter to Whewell, we find further objections to Harcourt's approach. The authority of such progress reports would come from the authority of the writer, not from that of the Association. He expected the meeting to be a mass of mediocrity, and, "A thousand mediocrities don't make up one excellence, nor can a thousand eyes looking through as many spyglasses see as well or as far as one with a first rate telescope." It would be foolish for such people to try to direct the research of creative individuals, who would normally lead the way themselves. "Perfect spontaneous freedom of thought is the essence of scientific progress."

Herschel especially disliked Vernon Harcourt's idea of designating particular men to handle particular research areas. "If any way could be devised more fertile than another in annoyance & disputes & in the long run more self-defeating, it would be the establishment of *scientific preserves* of the kind above mentioned. . . . No no, Science is a common on which every body has an equal right of occupancy and I trust that every inhabitant would rise up & demolish the first fence that should be erected across a single corner of it. It is often called a *republic* but such a society as is here proposed will make it a *democratic tyranny* with all the vices of the narrowest oligarchy."

Herschel's primary tone was, therefore, what Morrell calls attention to: individualism. It was, however, the individualism of the advanced professional, who fears the restrictions on scientific advance which the mediocrity of a large organization will produce. It is also as an advanced professional that he has a horror of being "represented" in dealings with the government. Not only would it be degrading to clamor for what should be automatically acknowledged, but also he will actually have more power if he stands alone and speaks as an individual. (And in Herschel's own case, this was probably true.)

Herschel did not, however, urge Whewell to avoid the meeting. His objections were to "the crude proposals contained in Mr. Harcourt's letter." He expected more evil than good from such a group; at the least, it would produce a great waste of time, and much involvement in quarrels on the part of participants. But if good men attended, constructed the organization sensibly, and made haste slowly, good might come of it—as long as it did not take on a specifically political character in the current excited state of the nation. Its intention to influence government might, unfortunately, encourage some members in this direction.

Herschel's letter was written, we remember from Foote's article, during the period of political agitation over the great parliamentary Reform Bill.

The third stage of Harcourt's proposals came as the principal item of business at the first session of the "General Scientific Meeting at York" (as the membership tickets read), on the afternoon of Monday, September 27, 1831, in the Theatre of the York Museum, before an assemblage of over 300 people.[57] Among those missing were William Whewell, William Buckland, and, of course, John Herschel.

By studying not merely the particular proposals, but also the emphasis Harcourt gave to each in his presentation, we will see that the opinions of the objectors were beginning to produce modifications. Indeed Harcourt candidly admitted this (or, rather, bragged about it) in the opening of his address.[58] I will note, as occasion arises, changes in the direction of Herschel's position, without assuming that the changes were caused solely by Herschel's letter to Harcourt or by any of his opinions which may have been transmitted by Whewell. My feeling is that Herschel's reactions represented, as much as those of any one person could, *the instincts of the professional scientist of the period.* I would expect that others had reacted much as he did to one part or another of Harcourt's proposals. Harcourt, as far as I can judge from this opening address on the "Objects and Plan of the Association," did not quite understand those instincts; but he understood the necessity of yielding to them.

A few matters of terminology are of interest. Brewster in his letter to the Yorkshire Philosophical Society in February had called for an association of "Men of Science." The official circular letter of July 12 announced a meeting of the "Friends of Science." However, the usual phrase at the meetings was the "cultivators of science." This followed the usage both of Brewster in his *Journal* (usually) and of Herschel in his *Preliminary Discourse.*[59] The phrase fairly obviously excludes the general public. Whewell in 1832 apparently felt this, and invited "all who are cultivators of science or interested in its objects" to the next meeting at Cambridge. Buckland tried out an alternative, "the improvers of science," but it did not catch on.[60] Eventually, as we know, Whewell introduced the word "scientist;" but that is not at all the same thing. It certainly excludes the interested Earl or devoted ex-Admiral. Therefore "men of science" eventually returned to popularity because, I suggest, "scientist" sounded too professional and "cultivators" a bit too genteel—more like flower gardeners than like Herschel's open-field farmers.

There were, however, clearly a special set of cultivators. They were "distinguished cultivators of science," or "men of scientific eminence," or even "masters in science."[61] These were the ones who had suggested that there be such a meeting; without such endorsement, the meeting would have had no right to convoke itself. Whether or not these men were our "professionals," I will leave to the reader to decide.

I note, however, that Harcourt never used the word "profession," even though (or perhaps because) Charles Babbage had used it in his *Decline of Science*. Harcourt did mildly advocate, as we shall see, full-time fully-paid men along the French model, but he did not call them professionals, or make any comparison to the professions of law, medicine, church, and military.

On, then, to the meeting.

Notes to Chapter Six

This is the first presentation of this material. It is a justification of the paragraph in an earlier article cited in note 13.

1. O. J. R. Howarth, *The British Association for the Advancement of Science: A Retrospect,* 2nd ed. (London, 1931), p. 42. The first edition of 1922 is identical with the second edition in the early chapters; I always cite the second edition because of the additional information towards the end.

2. Quoted in Howarth, *British Association,* p. 38.

3. Earl Fitzwilliam (formerly Viscount Milton), reported in *Lithographed Signatures of the Members of the British Association for the Advancement of Science...with a Report of the Proceedings* (Cambridge, 1833), p. 86. I am much indebted to Dr. M. J. S. Rudwick for my copy of this fairly rare publication.

4. George Foote, "The Place of Science in the British Reform Movement, 1830–1850," *Isis, 42* (1951), 208, 192.

5. Foote, "Place of Science," pp. 199, 200.

6. Foote, "Place of Science," pp. 200, 208. The Reverend William Vernon became William Vernon Harcourt in January 1831, just in the middle of our story, when his father Archbishop Vernon of York, succeeded to the Harcourt estates. He was thereafter referred to sometimes as Mr. William Harcourt, sometimes as Canon Harcourt (he was a canon of York Minster) but more often as Rev. W. Vernon Harcourt. His son, the well-known politician, was known as Sir William Harcourt.

I prefer to call our man Vernon Harcourt, but I use his name so often in this paper that it is only sensible to shorten it to Harcourt. This has contemporary justification; when beginning a discussion of, say, electricity, one might well refer to "Clerk Maxwell," and then shorten it to "Maxwell" for the rest of the paper. "Maxwell's equations" therefore do not bear the name of their inventor's family (his father was born Clerk) but of a small estate in Scotland his father inherited.

7. L. Pearce Williams, "The Royal Society and the Founding of the British Association for the Advancement of Science," *Notes and Records of the Royal Society, 16* (1961), 230 and note 37.

8. Williams, "Royal Society," pp. 221, 225, 223, 224.

9. Williams, "Royal Society," pp. 225, 228.

10. Williams, "Royal Society," p. 222.

11. The letter is printed in Howarth, *British Association,* pp. 13-14. A. D. Orange, "The British Association for the Advancement of Science: the Provincial Background," *Science Studies, 1* (1971), 316 n8, refers to such preliminary discussions.

12. L. Pearce Williams, *Michael Faraday* (New York, n.d.), pp. 354-355. If Faraday was a charter member, it was in a most Pickwickian sense. Faraday, along with 200-odd other individuals, was invited to the first meeting, and politely declined (*BA Report 1 and 2,* pp. 19, 21). Did he nevertheless send in a £1 membership fee? It does not seem a very important question.

   See also Harcourt to Faraday, 5 September 1831 (after Faraday's refusal) in *Selected Correspondence of Michael Faraday,* ed. L. Pearce Williams (Cambridge, 1971), I, # 116. It shows Harcourt the politician, not averse to stretching a point a bit ("I am in correspondence with Mr. Herschel," meaning that I have written him and have not yet received his refusal to cooperate), but determined to involve Faraday in some way or another.

   Williams bases his claim that Babbage lost the election, both in article and book, on just *one* letter, from George Harvey of Plymouth to Babbage, which shows that Babbage wrote telling him "it was not at all necessary to come up." But the rest of the letter as quoted by Williams ("Royal Society," note 37, p. 233) does not imply that there were many other people in the same category. It implies that a significant number of Herschel supporters lost their nerve at the last minute, and Harvey wants to make sure that people know *he* was not one of those; he, at least, has the excuse of Babbage's statement to him.

   In the article, Williams presents his speculation in a footnote: "It is possible that the election was lost because Babbage.... It would appear that Babbage had written to a number of Fellows who had planned to come to London especially to vote and told them not to bother." In the text of the book, still on the evidence of the one letter *from* Harvey, these statements have become, "The campaign, however, was lost by Babbage.... he wrote to a number of Fellows who resided some distance from London that they need not come to London since Herschel's election did not seem in doubt."

13. Walter F. Cannon, "History in Depth—the Early Victorian Period," *History of Science, 3* (1964), 24-25. The account that inspired me is in J. C. Shairp, P. G. Tait, and A. Adams-Reilly, *Life and Letters of James David Forbes* (London, 1873), pp. 75-79.

14. Nathan Reingold, "Babbage and Moll on the State of Science in Great Britain—A Note on a Document," *Brit. Jour. Hist. Science, 4* (1968), 58. Reingold's footnote ends, "Although not explicitly on this topic, Walter Cannon's articles...are necessary background." I appreciate the reference; but the reader must be a careful one to realize that the antecedent of "this topic" is "the Babbage-Moll exchange." If he is not care-

ful, he will gain the impression that I had never said anything explicitly about the subjects of Reingold's first sentence—including the founding of the British Association.

15. *Victorian Science: A Self-Portrait from the Presidential Addresses of the British Association for the Advancement of Science,* ed. with an intro. by George Basalla, William Coleman, and Robert H. Kargon (Anchor Books, Garden City, N. Y., 1970), pp. 7, 8; see also p. 27. [Hereafter cited as BCK].

16. BCK, p. 28; Howarth, *British Association,* p. 27.

17. BCK, p. 10.

18. BCK, pp. 10–11.

19. Howarth, *British Association,* p. 39.

20. BCK, pp. 15, 17.

21. J. B. Morrell, "Individualism and the Structure of British Science in 1830," *Historical Studies in the Physical Sciences, 3* (1971), 184, 187, 201, n37, 201.

22. Morrell, "Individualism," pp. 202–203. "Northern provincial-minded" is my phrase, not his; I invented it so as to include such persons as Murchison, in London but a Scot; and Brewster, in Scotland but not of the University of Edinburgh set. "Metropolitan" in contemporary usage usually meant only London and London-centered; I use it for the London-Cambridge axis. Morrell in this passage specifies not only London and Cambridge, but also Oxford (except Daubeny) and the Universities of Edinburgh and Glasgow as hold-outs.

23. Morrell, "Individualism," p. 201 n37.

24. Morrell, "Individualism," pp. 190–191; his statement that government support was "possibly niggardly" is explicit enough; but his comparison is to France and the Germanies, not to Great Britain in 1780 and 1880.

25. Morrell, "Individualism," p. 186; citing Geoffrey Millerson, *The Qualifying Associations* (London, 1964), pp. 10–12.

26. Orange, "British Association," p. 326. See also Orange's "The Origins of the British Association for the Advancement of Science," *Brit. Jour. Hist. Sci, 6* (1972), 152–176; and his *Philosophers and Provincials* (York, 1973).

27. James South, *Charges Against the President and Councils of the Royal Society,* 1830, p. 13 (2nd ed., 1830); bound as #8 and 9 in "Debates and Dissensions of the Royal Society" (Royal Society Library). (On other contributions to the controversy, see Foote, "Place of Science," *passim;* Reingold, "Babbage and Moll," *passim;* Orange, "British Association," p. 319.) South's *Charges* were first published just before the contested Royal Society election.

28. Herschel to Babbage, 22 May 1830; to Fitton, 18 October 1830 (John Herschel Papers, Royal Society).

29. Herschel to Rev. T. J. Hussey, 2–3 August 1831 (John Herschel Papers, Royal Society).

30. Herschel to Whewell, 20 September 1831 (John Herschel Papers, Royal Society).

31. Herschel to Daniell, 24 November, 29 November 1831 (John Herschel Papers, Royal Society). I have not seen Daniell's letter or checked his private opinion.

32. Herschel to Editor *Edinburgh Journal,* 25 November 1831; Herschel to Brewster, 25 November, 8 December 1831; Brewster to Herschel, 17 December 1831; Herschel to Brewster, 1 January 1832 (John Herschel Papers, Royal Society).

33. Brewster to Herschel, 19 November 1832, Herschel to Brewster, undated rough draft

(John Herschel Papers, Royal Society). The other letters from Herschel I cite are carbon copies, and may be presumed to have been sent.

34. "Testimonials in favor of James D. Forbes...for the Chair of Natural Philosophy in the University of Edinburgh," (Roderick Murchison Pamphlet Collection, IV, Geological Survey Library).

35. "Copies of the Decline of Science presented or delivered to Mr. Babbage's Order," Babbage Manuscripts (Science Museum, London). The U. S. copies went to the American Philosophical Society, the New York Philosophical Society, Columbia College, Boston Library, Philadelphia Library, New Orleans Library, Mr. Hassler, Mr. Bowdich [Ferdinand Hassler, Nathaniel Bowditch].

36. Cf. Morrell, "Individualism," p. 203.

37. Babbage to Quetelet, 24 December 1831, quoted in Reingold, "Babbage and Moll," p. 59; Herschel to Dr. Somerville, 16 September 1831 (John Herschel Papers, Royal Society).

38. Orange, "British Association," p. 317; Howarth, British Association, pp. 6–12.

39. See the President's remarks, in "Yorkshire Philosophical Society and New Museum" (meeting of February 2, 1830), Phil. Mag., n.s. 7 (Jan.–June 1830), 216–217; and Orange, "British Association," p. 320.

40. Harcourt to Babbage, August 1831, as quoted in Orange, "British Association," p. 318.

41. Morrell, "Individualism," p. 203.

42. Quoted in Howarth, British Association, p. 6.

43. Quoted in Howarth, British Association, pp. 13–14.

44. Orange, "British Association," p. 316.

45. Howarth, British Association, pp. 260–261.

46. Howarth, British Association, pp. 261–262; cf. Morrell, "Individualism," pp. 203–204.

47. "Yorkshire Philosophical Society and New Museum," pp. 216–217; Orange, "British Association," p. 328.

48. The announcement of July 12, and Phillips's remarks, are given in British Association Report, 1 and 2 (1831–32), pp. 18–19. I have not seen the original of any one of the recruiting letters. The portions of the one to Babbage quoted by Orange, "British Association," p. 318, and the portions of the one to Herschel quoted in his letter to Whewell, are substantially identical; but this does not show that all of the letters were identical on lesser points. Hence I refer to them as letters, rather than as a letter.

49. Harcourt to Charles Babbage, August 1831, as quoted in Orange, "British Association," p. 318.

50. Howarth, British Association, pp. 152, 148, 259.

51. Howarth, British Association, pp. 231–232, and Appendix I: Classified List of Grants Made by the British Association in Aid of Research, 1834–1931.

52. J. W. Clark and T. M. Hughes, The Life and Letters of the Reverend Adam Sedgwick (Cambridge, 1890), II, 123.

53. Morton Grosser, The Discovery of Neptune (Cambridge, Mass., 1962), p. 142.

54. Herschel to Vernon Harcourt Esq., 5 September 1831 (John Herschel Papers, Royal Society).

55. Herschel to Whewell, 20 September 1831 (John Herschel Papers, Royal Society).

56. John Herschel, *Preliminary Discourse on the Study of Natural Philosophy* (London, 1831), p. 352. In the full context, it is clear that Herschel had in mind two separate types of works: elementary works (we would say "textbooks") *and* the digests for research workers. It is a bit hard to make this clear in a short quotation, so I have inserted the [also].

57. The ticket, reproduced as the frontispiece of Howarth, *British Association,* reads "September 26," but the *BA Report 1 and 2,* p. 15, says September 27 at 12:30 p.m., with 353 tickets issued. Perhaps they decided to avoid a Sunday meeting.

    I will note here that, according to Adam Sedgwick at the third meeting, Harcourt was responsible "for the composition of the first Report" (*Lithographed Signatures,* p. 101). We may assume, perhaps, that he included such material as would in his judgment advance the cause of the Association.

58. *BA Report 1 and 2,* pp. 21–22.

59. *BA Report 1 and 2,* pp. 18, 16, 17, 19, 22, 103; the official "Objects of the Association" modified it to "those who cultivate Science" (p. ix); Herschel, *Preliminary Discourse,* p. 349; Orange, "British Association," p. 315.

60. *BA Report 1 and 2,* pp. 103, 98.

61. *BA Report 1 and 2,* pp. 17, 18, 34.

Chapter Seven

The Founding of the B.A.A.S.

The first meeting of the new British Association for the Advancement of Science, on Monday, September 27, 1831, at 12:30 p.m., was opened by the president of the host organization, the Yorkshire Philosophical Society. The Reverend William Vernon Harcourt had recently stepped down from this post, which he had held virtually since the beginning of the Society. The new president was a local notable, currently Viscount Milton, an M.P. for Yorkshire, soon to become Earl Fitzwilliam, and eventually a Knight of the Garter. Quite a notable indeed.

Milton made it clear that he was not qualified to tell "cultivators of science" what to do, and then proceeded to do it. They should impress on the government that the pursuit of science is "not confined to the metropolis," so that the government will give "every proper stimulus" throughout the empire. In particular, they should influence the government to remove legal obstacles to science, such as the regulations on the manufacture of optical glass. Government must, to be sure, undertake some investigations which are on so great a scale as to be beyond individual enterprise. At the Association's meetings in different parts of the country, "men of science" will meet and exchange ideas. "They will state the advances which have been made in their own respective spheres of action, and also what the deficiencies may be. Thus not only will an extraordinary impulse be given, but the individuals and the Societies taking part in the Meetings will learn what parts of science they can cultivate with the greatest utility, and will give their researches the most advantageous direction."[1]

Short, to the point, and not at all naive. This was what I shall call the "weak form" which the Association might take; and Herschel could have approved of every statement.

Harcourt and John Phillips, chairman and secretary of the Committee of Management (and also vice-president and secretary of the Yorkshire Philosophical Society), then gave details of how the meeting came about. It was made plain that the original promoters had asked the Society to work out the actual plan for a permanent Association, that the council of the Society had approved of the plan, and that, in short, what

Harcourt was about to say was *the* authorized statement and proposal.[2] "Mr. Harcourt then commenced his exposition of the OBJECTS AND PLAN OF THE ASSOCIATION."

First he proposed three objects:

[1] To give a stronger impulse and more systematic direction to scientific inquiry, [2] to obtain a greater degree of national attention to the objects of science, and a removal of those disadvantages which impede its progress, and [3] to promote the intercourse of the cultivators of science with one another, and with foreign philosophers.

These statements became the official "Objects of the Association" printed in the front of the Report, with two minor and three interesting changes.[3]

1. Objects 2 and 3 were reversed—this gave them the order in which they were actually treated in the speech.

2. "The cultivators of science" became "those who cultivate Science in different parts of the British Empire"—thus adding Milton's idea of empire.

3. "A greater degree of national attention" became "a more general attention." This removed any implication that the attention of the national government was being specified.

4. "Those disadvantages which impede its progress" could be construed to mean anything. It was replaced by "any disadvantages of a public kind, which impede its progress," thus narrowing it down to Milton's position.

5. All of these objects were prefaced by the statement, "The Association contemplates no interference with the grounds occupied by other Institutions." This was quite in accord with Harcourt's speech. It committed the Association to deliberately avoid any kind of competition with the Royal Society.

It was the "national attention" object which was nibbled away at by the other members of the Association.

Harcourt, however, devoted most of his argument to "the first and most important of these objects" on which "some differences of opinion may exist." He gave eighteen paragraphs (and one enormous footnote) to the need for a new Association "to give a stronger impulse and more systematic direction to scientific inquiry." He gave only one paragraph to the promotion of intercourse, and ended his argument with four paragraphs on "national attention."

The opening sentence of his argument was an overt repudiation of the declinists: "I do not rest my opinion, Gentlemen, of this want upon any complaint of the decline of science in England. It would be a strange anomaly if the science of the nation were declining, whilst the general intelligence and prosperity increase." He then immediately qualified his position so as to allow him to ask for everything which the declinists had asked for: "There is good reason, indeed, to regret that it does not make more rapid progress in so favourable a soil, and that its cultivation is not proportionate to the advantages which this country affords, and the immunity from vulgar cares which a mature state of social refinement implies." Babbage and Brewster, it would seem, had made the declinist position politically impossible even for those who agreed with their objects; and Harcourt went on to praise the lately deceased heroes whom the anti-declinists had used as evidence: Davy, Wollaston, and Young.

Then he combined ideas from his previous formulations. The new Philosophical Societies now made available a multitude of willing hands, if someone would only tell them what to do. The Association should concentrate on "pointing out the lines of direction in which the researches of science should move," in indicating the most pressing particulars of problems and data, and "in assigning to every class of mind a definite task." (By "assigning" he meant "suggesting," but it has an ominous sound.) No other organization "undertakes to lend any guidance to the individual efforts of its members." The Royal Society was founded to do just this, as an embodiment of Bacon's *New Atlantis*—and here Harcourt went into a long historical analysis stemming from his 1830 paper to the Yorkshire Philosophical Society.

He emerged with the obvious conclusion that the Royal Society at the present time, although it publishes, promotes, and advises, "scarcely labours itself, and does not attempt to guide the labours of others." Hence new special Societies have been formed to work for particular branches, "and by degrees the commonwealth of science is dissolved." This tendency will eventually retard science as a whole, if not countered by a way to bring the great men in individual subjects together to consider "the one great system of nature" as a whole. Harcourt's phrase for these people was "men who have exercised great and equal powers of mind upon different pursuits."[4] I think it is fair to translate this as "expert specialists." Harcourt's argument was an effective answer to Herschel's scepticism concerning general societies, for Herschel himself in his *Preliminary Discourse* had emphasized that "no natural phenomenon can be adequately studied *in itself alone,* but, to be understood, must be considered *as it stands connected with all nature.*"[5]

Furthermore, said Harcourt, the specialized Societies cannot employ to good advantage the "humble labourers" in the local Philosophical Societies; and already there are nine such local societies in Yorkshire alone. Harcourt particularly mentioned geology (the Geological Society of London had had the idea of organizing the results of the work of "humble labourers" almost from its foundation). But the labourers would be useful in natural history also, in meteorology, and indeed in every science founded on observation or on simple experiments. Harcourt did not cite Herschel on this point, but in the *Preliminary Discourse* Herschel had called for such "industrious individuals" especially in geology and meteorology but not limited to those subjects. Herschel had said:

> There is no branch of science whatever in which, at least, if useful and sensible queries were distinctly proposed, an immense mass of valuable information might not be collected from those who, in their various lines of life, at home or abroad, stationary or in travel, would gladly avail themselves of opportunities of being useful.[6]

It would be foolish to insist that either the earlier Geological Society hopes or the recent plea by Herschel was the source of Harcourt's idea; after all, he had read Bacon, as had, probably, all of these other men. But Harcourt could have called attention to the fact that he had designed an organization well-suited to carry out Herschel's own desires in this respect. He did not; he was saving his quotation from Herschel's *Preliminary Discourse* for a more debatable point.

Harcourt's own reference, then, was to Bacon. We now have, he said, in the local societies the manpower to carry out his scheme. The great men have crowds of ideas, which they do not have time to carry out themselves. If they will tell the "inexpert inquirers" precisely what is wanted and what methods to use, these latter will get better with experience and eventually turn out something useful. As for the "experienced in science," they will be flattered by being asked to do something. Even the greatest minds like to be asked—remember Newton.

As a historian, I find the most interesting feature of Harcourt's argument to be his great emphasis on Bacon, most of the details of which I have omitted. This was not "Baconianism," that is, a mere emphasis on fact-collecting; in a long footnote Harcourt demonstrated that not only Boyle but also Descartes was quite indebted to Bacon.[7] Bacon gave Harcourt the best of English precedents for his scheme; he could then bring in the French models so emphsized by the "declinists" casually, as a possible addition.

Harcourt did not omit the phrase Herschel had objected to, "assigning to every class of mind a definite task," but his emphasis was on the great men doing this for the willing workers. The great men themselves were merely to be tempted with the honor of being solicited to undertake an inquiry. Hence Herschel's fears about directing real scientists were unfounded. It was rather: if you want manpower to carry out all of your ideas, join us.

What do the humble labourers get out of the arrangement? Harcourt made that clear, but did not put it in these terms: they got an education in scientific methods working under the best and most enterprising men on real projects, and they paid no tuition for this advanced education.

On his second object, that of promoting intercourse among the cultivators of science, Harcourt said, "I shall content myself with remarking, that nothing can be better calculated to prevent those interferences, and reconcile those jealousies which sometimes disturb the peace of philosophy, than the mutual and amicable communion of such a Meeting as this." This sounds like a direct rejection of one of Herschel's arguments, although anyone involved in the Royal Society recently might have made an argument similar to Herschel's. It also sounds like a warning to any notable disturbers of the peace in the audience not to do it here.

On his third object, that of national attention to science, I think that Harcourt knew he was on dangerous ground. He seconded Milton's suggestion of the reform of fiscal laws, with heart-warming attention to the cost of museum exhibit cases because of the tax on glass. He added in Brewster's objections to the patent laws. Then he said, "With regard to the direct national encouragement which is due to scientific objects and scientific men, I am unwilling to moot any disputed or disputable question." Then he proceeded to do so.

Those who work "in the direct service of the country" should have good salaries, not merely enough to get by on. Abstract science would gain from "a liberal public provision." Perhaps, as Playfair has said, men should be supported in other sciences as well, since it has worked so well in France to give her a superiority in the mathematical sciences—but he would not decide whether Playfair was correct. Harcourt had, earlier in his speech, repeated Bacon's plea for a "whole man;" here he deliberately introduced the French model instead, carefully taking it from the respected John Playfair, not from younger admirers of the French.

Harcourt was indeed on debatable ground when he implied that a "liberal public provision" might act as an inducement for the best men to devote all of their time to science. At the second meeting at Oxford, he was directly answered (as I see it) by Charles Daubeny, in acknowledging the thanks of the meeting for having arranged for the meeting to be held there. His own love of science, said Daubeny, "implies in my case no extraordinary merit," since as professor of chemistry he made a satisfactory living from it. "It is to those only who have pursued such studies without partaking of the advantages derived from academical institutions, or that patronage of Government which in other countries supplies their place, to whom the praise is due of a high degree of disinterestedness in preferring the attractions of philosophy to those of emolument."[8]

At a meeting at which John Dalton and William Phillips were being honored, this was very appropriate. It also represented Herschel's strong opinion; and it is the kind of opinion which may have helped Faraday to give up consulting work and concentrate on pure science. And it is, by my definition given in the last chapter, a very professional statement: not that the professional scientist need renounce making a living, but that he should reserve his special praise for those who work more at science rather than making a better living.

This set of attitudes—which, I submit, are by no means dead in professional circles—as well as the problems they engendered are illustrated so nicely by a letter of Thomas Macaulay to Lord Lansdowne in 1856 that I will insert a long excerpt here. Macaulay was proposing that the British Museum (of which he was a trustee) have a separate superintendent for the Department of Natural History, which was in danger of being neglected in favor of the books and the sculptures. He proposed the leading anatomist Richard Owen for the position, and went on to say:

> I must add that I am extremely desirous that something should be done for Owen. I hardly know him. His pursuits are not mine. But his fame is spread over Europe. He is an honour to our country, and it is painful to me to think that a man of his merit should be approaching old age amidst anxieties and distresses. He told me that eight hundred a year, without a house in the Museum would be opulence to him. He did not, he said, even wish for more. He seems to me to be a case for public patronage. Such patronage is not needed by eminent literary men or artists. A poet, a novelist, a historian, a painter, a sculptor, who stood in his own line as high as Owen stands among men of science, could never be in want except by his own fault.[9]

What merits attention is: "he did not wish for more." Take care of your science, and the Trustees will see to it that you have enough to live on. Scientists do not expect to make money like painters and writers do. Einstein should *not* make a killing with a best seller on his Special Theory of Relativity; that is for popular authors to do.

(Incidentally, Owen got the job. During his tenure the separate Museum of Natural History was produced, of which he is therefore rated as the founder, much to the chagrin of some modern Darwinians who inhabit it.)

Returning to Harcourt's arguments for national attention to be paid to science, he continued by saying that we in England have certainly given public support to astronomy, and that is why our astronomical observations are so good. "And I may be permitted to think with Mr. Herschel, that 'it may very reasonably be asked, why the direct assistance afforded by governments to the execution of continued series of observations, adapted to this especial end, should continue to be, as it has hitherto almost exclusively been, confined to Astronomy.'" A hit—a palpable hit! The passage was from Herschel's *Preliminary Discourse,* and Harcourt presented it exactly in context.[10]

The final argument was also a good one. After Humboldt had shown me the natural collections in the Louvre, said Harcourt, "I felt, I confess, no elation of national pride when I recollected the state of the British Museum." Just as there are scientific enterprises too large for any but a nation to undertake (as Viscount Milton had said), so there are scientific establishments for other subjects than astronomy which only a nation can afford. For that matter, even our diplomats would benefit from more accurate scientific studies of our distant possessions, said Harcourt.

This, then was Harcourt's presentation of the objects of the new Association. In my terminology, it is almost exclusively a plan for the professionals to meet, enjoy one another's company, and decide what the other ranks should be doing. The other ranks will be rewarded by being permitted to be close to, and even work under the direction of, the professionals. In my analysis, most persons joined Lit & Phil societies as a move towards gentlemanly status. By becoming subject to professional scientists, they would gain more status. Therefore if Harcourt's scheme worked, it would mean that professional scientists did indeed have that higher social status which (I think) they were claiming in the contested Royal Society election of 1830.

Harcourt did not put it quite that way. What he said was:

Let philosophy at length come forth and show herself in public; let her hold her court in different parts of her dominions; and you will see her surrounded by loyal retainers, who will derive new light and zeal from her presence and contribute to extend her power on every side.[11]

Which, I submit, says just what I have said, only without using sociological metaphors.

This does not mean that the Association was founded by professional scientists. If Harcourt was a typical founder (and that can be argued), it was founded largely by the local cultivators of science, who were looking for professionals to be subject to. There had to be some professionals at the first meeting, of course: if the plea of the "loyal retainers" was that they needed to be told what to do, there had to be someone there to tell them.

Harcourt's emphases are understandable along these lines. He could not leave out government support of scientific projects, or financial support for individual scientists, because some of the professionals present were in favor of them. On the other hand, he well knew that some other professionals *not* present opposed them. So he reduced the idea of government support to limited areas which someone noteworthy, and not present, had supported: Playfair, Herschel, Humboldt.

Now let me review the organizational structure that was set up, to see if it fits this analysis. Harcourt's justification of the Objects of the Association was not permanently binding on anyone; but the structure was lasting. It was almost exactly as he proposed in the remainder of his address, with one amusing change, and one important addition which it is surprising that he omitted. And Harcourt was able to slip back in, albeit in a weak form, some of that direction of professional research which he carefully avoided in explaining the Objects.

The general unaffiliated and unscrutinized public was not to be permitted to join the Association, only those who had already shown a sign of gentlemanly interest. This is the significance of the early debate over the admission of women to the sessions. William Buckland as president in 1832 asserted that this "would at once turn the thing into a sort of Albemarle-dilettante meeting"—that is, a popular affair like the lectures at the Royal Institution on Albemarle Street in London. Really public lectures were added to the program but were distinct from the sessions for members—although women did continue to filter into the sessions.[12]

Harcourt proposed that "all *Members of Philosophical Societies* in the British empire" be permitted to join. He presented this as a liberal provision, renouncing the model of the German *Naturforscher,* since the Association had more extensive and more specific goals than did the German group. "A public testimonial of reputable character and zeal for science is the only passport into our camp which we would require." This was too uncontrolled a provision for his audience; debate ensued; and the simple qualification was changed to: all members of "Chartered Societies," all officers and council members of Philosophical Societies; and all members of Philosophical Societies *who were recommended by their Council.*[13] I suppose this was so that, if some disreputable person had managed to get into a Society, his superiors could spot him and keep him out of the Association. Naturally such a provision would have been an insult to members of real scientific ("Chartered") societies, such as the Royal Society.

As is usual, all members at the organizing meeting were given membership without qualification (it is difficult to get a harmonious majority on future restrictions otherwise). And finally, "persons not belonging to such Institution" could join if recommended by the General Committee of the Association. I do not know how this provision was used; it could, of course, be used to swamp all of the other restrictions. An obvious use of such a clause is to admit distinguished visitors, and foreigners. The Association seemed rather proud of its foreigners at the third meeting, and included their names on the List of Members, in italics.[14]

Internally, there was no intention of making the Association an egalitarian organization. Two higher types of people than the general run of membership were recognized. the "General Committee" was to have the "governing or executive power," and it was limited to "those who appear

208

to have been actually employed in working at science." They were defined as: all members present at a meeting "who have communicated any scientific paper to a Philosophical Society, which Paper has been printed in its Transactions, or with its concurrence."[15] This, as I mentioned earlier, was not a very stringent requirement even for a baronet or a bishop; but it shows the Association did not assume that all its members would work at science. Those who did, I think, are the active "cultivators of science," as opposed to the "friends of science" on the lower side and the real "scientists" on the higher side. At any rate, that is how I shall use the phrase "cultivators of science" hereafter.

Although the General Committee was the group of final appeal, it met only during the annual meeting and had little executive power. This was intentional. It was the special committees (or sub-committees; later, section committees) for the various sciences, "which the general committee shall appoint, placing severally on each those members who are most conversant with the several branches of science," which were to be in continued session between meetings, and which were to select points in their science needing inquiry, engage persons to investigate them, apply for assistance to other scientific groups, and "attend especially to that important object of obtaining Reports in which confidence may be placed on the recent progress, the actual state, and the deficiencies of every department of science." It was therefore the subject committees, said Harcourt, which were to be "the instruments by which, through the medium of the General Committee, the impulse of the Association must be principally directed.[16]

It was through these committees that the experts were to control all of the work of the Association. It is noteworthy, as Howarth points out, that each committee was autonomous; there was no coordinating committee of experts as a whole.[17] *Therefore these committees met Herschel's specifications in his letter to Harcourt:* a group for a particular subject, not controlled from above, in which *"the whole body* can form an opinion" of any proposed projects. Furthermore, without surrendering his desire to have a system "of advancing science in *determinate lines of direction,"* Harcourt stressed above all the progress reports which Herschel had praised and Whewell had proposed. These, said Harcourt, could be started at once. Requests to researchers to undertake particular researches could also be made now. If the funds of the Association ever permitted, it would be time to take the next step. That would be (here Harcourt switched directions) offering "PRIZES for particular investigations." Prizes had worked for chronometers, he said; why should they not work for science?[18]

The officers of the Association were to be changed annually, in part to limit the burdens of the positions (here again is a response to Herschel's position), and in part "to extend the interest which they may be supposed to kindle." There might be, said Harcourt, one or more permanent secretaries for administration; this was duly arranged at the next meeting with, not surprisingly, the Rev. William Vernon Harcourt being made General Secretary. The presidency was used, not surprisingly, directly for promotional purposes in the first decade. Howarth has noted that among the first ten presidents were a duke, an earl, two marquises, and a viscount (5 out of 10).[19] The viscount soon became an earl, and the earl eventually became a duke (indeed, the Duke of Devonshire); moreover, Sir Thomas Brisbane, K. C. B., was a career soldier and administrator (governor of New South Wales, 1821–25), soon to be a General and a G.C.B. So the social standing of six of the first ten presidents was quite high. Of the three scientist-like presidents I will discuss the reasons for choosing Buckland and Sedgwick shortly; the third was Bartholomew Lloyd, provost of Trinity College, Dublin, and president of the Association for its 1835 meeting in Dublin. the tenth man was the Rev. William Vernon Harcourt.

There is clearly something missing in the "Table of Organization" proposed by Harcourt and adopted at the first meeting. I wonder if he left it out deliberately, in order to emphasize the importance of the autonomous subject committees. At any rate, it was added at the second meeting: a council. This, then, was where the normal executive power was to lie; and it would be interesting to know who proposed it, and what discussion there was. Control of the Council was the usual way to control a Society; and it may have been a touchy subject after the Royal Society fights. It occurred to me to see if the scientists dominated this body to the extent that I presumed they would dominate the subject committees.

A quick study of the list of members of the Association after the second meeting, and the personnel named at the meeting, gave results that surprised me a bit. Of course, the results are not conclusive for the whole early period of the Association, but they are suggestive. Of 696 names on the list (I imagine some had not paid their dues),[20] I rate 56 as professional scientists and engineers. Of course my standards are quite high: teachers of science like Thomas Hope at Edinburgh and Bartholomew Lloyd at Dublin, however good, are excluded. On the other hand I include the son, Humphry Lloyd; although he had not yet done much, he was going to almost at once. The same is true of James F. W. Johnston.

My professionals dominated the Council even more thoroughly than the names on the Council list indicate, because the three trustees and the officers were on the Council *ex officio*. The result I did not expect is that the professionals did not dominate the subject committees so thoroughly, and indeed were outnumbered on two of them. I rank the groups, from greatest to least percentage of professional scientists as members, in the following order:

Officers and Trustees
Council

Committee II: Chemistry, Mineralogy, Electricity, Magnetism (11 pro's, 6 non-pro's)
Committee I: Mathematics and Physical Science (12–10)
Committee III: Geology and Geography (8–11)
Committee IV: Zoology, Botany, Physiology, Anatomy (5–11)

Of course these lists are of persons appointed; being listed does not prove that the person accepted, or ever attended a meeting. But the lists represent the desire of the General Committee, and that is what I want to get at.[21]

Even if my decisions about the status of individual members should be challenged, I imagine the general trend of my results will remain. The trend leads to some nice questions. Did the General Committee desire to have geology and biology represented less by professionals and more by "cultivators," or were they simply short of professionals in these subjects? If the latter, was this because the Association did not appeal to them, or because there was a nationwide shortage? If, as I prefer to think, the latter explanation is correct, then we can speculate that British geology had its Golden Age and led the world (in the 1830's and 1840's) when there weren't many professionals available; and that as more money, time, and personnel were invested in the subject, British geology gradually ceased to lead the world.

This is Parkinson's Law applied to the development of science. I wonder if it is true of other times, other places, other sciences. If so, it suggests that increased professionalization is not caused by the need to improve science. Perhaps it is caused by the need of society to find jobs for more workers.

At any rate, my figures for Committee IV enable us to predict that the really great scientific work in Britain in the 1830's and 1840's should have been done in biology. A rational procedure which unexpectedly yields a theoretical result which is so strikingly confirmed by the evidence must have some validity. (Charles Darwin's great creative period was the late 1830's.)

One general cultural observation can be made about the 1832 membership list: it does not contain many names which I have encountered in British history outside of the history of science and technology. The only names which struck a chord in my memory were: George Birkbeck, of the mechanics' institutions; the sculptor Francis Chantrey (who was a cultivator of geology); the historian Henry Hallam; Rowland Hill of penny-post fame; Bishop Blomfield of London; and Sir Robert Peel.

Both Archbishops belonged; York (Harcourt) more or less had to, one would imagine. It would have been churlish for the Bishop of Oxford not to have been a member at the Oxford meeting; and there were two more bishops on the list (Exeter and Cloyne) for a total of four. I was glad to encounter my old friend the Reverend Edward Stanley (a bishop-to-be) as a member this early.

The organization of the Association, and its early choice of leaders, confirms the analysis of its founding which I gave above. Ultimate authority rested with the active cultivators of science (the General Committee). They appointed as expert scientists as they could get to do whatever the scientists thought best (the Council, the subject committees), in the belief that the active and passive cultivators would benefit from the results (going to the sessions, reading the Reports, being told what they could do that would be of use to science). All of the proposals for directing professional research, subsidizing research by salaries or prizes, or putting pressure on the government were left without mechanisms or without funds. If the professionals chose to exert themselves along these lines, something might be done.

The experts were themselves so split up into autonomous committees which were given so much authority that there was no one group which could claim to be "the voice of science." The Council may have been made up of scientists, but it was an administrative organ; the prestige of being the locus of science was definitely conferred on the section committees. No one could manipulate this structure to make it anything but a "weak form" of organization; and that is just what professionals like Herschel wanted.

In short, the cultivators of science had set up, or at least approved of the setting up, of an organization whereby they would invite professional scientists to dominate them for their own good. It was, next, necessary to convince the professionals that the professionals would enjoy coming and dominating them.

In a passage which Morrell cites in his article but does not have space to quote, Roderick Murchison, in 1831 the young Scots president of the Geological Society of London, recalled the situation. He had appealed to all of his scientific friends, but most of them were sceptical of the proposed Association. At York, "the feebleness of the body scientific was too apparent. From London we had no strong men of other branches of science, and I was but a young president of the geologists; from Cambridge no one, but apologies from Whewell, Sedgwick, and others; from Oxford we had Daubeny only, with apologies from Buckland and others....We were but a meagre squad to represent British science."[22]

On the basis of the scanty and unsatisfactory evidence I have, it seems to me that the active men at the first meeting were not all "provincial" but, as Murchison's statement intimates, were ones who felt themselves to be of second-class status. (Of course, some of them such as Brewster *resented* this, since they considered their science to be first class; but they felt it nevertheless.) The ones I have identified are:

The professional Scots: Brewster, James F. W. Johnston, John Robison (the son, not *the* John Robison), J. D. Forbes.

The York men: especially John Phillips and Harcourt.[23]

Roderick Murchison: a young Scot, president of the Geological Society of London. (George Greenough, a leading founder of the Geological Society, was also present.)[24]

Charles Babbage: definitely metropolitan but often at odds with his peers. (Babbage's rival as a man difficult to get along with, Sir James South, sent an astronomical note to be read at the meeting.)[25]

Charles Daubeny: not a leading chemist, not from the science-oriented university, and not the leading scientist at his own university.

William Pearson and Sir Thomas Brisbane: both good private astronomers, Pearson indeed a co-founder of the Astronomical Society of London and Brisbane the private founder of the Paramatta Observatory while governor of New South Wales—but not dominant astronomers like Francis Baily, John Herschel, and George Airy.

William Scoresby: master mariner in the Greenland trade turned cleric after his wife's death; active in Arctic geography, meteorology, and magnetism..

"Old Dalton": quasi-dictator of the Manchester Lit & Phil, consciously embodying the myth of the rustic provincial genius—a vignette from Gray's *Elegy*. (The determination of the kind of people who were at the first two meetings to establish authoritatively the myths of Dalton, and of William Smith as the "Father of English Geology," is quite apparent.)[26]

To describe such people, "provincial" was good enough as a first approximation, but it is a bit hard to stretch it geographically to include the president of the Geological Society of London *and* a vice-president of the Astronomical Society of London (Pearson) *and* Oxford University *and* Edinburgh (which I have never considered to be a provincial town) *and* Charles Babbage. I could explain away any one or two of these, but not all of them. The Association's origin was *not* from "metropolitan" scientists; its main impetus *was* from Scotland and the more-or-less northern provinces of England; but its initial appeal was to *persons in the provinces of science,* as well as to those who lived in areas geographically protected from London. Some of the initial enthusiasts, as I said, resented their second-class status. Others knew that they were only middling-good, but wanted to do something anyway. Still others (e.g., Murchison, Forbes, John Phillips) were on the way up but still had their first-class reputations to establish.

How, then, and by whom were the much higher-ranking scientists of the second and especially the third meeting persuaded into coming? There is some evidence already available as to how this was done for one or two key figures: by first getting Buckland at Oxford, for example, and then (as Morrell points out[27]) having Buckland and Murchison work on their fellow-geologist Sedgwick of Cambridge. Or by convincing William Whewell; and through him reaching a number of Cambridge scientists, and perhaps even his friend John Herschel. The route from York (1831) to Oxford (1832) to Cambridge (1833) was, perhaps, an indirect way of opening a path from London to the Association, since the "decline" controversy had closed the direct route.

Buckland was an obvious person to get to first, but not because there was a dominant Oxford school of geology (no such group existed in 1831), and not because he personally could give the Association an "effortlessly superior" social tone. As a matter of fact, he was too rough-cut and enthusiastic for many Oxford tastes, and even for many Geological Society tastes. His social contribution to the 1832 meeting, as president, was rather that of "diffusing sunshine through all its proceedings" (to use Adam Sedgwick's phrase) while a good bit of needed organizational work was being done.[28]

Buckland was an obvious person to try for, because the Oxford chemist Charles Daubeny was an active worker at the first meeting.[29] If he could lure his colleague into participation with the offer of the presidency, the two of them would then represent enough weight in the Oxford scheme of things (according to my study of that scheme) to obtain a reasonable welcome for the Association. The recruitment of Baden Powell, the Savilian Professor of Geometry, gave them a plausible local working group of three substantial professors. It also almost assured the hostility of the poetry professor, John Keble, although the opposition between the theologically radical Powell and the conservative Keble did not break out into violent denunciations (so far as I know) until the time of the Hampden controversy a few years later.

Buckland was an especially obvious person to woo because of his close connection with Yorkshire science. The York Museum and Society had been founded in 1823 specifically to take care of the fossil bones and other relics found in Kirkdale Cavern; and Buckland had been the outstanding scientific investigator of the cavern. His well-known *Reliquiae Diluvianae* of 1823 was the result largely of these studies.[30] It might even be said (but this is a speculation) that Buckland was in this indirect manner a co-founder of the Yorkshire Philosophical Society. He was hardly in a position to resist a little friendly persuasion from the men at York who were the keepers of the archives documenting his *chef-d'oeuvre.*

Additional factors were that Buckland had travelled in Germany and was aware of the *Naturforscher* meetings which had suggested the idea of a British peripatetic group. He was the only Oxford scientist (almost the only Oxford scholar) who was internationally well-known. He was, in Oxford terms, a moderate liberal, between the extremes represented by Powell and Keble. The organizers thought it would be a good idea for the Association to go south for its second meeting. Daubeny, presumably, thought it would be good for science at Oxford to have the Association there. With Cambridge still sceptical, what better combination than Buckland and Oxford (we may ask), if Daubeny could bring it off? Daubeny guaranteed at the first meeting that he could.[31]

The same tactic—of making the better scientist but less enthusiastic supporter president, and putting the more active supporter in charge of arrangements—was used the following year also, with Adam Sedgwick as president at the Cambridge meeting, and William Whewell (together with the non-controversial J. S. Henslow) as local arranger. Sedgwick was an exemplary professional scientist: that is, he came down with all sorts of incurable ailments during the winter which could only be alleviated by extensive field-work all summer; he entertained his colleagues at professional society meetings whenever he gave a paper; he hated the actual writing of monographs; he gave excellent lectures to students, not noticeably pushing his own work or special interests; and he was very, very good.

Another way of recruiting scientists was to ask them to write one of the progress reports which Whewell had suggested and Harcourt had enthusiastically adopted as a project for the Association. It was a tempting offer, as Harcourt had said, to be asked to be the judge of all recent work in your field. Once you had exerted enough effort to prepare the report, it would be hard to be stand-offish about the organization. These attractions might not suffice for the top scientists (Herschel, Faraday) but they would for middling ones (Conybeare, Powell), for young ones with a new subject to promote (Lubbock on the tides), and young men out to establish their standing (Airy, Forbes). The early assignment of reports, it seems to me, certainly functioned as a recruiting tool, and the only question in my mind is the extent to which this was done deliberately rather than instinctively. The pattern of assignments—some to Edinburgh and the provinces, one or two to Oxford, many to Cambridge, a few to London—is not explicable by the quality of scientists in those places; London is grossly under-represented. Nor does the whole pattern seem to me explicable by the influence of one man, William Whewell at Cambridge, although that may explain the over-representation of Cambridge. The pattern does suggest, to me, the route of the Association. Edinburgh and the provinces met in York; from there, as I said, direct access to London science was barred; the way thither led through Oxford to Cambridge.

There is much more that could be done along these lines by someone who had more data than I do. I will simply go on to note various incidents in the proceedings which help to make my analysis more convincing. In spite of the large opening-day crowd at York, after the meeting the Association had just 96 dues-paying members. At Oxford the next year the University very nicely held a convocation and gave honorary degrees (D. C. L.) to four scientists. Very diplomatically, they were David Brewster (Edinburgh, "declinist"); John Dalton (Manchester, provincialist); Michael Faraday (London, "anti-declinist"); and Robert Brown (London, organizer for Herschel in the Royal Society election). Sedgwick and other Cantabrigians were won over; after the meeting there were 583 dues-paying members; John Herschel had joined by 1833;[32] and the third meeting at Cambridge, with Sedgwick as president, was the one at which the Association had attracted a good sample of the "metropolitan" scientists and could claim thereafter to be fully representative.

(I will pause here to explain that I use "metropolitan" to mean the London-Cambridge circle rather than, as it meant in contemporary reference, London and its environs alone. Otherwise, one could claim that William Whewell was not only a provincial but even the representative of a provincial Lit & Phil, since he was secretary of the Cambridge Philosophical Society. But this does not seem to me to be a very useful way of looking at either Whewell or the Cambridge Philosophical Society. Actually, I think that my usage is only justifiable from the 1820's on, when Herschel and Babbage and other members of the Cambridge Network had become established in London.)

At the beginning of the Cambridge meeting, there were 688 members; at the end of the week, 1377, as the treasurer happily announced.[33] Not all, of course, were "professionals;" I doubt that there were that many professional scientists in the British Isles. If it had been the aim of the founders to create an association "run for professionals" (in Williams's phrase), they were clearly in trouble. But actually they were clearly jubilant. It seems that John Dalton and John Herschel were the sentimental favorites at the meeting,[34] and no wonder: the recognized leader of London professional science had come to join the arch-representative of provincial science. There was an additional triumph: special note was taken of the (late) arrival of the Reverend Dr. Thomas Chalmers, who was then put on the Council. The adherence of Britain's leading Evangelical preacher was a noteworthy event indeed.[35]

One matter that caused debate was just how much professional science the cultivators could take. The Reports were a great success (the printed volume for the first two meetings had just been distributed); the Irish astronomer T. R. Robinson even said that "when the Association was first proposed he did not think it would come to much; but the Reports had convinced him how much he had been deceived." But did they have to be delivered at the general meetings? The first president of the Association, now Earl Fitzwilliam, proposed that abstruse speculations be kept to the sections. He had not been able to follow some of the Reports, and he thought the general meetings needed topics suited "to the minds of those who came to assemblies of this kind, not bringing much of science with them."

W. R. Hamilton, the other Irish astronomer, opposed him; Hamilton said that attempts to speak laymen's language were good for mathematicians like himself. Robinson went further; he felt it was good for "young aspirants" to encounter these "abstruse mysteries" they could not understand, so they would realize that they ought to learn mathematics.

Fitzwilliam had the last word. He agreed with Hamilton: popular presentations in laymen's language by experts was what he wanted. He cited George Peacock, who had given only an abstract of his report on algebra, not the whole thing. And he praised William Whewell's presentation of the subject of the tides, which had delighted even the ladies in the audience.[36]

That the Association was shaped into something rather different from what some of its original proposers had wanted did not go unnoticed. Reviewing the work of the first three meetings, David Brewster felt that he had lost out to Whewell and Whewell's Cambridge allies.[37] But, if my analysis is correct, blaming the course of events on one clique of scientists or another pays too little attention to the cultivators of science. It was they, ultimately, who chose which masters to serve; and if the masters they wanted refused to come into a strong organization, then it was up to the organizer to come up with a weak one. Even then, if the masters fed them too rich a scientific diet, they were capable of making their discomfort known, as Fitzwilliam's objections at the third meeting indicate.

Finally, let me sketch what I think are the most interesting positions to hold on the various topics that our collection of professional historians have debated. Note that I consider these as suggestions; none of them, perhaps, has been satisfactorily demonstrated in overwhelming fashion. In particular, I am open to the criticism which I made of Williams: if his position was too Babbage-centered, perhaps mine is too Herschel-centered. My defense is that Herschel *was* at the center of English science, and Babbage was not.

1. The "reforming" energy in English society manifested itself in the

founding of one-subject metropolitan scientific societies and in provincial Lit & Phil societies. Like the parliamentary reform agitation of 1830–32, both of these movements had something to do with increasing the social status of "new" men. The one-subject societies did so by helping to make the sciences more professional (as I have defined the word in the last chapter). The Lit & Phils did so by associating "merchants, manufacturers, medical men, bankers, solicitors, land-agents, tutors, dissenting clergymen" (to use Orange's characterization) in an older pattern of activity suitable for gentlemen. Orange attributes to these men an "appetite for natural knowledge;" I prefer, more cynically, to attribute to them an appetite for a gentlemanly pursuit.[38]

That the founding of local societies, at least, was directly related to socio-economic matters can be seen from the data Howarth gives as to the date of founding of 123 local societies which were eventually corresponding societies of the British Association.[39] Here is the chart, with my characterization:

Period	No. of Societies	
1790–1800:	1	
1801–1810:	2	
1811–1820:	2	The movement toward Victorian prosperity
1821–1830:	5	
1831–1840:	8	
1841–1850:	5	The hungry forties
1851–1860:	16	
1861–1870:	19	Victorian prosperity
1871–1880:	21	
1881–1890:	16	
1890–1900:	14	The waning of the Victorian era
1901–1910:	10	
1911–1920:	4	

The absolute figures are meaningless, since many local societies perished; but the relative figures are magnificent. This is the profile of the British 19th century which is taught by professors of ordinary history who have never even considered the history of science or scientific institutions. If you do not like economic terms, you can make it "the height of Victorian culture," etc., and it comes out just as well. As a matter of fact, this chart can be used, together with economic figures, to define what you mean by the height of Victorian culture.

2. All changes in scientific institutions can be said to be, in a sense, expressions of dissatisfaction with the Royal Society. Joseph Banks sensed this as early as 1807 when he opposed the formation of a serious, independent Geological Society. But such a statement does not say very much except what Harcourt said clearly: there was more interest in science than the Royal Society could deal with. Taken by itself, a statement that people were dissatisfied with the Royal Society does not explain much. Note that the "more" means more persons in the absolute sense; it says nothing about a proportion of the (increasing) population, or more intensity of interest, or a larger role in the total high culture. It might not be impossible to demonstrate "more" in all of these latter senses, but no one has attempted to do so. If one were to make the attempt, one might try to find a direct proportion between this "more" and the increase of a certain educated section of the middle class in all of these ways; one might correlate it to, say, the number of girls learning to play melodies from Mendelssohn, combined with the number of persons in England (excluding Wales) singing Handel's "Messiah" in choruses. But this is only a first approximation to a working hypothesis.

3. Given this numerically increased interest in science, from this point the developments which led to the founding of the British Association were largely distinct from the efforts to reform the Royal Society. Their sources were the increasing scientific pretensions of Edinburgh and some English provincials; and the increasing desire to have something better, among many members of local Lit & Phil societies. The thought-experiment to perform here is to ask: did Brewster need Babbage? Did Brewster propose a new society largely because the beloved old one could not be reformed? (To me, this seems a little like asking if Lenin would have been satisfied with Kerensky's regime, had it worked.)

4. The struggle within the Royal Society should not be called in any simple sense a struggle between "scientists" and "amateurs." It was a conflict between an older style in the life of science, as represented by Humphry Davy (and then Davies Gilbert) trying to carry on the tradition of Joseph Banks, meeting a new more "liberal" style (as the reformers thought of it) of which William Wollaston had been the senior London figure in the decades before 1830. It can crudely be called a conflict between gentlemen and professionals who thought they were gentlemen too.

It was not, however, an affair of young Turks, as it sometimes seems to be when the role of Charles Babbage is emphasized. How John Herschel came to be the reform candidate for president, as evidenced in his correspondence, shows the kinds of people involved. Although Babbage felt him out on the subject on October 15, 1830, Herschel shrugged off the suggestion. It was William Fitton, a well-respected senior figure, who manoeuvred Herschel into running, with the help of Robert Brown (whom Herschel had suggested to Fitton as a good man to help him in his anti-Gilbert efforts) and Dr. William Somerville, one of Herschel's personal friends. Somerville was a gentleman but not really a scientist, I believe; at any rate he is best-known today as the husband of Mary Somerville. When Herschel finally agreed to stand for election because Gilbert's antics were, he thought, becoming *too* intolerable for self-respecting people to submit to, on November 26 he asked two men, Babbage and Francis Beaufort, the well-respected hydrographer of the Navy, to see Fitton and each other; the three of them would decide how to manage the affair.[40]

I would say that Herschel chose Fitton to represent the senior men of the Society, as well as being the prime mover; Babbage to represent the young Turks; and Beaufort to represent the men of the armed forces. I don't know that he thought in just those terms, but, at any rate, that is how it came out.

5. So far, no one has shown that the loss of the Royal Society election of 1830 sent reformers fleeing north to found a new association. The evidence seems to be just the reverse. The effect of the election on one reformer, Herschel, was to make him suspicious of all general science societies. Did this same scepticism extend to others in London, such as the friends who ignored Roderick Murchison's appeal to attend the first meeting?

A more interesting question is: were the one-science societies in London the chief beneficiaries? Did scientists turn more to them because of disgust with the Royal Society? This would be a rather tricky problem to investigate (unless the movement were overwhelming) because some of these societies had their own positive reasons for growth in the 1830's, in particular the Geological. The Linnean and the Astronomical might be good starting-points.

6. The decline of science controversy, according to the available evidence, seriously handicapped the founding of the British Association; and it took fairly skillful manoeuvring for about two years after 1831 to overcome its influence. To get to Herschel, for example, the steps were first to make much of Whewell at the first meeting, then to make much of Brown and Faraday at the second meeting, then to set up a session which Herschel would enjoy (on the aurora borealis) as the first session of the third meeting.

7. The Association therefore came into being without being caused by particular events in London or in the Royal Society. The impetus from Edinburgh met men looking for a justification for starting something, in York. The force behind the founding of the Association was the will-to-power of the cultivators of science, who needed something to attract eminent men of science in order to improve their own status.

As symbolized by people, the credit given at the third meeting seems fair: proposing the Association was David Brewster's greatest experiment; the actual Association was Harcourt's organization.[41]

8. The movement coalesced into an Association at York not by chance, but because York and William Vernon Harcourt had just the right qualities to make it succeed.

Harcourt himself was a product of the new "liberal" spirit, having been encouraged in chemistry by both Wollaston and Davy; and his particular interest, eventually, was optical glass—that is, the subject on which Faraday and Herschel had cooperated for the Royal Society in the 1820's.[42] But also he carried with him the prestige of York Minster, of which he was a canon, and the prestige of the Archbishop of York, of whom he was a son. The city of York was undergoing a cultural efflorescence of sorts in the 1820's, and its most obvious center was the cathedral. Its dean, William Cockburn, encouraged the use of its library, and was involved in the well-noted Musical Festivals which began in 1823[43]

Cockburn and Harcourt were rivals (this is the same Dean Cockburn who attacked the Association later),[44] and Harcourt's cultural contribution was the Yorkshire Philosophical Society, which he erected on the bones from Kirkdale Cavern. York's old social life was decaying in the 1820's, but this was the life based squarely on the gentry. Efforts such as those of Cockburn and Harcourt were building up a new, more middle-class set of activities. Historians sometimes date York's cultural renaissance from the prosperity brought by the railroads in the 1830's; but it had begun fairly clearly in the '20's and thus was well enough along to produce its most spectacular success in 1831.

I bring this material in to help explain why it was York, rather than a more prosperous town—say, Leeds—that proved to be a suitable starting-point for a new Association. Harcourt had back of him, not only a successful local society, but also the social prestige of York (if faded, still superior to that of a mere industrial town) and of the Archbishop of York. York was, we might say, an "obvious" place for an aspiring set of provincials to meet. Less gentlemanly, perhaps, than it had been, it was still more gentlemanly than Leeds.

9. The hopes of various founding fathers for a strong form of organization, which would have a large share in directing the main lines of scientific activity, and in influencing the government to provide money and other benefits for science, were thwarted step by step by the opposition of professional scientists who did not believe that science would thrive under such conditions. Their weapon was the boycott; it was largely the scientists who did *not* attend who constituted this opposition.

10. The result was a body which served professional scientists in a way which does not fall easily into the patterns of activity which histo-

221

rians so far have emphasized under the title of "the professionalization of science." But this way also does not easily fit the ideas of scientists addressing a general audience, or of trying to mould public opinion. I think, however, that it does fit into the pattern which we *ought* to describe as "the professionalization of science," and that this way (I have called it "providing congenial arenas for debate") is indeed a most important part of the professionalization pattern for scientists and scholars, whether or not it is important for technicians, clerks, and civil servants.

A common complaint about Royal Society meetings was that, after a long paper was read, no discussion or debate was allowed. The Geological Society of London flourished precisely on its free-wheeling debates. At the British Association, one could debate with one's peers *before a large, admiring, quasi-informed audience of cultivators of science.* No further explanation of its popularity among scientists seems necessary.

This does not rule out the function of getting science accepted as deserving a larger place in educated culture. (I have my doubts as to whether this was ever accomplished, however.) It is part of professionalization to get one's profession accepted as such by the part of the community that counts (or, as we sometimes say, by "public opinion"). It was perhaps, an especial feature of the Association that the process of professional debate and discussion (which the professionals used to professionalize one another, and young aspirants) was also the public attraction which made the greatest impression on the part of the community that counted. At least that is my guess; but if you think the more popular sessions and public lectures had equal or greater impact, at least they did so in large part because they were summaries given by the experts themselves.

In spite of these functions it served in the professionalization of science, however, the Association was not set up either by professionals or for professionals. It was set up by cultivators for cultivators; it was run (after the first meeting) by professionals, because that was the way you could get other professionals to attend; and the attendance of the professionals was what the cultivators wanted to benefit from themselves.

So it can be seen that in this chapter I have done, at length, three things that a professional historian finds it pleasant to do:

I have decided that I was basically right all along.

I have introduced my man, Herschel, as an important figure while minimizing the importance of someone else's man, Babbage.

I have suggested that what is needed is a lot more research by other historians along the lines I have suggested.

One can hardly ask for more.

Notes to Chapter 7

1. *BA Report 1 and 2*, pp. 15–17.

2. *Ibid.*, pp. 17, 18.

3. Compare *ibid.*, p. 22 to p. 41 and then to p. ix. It is not made clear in the *Report* by what procedure the second version (p. 41) was modified into the final version (p. ix).

4. *BA Report 1 and 2*, p. 28.

5. Herschel, *Preliminary Discourse*, p. 259. Italics in the original.

6. Herschel, *Preliminary Discourse*, pp. 133–134; see the interesting article by M. J. S. Rudwick, "The Foundation of the Geological Society of London: Its Scheme for Cooperative Research and Its Struggle for Independence," *Brit. Jour. Hist. Science, 1,* (1963), esp. pp. 333–336, 354.

7. *BA Report 1 and 2*, pp. 24–25.

8. *BA Report 1 and 2*, p. 108.

9. George O. Trevelyan, *Life and Letters of Lord Macaulay,* 2nd ed. (London, 1901), p. 659.

10. *BA Report 1 and 2*, p. 34; Herschel, *Preliminary Discourse*, pp. 213–214.

11. *BA Report 1 and 2*, p. 30.

12. Howarth, *British Association,* pp. 16n, 99.

13. *BA Report 1 and 2*, pp. 35, 41–42, ix (my italics).

14. *Lithographed Signatures,* p. 111; *cf.* pp. 80, 109.

15. *BA Report 1 and 2*, pp. 35, ix.

16. *BA Report 1 and 2*, pp. 36–37.

17. Howarth, *British Association,* p. 254.

18. *BA Report 1 and 2*, pp. 38–39.

19. Howarth, *British Association,* pp. 107, 292–294.

20. There were not this many by the treasurer's report of October 1832 (*BA Report 1 and 2,* p. 124); I make it as 583 then. There were only 688 at the beginning of the third meeting (*Lithographed Signatures,* p. 82). From 1833 on, one was a member for two years before being ousted for non-payment of dues (*BA Report, 3* [1833], p. xxxv).

21. *BA Report 1 and 2*, pp. 111–114, 607–624.

22. Quoted in Howarth, *British Association,* p. 34.

23. Orange, "British Association," p. 325, gives the names of some local cultivators, whom I do not know anything about.

24. *BA Report 1 and 2*, p. 83.

25. *BA Report 1 and 2*, pp. 87–88.

26. *Cf. BA Report 1 and 2*, pp. 99, 100. "Old Dalton" was Murchison's phrase, quoted in Morrell, "Individualism," p. 202.

27. Morrell, "Individualism," p. 203 n44.

28. *BA Report 1 and 2*, p. 102.

29. *BA Report 1 and 2*, pp. 42, 83, 85, 88.

30. Orange, "British Association," p. 324. On the importance of *Reliquiae Diluvianae*, see my article "William Buckland" in the *Dictionary of Scientific Biography*.

31. *BA Report 1 and 2*, pp. 20, 108.

32. *BA Report 1 and 2*, pp. 100, 124, 615. I don't know the exact date or occasion of Herschel's joining. He is on the List of Members at the end of this volume; the List is undated, but the volume's "Preface" is dated 9 April, 1833.

33. *Lithographed Signatures*, p. 82.

34. *Lithographed Signatures*, pp. 71, 79. Both speeches are by Sedgwick, so my statement rests on the assumption that here, as on other occasions, he knew how to catch the mood of his audience and give it back to them in his speeches.

35. *Lithographed Signatures*, pp. 84, 106; *BA Report 3*, p. xxxviii.

36. *Lithographed Signatures*, pp. 86, 95–98.

37. Orange, "British Association," p. 316 n.

38. Orange, "British Association," p. 321.

39. Howarth, *British Association*, p. 95.

40. Herschel to Babbage, 15 October 1830; to Fitton, 18 October; to Beaufort, 26 November; to Babbage, 26 November (John Herschel Papers, Royal Society). The otherwise good narrative now available in Arthur Todd, *Beyond the Blaze* (Truro, n.d.), pp. 240–266, is still Babbage-centered.

41. *Lithographed Signatures*, pp. 79, 99.

42. A. V. Gardiner, *The Life of Sir William Harcourt* (London, 1923) I, 15.

43. I am much indebted, both for information and for guidance to the York Minster librarian, Mr. Barr, as well as to Miss Langley, and to the reference librarians of the York Public Library.

44. A brief account is given in Howarth, *British Association*, pp. 58–60.

Chapter Eight

History in Depth

The first half of the 19th century in Britain—and perhaps the second quarter more particularly than the first quarter—was one of the most formative, searching, and basically determinative periods in the history of science. By 1850, many of the great problems had been posed, many of the tools needed to solve them had been invented, and many of the men who were to solve them had been trained in a professional way which their elders had developed satisfactorily only in the preceding generation. To be sure, these elders had not relied only on British sources of inspiration. Lagrange in mathematics, Bessel in astronomy, Fresnel in optics were among the great stylists whom the early Victorians transplanted and transmuted.

There is very little satisfactory information in print about this turbulent period. In most fields, there is not only no satisfactory general work, there is also no particular monograph whose scope and accuracy can be trusted. One must often determine every factual statement from original sources, not only in one's central subject, but also in peripheral matters. Trusting one's predecessors is the single most common source of error in one's own writings.

In spite of the difficulty this situation causes in writing even a simple narrative accurately, I can point to two fundamental matters in which basic change took place between 1800 and 1850. The first is the matter of evidence. The mass of data to be explained was much greater by 1850. This accumulation was deliberate, not the result of some inherent sociological growth factor. In area after area, scientific spokesmen called for more data, and the scientific community laboured to respond.

We can, however, distinguish subjects at various different stages of development. Many monographs in the physical sciences reported exhaustive tests on a wide variety of different substances and different forms of substances; they are a very different kind of reading from Newton's *Principia.* As a person trained in physics, they first struck me as what I would expect from chemists. But I have since run across the same phenomenon in cosmic ray literature in the 1930's, and nuclear spectroscopy around 1950. When a subject catches on, but has a shortage of exact theories to verify or refine, scientists will try out experiments on anything that occurs to them. In the early 19th century, examples of what I mean are many of the papers on polarized light by David Brewster and John Herschel, Faraday's *Experimental Researches,* and Herschel's "action of the rays of the solar spectrum on vegetable colors" of 1842.[1]

In astronomy, we can distinguish several different impulses. The historians Dreyer and Turner report a general feeling prevalent around 1820 that it would be desirable to study every individual star in the heavens; I have no explanation for this.[2] In contrast, John Herschel wanted to chart and classify every nebula in the heavens, but his original motive is quite clear. It was to furnish overwhelming evidence for his father's disputed theories concerning the structure and evolution of the universe. If John did not do this, it seemed that no one else would. William Herschel's theories were being more or less pushed to one side because no one else had the telescopes to verify or extend his observations. John, that is, did not collect catalogues of nebular data for the sake of having catalogues. As he proceeded, he became fascinated by the sheer beauty of what he was observing; and he produced particular theories of his own; but the original motive was to test theories of the widest possible scope.

We cannot be sure, however, that a scientific project has a scientifically valid personal motive. Take Herschel's gigantic double-star catalogue. His diaries make it clear that this was something he did when he was sick and unable to do anything that required thinking. As he put it in a letter to Augustus De Morgan, it filled up the time, and was for him "like worsted work is for ladies."[3]

A fascination with the beauty of accumulating more and more detailed information is apparent in geology. The Geological Society of London was founded in 1807 precisely for this purpose, in reaction against premature speculation. But it was cosmological system-building in the styles of Werner and Hutton that were shunned. The men of the Geological Society were prolific in producing ordinary scientific theories and working hypotheses; and by the late 1820's, with Charles Lyell, Roderick Murchison, Adam Sedgwick, and others, geology's great age of theorizing had begun. These men could take for granted what constituted a proper research paper; that is, what a geological monograph should contain in order to mean something. This more fundamental problem had been worked out by their predecessors, of whom James Hall and John Playfair, William Smith and James Parkinson, William Fitton, Georges Cuvier, W. B. Conybeare, and William Buckland are only some of the better-known names for British geology.

The new theorists were themselves among the most active contributors to factual knowledge. By 1850, not only was the structure of the earth as described by geology approximating to modern form, but also, as in the collaboration of Murchison with Russian geologists, it was taking on modern geographical scope.

In terrestrial magnetism, in meteorology, in tidal observations, there was a great push in the 1830's to set up worldwide co-ordinated networks of observatories, and there was prevalent a belief that if the variations of, say, the atmosphere could be thoroughly charted for five or ten years, the laws of the variations could then be worked out. This belief, helped along by the formulation of Gauss's theory of terrestrial magnetism, was nevertheless based more upon self-confidence than on a thoroughly valid precedent. In astronomy, observations went back for a century or more.

To be sure, not all astronomers of the early 19th century believed that the earlier astronomical observations were very reliable. It took men of the stature of Bessel, with his discussion of James Bradley's results, and of John Herschel, with his use of Bradley's observations in computing double-star orbits, to prove that Bradley had indeed been one of the great astronomical observers. (The whole subject of observational accuracy from the mid-18th to the mid-19th century is one which in my opinion needs much more discussion. That is to say, I am not satisfied with the judgments in Edward Walter Maunder, *The Royal Observatory, Greenwich* [1910]).

It was only the triumphant discovery of Neptune which established both the credit of 18th-century astronomical observers *and* the validity of Newton's gravitation law. Before that, it seemed to some to be more scientific either to suspect earlier observations or to doubt the simplicity of the inverse-square law than to postulate the existence of a planet which had never been seen. Thus even Bessel for some time had doubts about the exactitude of Newtonian law—a failure of faith which his admirer Schumacher wished to suppress after his death.[4]

With, then, a belief in the ability of modern observers to furnish the right kind of data for modern theories, the advanced men of the 1830's wanted to set up physical observatories—we might call them geophysical observatories—on the model of astronomical observatories, devoted to continuous observations of the variables connected with the earth, as in meteorology and magnetism. These were distinguished from another kind of establishment, also desirable, to determine physical constants extremely accurately—what we might call a Bureau of Standards.[5]

It is not, therefore, surprising that the second quarter of the century saw the firm beginnings of co-operation among astronomical observatories. Nor is it surprising that two of the most influential English scientists behind the movement for physical observatories were astronomers. George Airy converted Greenwich not only into the premier fact-producing astronomical observatory of the world but into a physical observatory as well. John Herschel—who seems to have been the only British scientist whom Airy genuinely considered his peer—had as his special campaign the regular and systematic use of photography in astronomy, as contrasted to merely taking a portrait of the moon or sun. (Since Herschel also invented the glass plate negative, we can say that he was responsible for both the idea and the instrument of modern astronomical methodology.) A revived Kew Observatory was one result of Herschel's desire for physical observatories, and it was there that Warren de la Rue found a place to put Herschel's photographic scheme into effect.

The British Association's original proposal for the use of Kew is a fascinating potpourri of projects. Tracing each of the ideas back to individuals, and comparing them to the actual activities at Kew under the Association, would be a major way into mid-century British science. One major influence was the need to give Edward Sabine an area of magnetic research free from Airy's overlordship. The combination Herschel-Airy-Sabine (with Alexander Humboldt's backing) had enough prestige with the institutions involved—the Admiralty, the War Office, the Chancellor of the Exchequer, the British Association—to obtain *two* physical observatories, one at Kew and one as part of Greenwich, since it was apparent that the personalities of Sabine and Airy would not fit into one observatory.[6]

Another figure can be discerned edging into the picture. Robert Fitzroy invited a young naturalist, Charles Darwin, to go on the voyage of the *Beagle* and take care of that kind of observation, because his own interests were in the physical sciences. Back in England, he actually helped the building of the physical observatory at Greenwich, as we learn from the report of the Astronomer Royal in 1841:

> A small wooden house, the property of Capt. Fitzroy, R. N., which was carried by him, in the Beagle, in his circumnavigation of the globe, has been planted in the southern part of the Magnetic Ground for observations of the dipping needle, and any other observations which would be prejudiced by the action of the large magnets in the Magnetic Observatory.[7]

This deliberate search for more facts, and for a satisfactory charting of the differences of observations made in different localities (as in geology, biology, meteorology, tides, terrestrial magnetism) or in different substances (as in optics and crystallography, chemistry, electricity) was not "Baconian," if by "Baconian" it is meant that new scientific concepts and theories will arise or make themselves evident through the amassing of facts. Adam Sedgwick's Cambrian System did not emerge obviously from the collection of geological facts; which is why Roderick Murchison could deny that it had any independent existence at all. The myth that British scientists were "Baconian" in this sense dies hard; but it is reasonably difficult to list many good scientists who defended such a position seriously—at least I have never seen it done—although it is easy to list many who expressed an admiration for the writings of Francis Bacon. One explanation for the genesis of the myth has occurred to me: good scientists spent enough effort in *attacking* this "Baconian" idea that they may have given modern historians a belief in the existence of an even more powerful group being attacked.

I have remarked in earlier chapters on various distorting effects of the "Baconian" myth on modern historiography. Let me here face the issues squarely: who of the period advocated "Baconianism" such as modern authors refer to; and what was the development, the "new thing" if it was not "Baconianism"?

Persons who did use the "Baconian" position were not a powerful group of scientists but were, for example, clergymen who wished to defend the Bible against more modern cosmographies. Thus the Reverend Thomas Chalmers certainly did deny the validity of geological theories which cast doubt on the accuracy of early chapters of Genesis, and did emphasize the need to amass more and more facts, as (he believed) Bacon's philosophy dictated, before theorising so rashly.[8] This is not to say that Chalmers was a "conservative." He was one of the clergymen most interested in social amelioration in the 1830's. He knew how essential it was to collect facts in the moral sciences. Of course he did not need these data in order to find out for himself if his own Glasgow slum-dwellers were in desperate plight; he needed them to prove it to public opinion.

That persons like Chalmers exist, persons who are "conservative" on scientific and theological matters while being rather "radical" on social and institutional matters—being "radical" either by adhering to new economic theories, or by supporting older traditions than those of Peelite Conservatism—is one of the truths which needs to be brought home again and again to the historian who loves to extend political classifications to cover all parts of a person's beliefs. There is no such necessary sameness in an individual, nor even in an institution. In the 1830's, for example, the politically conservative *Quarterly Review* was considerably more "liberal" on scientific matters than was the Whig *Edinburgh* or the radical *Westminster*. (I note that, with Chalmers and the *Edinburgh* as examples, it is not safe to generalize about Scottish metaphysics vs. English empiricism. The Scottish intellectual scene is too complex for such easy distinctions to be valid.) And a "radical" may have been a believer in Ricardian economics.

Far from being a period of "Baconian" influence, the second quarter of the 19th century was the period when Idealism had its strongest impact on British science. This is a point often ignored by intellectual historians, some of whom seem to think that philosophy means Hegel and that, because the British did not take to Hegel, they were too unsophisticated to understand Hegel. The situation was rather that Hegel was read by key intellectuals who, because they had a standard of comparison in Kant and Schelling and Goethe, very often rejected Hegel. Indeed, the situation may have been that the more they understood Hegel, the more they rejected him. The most scholarly of the Germanists in England, Connop Thirlwall, translator of the great historian Niebuhr, wrote to a philosophic friend: "My own examination of certain portions of Hegel's work, which I had occasion to study, attentively, has impressed me with the deepest conviction, that he is, to say the least, one of the most impudent of all literary quacks."[9]

I suppose that in principle one could believe in "Baconianism" and also tend towards Idealism, but I have not encountered any examples in actuality. Tendencies are important in studying scientists, because many of them did not have a philosophical position consistent enough to satisfy modern analysts. Those who did often did not set forth their positions in their scientific monographs; one did not convince fellow-scientists by citing a German metaphysician. There was much more philosophy among scientists in private than in public.

Tendencies which were not overtly Idealistic but were at least in the direction away from common-sense "empiricism" were of importance for some key figures in the first half of the century. One of these tendencies was to welcome the atomistic ideas of the 18th-century Newtonian, Roger Boscovich. Today, atoms may not seem to be very Ideal; but they were unobservable thought-entities then. Moreover, Boscovich postulated that atoms are nothing material but are only a mathematical locus of forces.

I am not competent to judge whether or not Pearce Williams is correct in seeing Michael Faraday's actual course of research as being influenced by Boscovichian ideas; but in drawing attention to the importance of Boscovich for the British scene, Williams is surely right.[10] Faraday, William Rowan Hamilton,[11] and John Herschel[12] have been identified as admirers of Boscovich in the second quarter of the century; we may expect that others will be discovered. In the 1840's, Robert Ellis explained to the Cambridge Philosophical Society why Boscovich was attractive. He was an alternative for those who wanted to maintain a mechanical philosophy without resorting to billiard-ball atoms or to vague chemical "affinities."[13] Chemists in the 19th century often disbelieved in atoms altogether,[14] whereas physicists liked to have them around—perhaps it made them feel vaguely Newtonian. But some of them were coming to see the world even at this basic level as an interweaving pattern of forces; it need not surprise us that physicists soon were to find something of this kind real enough to be conserved.

We should not expect to find complete acceptance of Boscovich's scheme when we are trying to establish "influence" or "tendencies." We need to estimate both the *intensity* and the *completeness* of belief in a new idea. In an interesting article on the reception of Boscovich's ideas in Edinburgh before and around 1800, Richard Olson has shown some of the variations possible.[15] Scientists felt free to reject Boscovich's metaphysical postulates on the one hand, and many of his detailed applications on the other, while exploiting his central images of how atomic forces work. Thus John Robison liked Boscovich primarily (it seems to me) because Boscovich was an atomist. Robison believed that it was Newton who first suggested valid atomic ideas, and that it was Boscovich who first reduced them to a system. For Robison, Boscovich's "great principle" was "the discrete constitution of matter, and the non-existence of mathematical contact." He admitted that he could not cope with atoms as "unextended mathematical points," but he did not think this idea necessary for the system.[16]

John Leslie, in his well-respected *Experimental Inquiry into the Nature and Propagation of Heat* of 1804, rejected the notion that atoms are mere points with no size at all; but for him this did not invalidate the concept of atoms as force centers rather than material bodies. What such a position implies, as John Herschel later pointed out to William Rowan Hamilton, is that on the atomic level force is quantized.[17]

"Influence" can be transmitted by an author who disagrees in part or in whole with a theory, as well as by one who agrees, provided that his explanation of the theory is a prominent part of his presentation. Readers with a different tendency may agree with the explanation and not the author's refutation. Thus it is useful to note how much exposure Boscovich received in books which a 19th-century Englishman might come across.

Robert Schofield has pointed out Joseph Priestley's use of Boscovich; but it could be argued as to how many scientists in the 19th century read the books in which Priestley mentions Boscovich.[18] The same is true of Dugald Stewart's *Philosophy of the Human Mind;* Leslie's *Heat;* and Robison's *Elements of Mechanical Philosophy* of 1804. But we also find from Olson's article that there was a long article on Boscovich (by Robison) in the Supplement to the third edition of the *Encyclopaedia Britannica*. Boscovich was treated in Thomas Thomson's *System of Chemistry,* which had edition after edition from 1802 until it was split into four separate works in the 1830's; and Thomson's work was not read by chemists only, since it treated other subjects as well.[19]

And finally, the historical dissertations which formed part of the Supplement to the fourth edition of the *Britannica* were reprinted up to the 1850's, both as part of the encyclopedia and separately.[20] Dugald Stewart's dissertation on the "Progress of Metaphysical and Ethical Philoso-

231

phy" praised Boscovich more than it explained what he said; but John Leslie, in his "Progress of Mathematical and Physical Science," gave more space in his section on dynamics to Boscovich than to anyone else, including such famous names as Euler, Bernoulli, and D'Alembert. He concluded of Boscovich's atomic theory that, "Purged of its antiquated metaphysics and crude chemical notions, it would form the best introduction to general physical science."[21]

Given such a number of Boscovichian *loci* (and there are surely more to be discovered), we do not have to trace how any particular scientist could have come to hear of Boscovich at any particular time before accepting evidence that he did know of him. John Herschel, for example, may have read Thomson *and* Stewart *and* Leslie. Joseph Henry in the United States prepared lectures with the use of Robison's *System of Mechanical Philosophy.*[22]

It may, of course, be interesting to know how much a given scientist knew of Boscovich; but more interesting to me is what it was in the scientist's own mental attitude that made him accept that part of Boscovich which he did. Such a treatment of a scientist would need to be detailed; words like Idealism and "Baconian" usually are not enough. (After all, Boscovich was presumably neither.) In this respect, Olson's treatment of Leslie in his article is worthy of praise.

However, some scientists did try to develop a consistent philosophical position; and about them we can use such terms as Idealism, albeit with a great deal of care. Thus the influence of the anti-Baconian tendency of dominant theorists in the first half of the century can be seen in the thinking of James Clerk Maxwell. He is presented in his biography as having paid attention to three writers on the nature of science, John Herschel, John Stuart Mill, and William Whewell.[23] These were the authors of the best-known English treatises about science in the period of Maxwell's youth.

Herschel's published view of scientific method was that perhaps the best results come when the mind "leaps forward" "by forming at once a bold hypothesis," and his letters show that he put this into practice on all kinds of subjects and liked other people to do the same.[24] Mill in his *System of Logic* enjoyed denouncing "empiricism." I am not quite sure what this was, since it is hard to find anyone of the period who admitted that his position was "empirical"—it was quite a bad word—but I assume that Mill was denouncing the same dominance of the facts which modern writers have in mind when they speak of "Baconian." Whewell was an out-and-out Idealist; but as an admirer of Bacon (not "Baconianism") he was not exactly a Germanic Idealist. Whewell had certainly read and understood the Germans, but he was from Lancashire, so it is not clear whether Whewell found in Kant and Schelling inspiration or only support. A number of Lancastrians even to this day are pub-Idealists of a particularly stubborn type.[25]

Maxwell had also been exposed to Scots philosophers as a youth in Edinburgh. But in 1856 he chose to express his position in terms of his study of Englishmen. Putting Mill as representative of "low" philosophy (less Idealist), and Whewell as representative of "high" philosophy (more Idealist), he said: "I find I get fonder of metaphysics and less of calculation continually, and that my metaphysics are fast settling into the rigid high style, that is about ten times as far *above* Whewell as Mill is *below* him...."[26]

Once the myth of "Baconianism" is dispelled, we can speculate as to why scientists wanted to accumulate much more data. One possible reason is obvious: they were influenced by the example of the queen of the sciences, astronomy. Scientists knew how much observational work was behind the triumphs of Newton or the calculations of Laplace. Kepler was already available as a scientific hero in his role as respecter of Tycho Brahe's accuracy. The suspicions I mentioned above of some astronomers about the usefulness of earlier observations were generated because the new breed of astronomers were leading the way in still more accurate observations, more refined calculations, and more extended series of reductions and publications.

For such observers, instrumental limitations were unacceptable. Mathematics was the essential crutch with whose help they hoped to correct the mechanical uncertainties of even the best divided circle, of even the best transit mounting. Practical precision astronomy was coming to be dominated by practical mathematicians like Bessel and Airy, because only such men could use mathematics to search out, identify, and compensate for the almost inevitable shortcomings of mere mechanical craftsmanship.

The demands made on instruments by scientists with new standards of precision would be a good starting-point for a study of, say, the Munich instrument-makers. A telescope is sometimes represented in overly simple books as primarily an optical system; and in better treatments it is still the optical quality which receives most of the emphasis. But to satisfy such a man as Struve, a telescope had to have not only a superior lens system, but also a superior mechanical construction, a superior mounting, and a superior measuring device. Fraunhofer and the other great telescope-makers would become more interesting if all of these aspects were equally well treated.

And so would astronomers. When Lord Rosse had two parallel masonry walls constructed as supports for his giant telescope designed to outreach even the Herschels, they were made as fully Gothic walls, with arches, battlements, and small towers. This is almost unimportant unless you ask just what stability of mount was obtained. Was this part of a desire for better data? We do know that the desire for better data led to new institutions. Thus the Astronomical Society of London, although proposed earlier, was not founded until 1820 because it was not until the work of Bessel was available that the Society's co-founder Francis Baily could find the proper guide for his crusade. That crusade was to introduce new precision techniques throughout British astronomy and indeed throughout British physical science generally.[27] Through Baily's own work, through that of his disciples Richard Sheepshanks and George Airy, and through the committees on standards of the British Association, the crusade did eventually succeed. What did it do to styles of telescope mountings and domes?

In contrast, the newer sciences were woefully short of even acceptable data. In meteorology, for example, James Forbes's reports to the British Association emphasized the worthlessness of sporadic observations made by undisciplined observers. What were needed were systematic data obtained according to a standardized scheme.[28] Indeed, from its beginning the British Association was devoted to searching out the areas in which more research was needed and then in stimulating that research.[29] Its "Reports on the Progress of..." were one of the ways to do this.

If the first obvious matter in which basic change took place between 1800 and 1850 is the matter of evidence, the second is the manner of treating evidence. Mathematically-equipped scientists began to take over large areas of science with a relentlessness heretofore confined to physical astronomy. Mathematics became a method of discovery, as in the case of Neptune, and a method of invention, as in the analyses of lens systems by Herschel, Airy, and others.[30] Creative ideas from Faraday were turned into mathematical form by Thomson, and it was in that form that they had their most far-reaching impact through the work of Clerk Maxwell. Crystallography was concerned with reducing itself to algebra in Britain, to geometry on the Continent. Whewell in his tidal studies used graphical representation to distinguish the various components resulting from the various forces at work. With Fourier, the study of heat and advanced mathematics became inextricably tangled. With the work of Boole and De Morgan, mathematics even began to invade technical philosophy.

The mathematically expert scientists by no means contributed all of the best work after 1850, but they were coming to be regarded as the final arbiters—as the mathematically expert astronomers had been for some years. To be sure, this increased prestige for mathematics was not an unmixed blessing, as the confrontation of William Thomson and Thomas Huxley in 1868–69 on the age of the earth shows. Huxley was not able to meet Thomson's arguments without equivocation, since he did not really understand the mathematics involved, even after being coached by John Tyndall.[31] But, although confused, his conviction that Darwinian bio-topographical vision was a sounder guide than applied thermo-dynamic reasoning was, as it turned out, sound.[32]

The nature of these two great areas of change shows how little true it is that great changes in science are revolutions in scientific theory. By saying this, I am putting myself in opposition to what has been the reigning orthodoxy in the history of science for several decades. Nevertheless, I will assert that the great changes which make modern physical science so strikingly different from science of the 17th and 18th centuries came out of the deliberate exposure to many new phenomena, and the deliberate push to mathematise them, in the first half of the 19th century.

The theoretical revolution which resulted was made up not so much of theories as of the elements of theories. What needed to be invented were the things to be put into theories—fields of force, spectral lines, indeed the spectrum itself (for it was by no means evident in the 1830's that invisible rays are simple linear extensions of the visible spectrum.[33]) These elements could then produce theories; that is, they showed scientists the proper "objects" for their attention. These were the things which needed to be connected together, by theories. They were somewhat like the ellipses which Kepler invented to be the proper objects of astronomical theory. (It is not proper to say that Kepler "discovered" that the planets move in elliptical orbits, because, as William Whewell pointed out, as a matter of fact the planets do not move in elliptical orbits).

These new elements came into being as men sought out factual complexity and tried to find new ways of ordering their experience. Another one of these new ways was a new subject. As is discussed in Chapter Seven, "physics" as a mathematically equipped, centralized subject of which such earlier special subjects as mechanics, acoustics, optics, and heat were part (but chemistry, astronomy, and anatomy were not) came into being in the first half of the 19th century, with French science furnishing the *'elan vital.* And a third new way, as I have indicated, was the founding or re-shaping of institutions for copious regularized observations.

In the journal *History of Science* for 1962 and 1963, D. S. L. Cardwell and A. R. Hall used the term "intellectualist" to name the view that the history of science is to be explained in terms of the internal dialectic of the sciences, and the term "externalist" to name the view that general cultural, economic, and social states are the key to the history of science as well as to all other special histories. Hall suggested, probably correctly, that "externalist" explanations had been of little interest in the history of science in recent decades. My own denunciation, in the same journal for

1964, of such an opposition as an "outmoded simplicity," seemingly went relatively unheeded.[34] Students and teachers alike during the past decade have relished debating which alternative is the good one; and I have found myself suspected of giving aid and comfort now to one side, now to the other.

On the whole, the "externalists" have grown in strength. This was natural, as the "externalists" can study the history of a scientific institution without saying much about science, by treating it just like any other kind of institution. "Externalist" history as produced in the past decade has sometimes been a way for persons to become historians of science, without investigating anything distinctive about science or scientists.

Before repeating my 1964 demonstration that neither "intellectualist" nor "externalist" approaches are adequate, I want to clear up a linguistic matter which seems to have led to confusion in the debates which I have heard. I admit to being lax, though not confused, in my original usage. The word "externalist" automatically generates the word "internalist" as its opposite for most hearers. But this term is not equivalent to "intellectualist," which as defined above is *only one kind* of "internalist" explanation, and is not even equivalent to "intellectual." The rivalry of Airy and Sabine mentioned above was "internal" to science without being especially "intellectualist;" the same might be true of a fight between the College of Physicians and the Royal Society. The influence on scientists of Coleridge, or Anglican theology, is an "external" influence, but it is intellectual in nature. When Henry Guerlac says that "our problem is no different from that facing any historian of ideas, for I would hasten to agree that the history of science is primarily (but not exclusively) the history of *thought* about nature,"[35] he is supporting an intellectual emphasis which is contrasted to my emphasis on what new things scientists could see with new experiences and new instruments. Nevertheless my emphasis may be "internalist" if the new experiences and instruments were not given to scientists from outside but were developed in response to a demand within science itself; and Guerlac's may well include external intellectual influences.

Finally, an interesting problem is how an external factor becomes internalized, becomes accepted as part of the internal dialectic of a science. At one time a given factor may be external, and at another time internal. It is not possible to separate matters into external and internal by using categories of pure reason.

With this explanation given, let me return to my original argument. Change at the basic levels I have mentioned cannot be understood by internal intellectualist methods alone. Given the state of the sciences in 1840 or 1850, the great Victorian theories perhaps can be shown to follow by internal logic: that is, there were unexplained phenomena known which were powerful enough to compel some sort of theories, and scientists were available adequate to supply theories. But given the state of the sciences in 1800, and the number and nature of the scientists then available, the new knowledge and the new concepts introduced by 1850 do not follow. (Nor does it matter to the truth of these statements whether they are applied to British scientists or to scientists generally.) In general, once a revolution in interest and attention has taken place, the intellectualist method may be fruitful in deriving, say, Hertz from Clerk Maxwell; but the method does not explain how or why the revolution happened.

To take a specific example, it is clear that it was his involvement in the practical matters of surveying and geodesy which inspired Gauss to his work on abstract mapping.[36] He needed a method for using the flat triangles which are the direct result of geodetic observation to construct a three-dimensional surface, the curved exterior of the earth. It was not, therefore, an internal logic in mathematics which initiated this development in Gauss's career, although it may be that Dirichlet's and Riemann's further work can be explained on internalist grounds.

Indeed, the myth of Gauss as "Prince of Mathematicians" has so obscured his particular susceptibility to influences external to mathematics (the duties of his position, the needs of astronomers, the projects of Alexander Humboldt) that the really major importance of Gauss in physical science is often not so much overlooked as slurred over by historians of science. Actually, in the terminology I used earlier, it seems to me that Gauss was to Humboldtian science what Clerk Maxwell was to physics: the fully formed professional.

To take a more complex example: the attention of such well-known figures as Halley and Euler did not produce an outburst of attention to terrestrial magnetism in the 18th century. By outburst I mean the same amount of attention as that paid to geodesy, let us say. The externalist cause—the obvious interest of the subject in matters of navigation—did not work. One reason may have been that the persons most subject to externalist influence were focusing their attention on what might be considered a somewhat more abstruse, although more needed, matter in navigation. They were seeking for a way to apply astronomy to determine the longitude. E. G. R. Taylor has described various schemes proposed in 1714 and 1715 after a Parliamentary reward was offered. Some of them were very abstruse indeed.[37]

Progress was made in one method (Mayer and equal altitudes) because astronomy was sufficiently advanced (intellectualist). Progress was made in another (Harrison and chronometers) because craftsmen were sufficiently advanced (non-intellectualist). Deciding the relative merits of two such different approaches was as difficult then as assessing their re-

237

spective roles in the improvement of navigation has been to historians ever since; on the whole, Harrison has had the better propagandists. One can argue that the craftsmen were advanced because the great successes of astronomy had generated a new breed of craftsmen (intellectualist). One can also argue that there were already advanced craftsmen in England working in other fields, and that the absence of strict guild restrictions permitted them to migrate into fields where the money and prestige were good (externalist). It would seem sensible to do as Taylor does, and argue for both of these positions simultaneously.[38]

However, what one loses sight of in this kind of discussion is John Harrison himself. The inspection of scientific instruments of the period makes arguments about craftsmen in general suspect. John Harrison is no more interchangeable with Jesse Ramsden than Laplace is with Lagrange. To intellectual and non-intellectual, internal and external explanations we must add what might be called "the role of genius" but which I prefer to call "the uniqueness of the great stylist."

Yet even with this addition, we have not provided a basis for explaining why there *was* a great outburst of enthusiasm for the subject of terrestrial magnetism in the early 19th century. This enthusiasm led to the establishment of a network of magnetic observatories throughout Europe and Asia and, once the British government became involved, throughout the world, all coordinated to "Göttingen mean time."[39] It led to the Clarendon Laboratory being built at Oxford without any metal being used. It led to the instruments and theoretical work of Gauss relating to magnetism and then to the introduction of absolute measurements. It led to Ross's great Antarctic expedition of 1838–41.

And it led to the foundation of modern meteorology through John Herschel's suggestion that the same or similar observatories and the same or similar methods of simultaneous observation could be used to put that subject on a sound basis. (I am using the position of Napier Shaw in his *Manual of Meteorology,* that modern meteorology was founded when and only when a continuously existing large-scale network of observatories, making large-scale weather maps possible, came into existence.)[40] By the 1850's, meteorology was not only well on the road to being systematized but also, at least at Greenwich, automated with self-registering instruments as well.[41]

Taken as a whole, this magnetic enterprise was, to a contemporary historian, "by far the greatest scientific undertaking which the world has ever seen."[42] Clearly it is of interest for almost every area of the history of science. Yet the chief externalist motive, that of improving navigation, no more demanded a magnetic solution in this period than it had previously—or did it? And yet what intellectualist or internalist impulse could have set such world-wide forces as these into operation? The only comprehensive explanation I have seen so far is that Alexander Humboldt had remarkable powers of persuasion over people so diverse as the Emperor of Russia and C. F. Gauss. It was Humboldt's enterprise; but what theory can we use to explain Humboldt's power?

A really satisfactory study of this magnetic undertaking is an obvious way to develop a methodology, and stemming from it a system of explanation, which would outmode the intellectualist-externalist simplicity. Such a study would need to investigate the relation of very different sources of inspiration in one man, as with Gauss. It would need to investigate the ramifications of how externalist causes are embodied, as with the British Admiralty's love of scientific voyages which would put at least part of the huge mob of officers left over from the Napoleonic wars back on full pay. Was this one of the first official recognitions of structural unemployment? (The idea of the unemployed warrior as social *victim* is to be contrasted to the notion, as old as the First Crusade, of the unemployed warrior as social *danger.*)

The study of voyages would need to consider whether a qualitatively new situation was created in the 19th century, or whether there was merely a steady expansion of 18th-century developments in scientific voyaging. Jacob Gruber's important article, "Who was the *Beagle's* Naturalist?" suggests some of the problems historians should face with respect to the scientists involved.[43] And what was the attitude of naval personnel? Did midshipmen beg for the chance to go on expeditions favored by the Royal Society in the 19th century, as young Horatio Nelson begged to go with the *Racehorse* and the *Carcass* in 1773 on their voyage to make scientific observations within the Arctic Circle? To be sure, this was not because the young man was interested in science, but because he wanted to serve under the captain of the *Carcass,* Skeffington Lutwidge.[44] Did the Admiralty in the 19th century choose captains for its scientific voyages whom young career officers were equally eager to serve under?

The study would need to be comparative, and take into account transit-of-Venus enterprises, from which the magnetic undertaking differed considerably, as well as Admiralty surveying voyages, from which Ross's Antarctic expedition differed decisively. And yet Ross's voyage did have externalist similarities with other voyages; and on board was young Joseph Hooker, for whose biological education this voyage was important in some of the ways that the voyage of the *Beagle* was for Charles Darwin or that the voyage of the *Rattlesnake* was for Thomas Huxley.

When some such studies are done, we can move on to really large-scale comparative problems. Why, for example, in spite of all this international cooperation among scientists, was it not the scientists but the *anti-slavery propagandists* who successfully introduced the international congress, a form of activity which scientists seem to find particularly congenial today? The first such regular scientific meeting (I am not referring to one-time conferences called for a particular purpose) was that of geodesists, from 1864—the Europaische Gradmessung—although it did not call itself international until 1886. From Quaker Christianity to Christianity in the British establishment to the meetings of theoretical physicists—I cite this example to show how diverse are the origins of even the professional parts of the life of the scientist. But surely the life of the scientist has something to do with the life of science? George Stocking has traced organized English anthropology from the same moral source: from the Aborigines Protection Society to the Ethnological Society of London to its rivalry with the newer Anthropological Society of London to their merging in the Anthropological Institute of Great Britain and Ireland.[45]

If we consider another development of interest in early 19th-century Britain about which there is more accurate information available, the methodological problems may become clearer, although not necessarily simpler. The interest in social statistics involved, among other factors:
1. a heritage from the Enlightenment (as with Laplace), and more especially from the Physiocrats.
2. a realisation that great social changes were taking place which no one understood as a whole—it is, for example, a chief message of Charles Babbage's *On the Economy of Machinery and Manufacturers* of 1832 that industrialists do not understand what is happening in industry;
3. a conviction that social reform was needed but could not be justified without evidence—the position of Thomas Arnold;
4. a desire to refute Ricardo and a general dissatisfaction with the deductive school of social theorists which he (and some would have included James Mill, and even Bentham) represented.
The first three factors are fairly obvious to the student of general British history in the 1830's, but the fourth is still in need of emphasizing. Even so competent a work as Eric Roll's *History of Economic Thought* does not make the connection between Thomas Malthus's methodological argument with Ricardo in the 1820's; the attacks on Ricardianism by Cambridge men such as Richard Jones, William Whewell, and Babbage; the founding of statistical societies; and the downfall of Ricardian theory dating from W. S. Jevons's work beginning in 1862 (*The Theory of Political Economy*, 1871) and Alfred Marshall's work beginning in 1879 (*Principles of Economics*, 1890).[46] The connection was, however, sensed by a Ricardian such as John Stuart Mill, who in his *System of Logic* attacked Laplace's probabilistic reasoning on several points. Unfortunately, he soon found out from John Herschel that his own reasoning on these points was often at fault.[47]

Interestingly enough, Jevons was an admirer of Herschel, and he drew his analogy as to the nature of his own particular new laws from Herschel's "Meteorology" article in the eighth edition of the *Encyclopaedia Britannica*. Economics, he went on to say, can be "as exact as many of the purely physical sciences; as exact, for instance, as Meteorology is likely to be for a very long time to come."[48]

Dissatisfaction with deductive social theory was also, as J. W. Burrow has observed, basic to the foundation of the Ethnological Society, which from 1848 dealt with the diversity of mankind by tracing the historical connections of its varied races, and the rival Anthropological Society of 1863, which sought general laws based on comparative study—as in comparative anatomy. For the founders of the Anthropological Society, says Burrow, "the common enemy was what they regarded as the a priori theories of the utilitarians and the economists."[49]

It seems clear that the advocacy of a mathematical statistical approach was based on more than the 18th-century orthodoxy that probabilities were the best that could be arrived at in the "moral sciences" (that is, all studies in which human psychology played a role—including political economy and the historical evidence for Christian miracles as well as ethics). This orthodoxy had permitted the influence of Smith, Ricardo, and Bentham. One new factor in establishing the validity of a statistical approach was the prestige of the theory of probability in its role as trusted assistant to the queen of the sciences. It was the men of theoretical astronomy, Laplace, Gauss, Quetelet, John Lubbock, John Herschel, who certified as to the validity of a statistical approach to knowledge. Thomas Galloway, professor of mathematics at Sandhurst and an actuary for the Amicable Life Assurance Office, even used such an approach in trying to determine the proper motion of the solar system in space. I have not checked Gauss and Galloway, but the other four men also advocated the use of this approach in social studies.[50] The Belgian mathematician and astronomer Quetelet was perhaps the most extreme proponent of reducing the moral sciences to the exact sciences. He even believed that in his concept, the "average man," he had discovered the analogue to the centre of gravity in mechanics. And he believed in the validity of the analogy so much that he used it to interest other scientists in his work.[51]

If astronomy gave prestige to the theory of probability, then practical statistics found its chief economic use in life assurance companies. It was in the first third of the 19th century that the two traditions squarely joined in Britain, as the companies turned to the theorists for guidance. Such mathematicians as Thomas Young, Charles Babbage, Frances Baily, Augustus De Morgan, and Thomas Galloway showed the businessmen how to make the businesses scientifically secure. In Babbage's case, the contrast in reliability between men of science and men of business was extreme. Whereas

Babbage's annuity tables were used not only in England but also on the Continent, the combine which had involved him in the subject was so insecurely financed that it never came into corporate existence.[52]

By 1832 the government also was feeling the pressure for more accurate knowledge, and in that year the statistical office of the Board of Trade was founded. One of its chief backers was C. E. Poulett Thomson, who was, significantly enough, M. P. for Manchester. He was also brother to G. J. Poulett Scrope, M. P. for Stroud and geological friend of Charles Lyell.

Social statistics as a distinct entity with its own reasons for existence, separate from the desires of astronomers or life assurance executives, can be said to have fully come to England in 1833, with the founding of the Statistical Section of the British Association. This was followed directly by the founding of the Statistical Societies of Manchester (1833) and London (1834) and then by Birmingham (1835), Glasgow (1836), Bristol (1836), and Leeds, Liverpool, and Ulster (1838).[53] The British Association event was made possible by the conjuction of five individuals at that meeting, and their presence was by no means accidental or uncaused. It was caused by the widespread hope, or spirit of the possibility of doing something, that William Vernon Harcourt, John Phillips, and David Brewster had added to the British scientific scene by founding their new body at York in 1831. We note that a similar spirit had already produced a statistical society in Saxony (1830), and a geographical and statistical society in Mexico (spring of 1833).... The geographical spread suggests a Humboldtian spirit.

The five men were: Quetelet, come over from Belgium to reveal his new theories to his mentors and judges; Charles Babbage, fresh from his exposé in the *Economy of Machinery and Manufactures* of how little British economists (or indeed British businessmen) knew about what was going on in industry; the ailing Thomas Malthus, who had long called for better statistical knowledge; Richard Jones, the anti-Ricardian economist who was to succeed Malthus as professor at Haileybury, the college of the East India Company; and Adam Sedgwick, the geologist friend of Babbage and Jones who was no lover of Ricardo or Bentham or any other deductive theorist, and whose co-operation as president was needed if a section was to be added to the Association by a simple *coup*.

Jones was the entrepreneur of the movement. He had gotten the help of John Lubbock in working out a more exact mathematics for population figures as early as 1829.[54] In his Inaugural address as professor of political economy at King's College, London, in February, 1833, he advocated a statistical society for England; and he and his friend William Whewell had discussed strategy several months before the British Association meeting. It was he who collected the small group in his rooms at

Cambridge in June during the meeting. A decision was reached; Babbage acted as spokesman to the general meeting; and Sedgwick knew how to carry off any situation in front of an early Victorian audience with his own mixture of humour, eloquence, and an affirmation of his belief in God and Truth. He assured the meeting that the section would be devoted to mathematics and facts, not social rabble-rousing.[55] The *coup* was accepted; the new section prospered; by the end of the decade the Reverend Thomas Chalmers was a prominent member—and William Whewell complained that the section *was* being used for social rabble-rousing.

The British Association group agreed to form a society with regular meetings, as well as the BA section with one meeting a year. Again Babbage was the public arranger, with Malthus and Jones as principal supporters; and the London Statistical Society came into being in the spring of 1834. It had more than 400 members by the end of the year, with enough prominent politicians (including several future Chancellors of the Exchequer) to give "a readiness of access to Governmental circles which was invaluable."[56]

Just as an impression based on inadequate information, I would guess that the London Statistical Society became a vehicle for proper persons to exert proper influence on government, and the Statistical Section of the British Association became a vehicle for other persons to try to exert improper influence on government.

The Manchester Statistical Society was not so much derivative from the British Association event as just subsequent to it. Manchester was in an institution-forming mood in the 1820's. In addition to four newspapers (including the *Guardian*), there came into being, to supplement the older Manchester Literary and Philosophical society (1781), a Natural History Society, a Mechanics' Institution, the Royal Manchester Institution, and the Royal Medical College. To this increasingly public-spirited area in 1829 came William Langton, who helped set up the Manchester and Salford District Provident Society, and who suggested to his fellow-secretary James Kay in 1833 the need for a statistical society. Kay, who was to make his name as one of the greatest of public health reformers, presumably needed little convincing; and there were public-spirited businessmen in Manchester to give the necessary support. The founders, not surprisingly, "were less interested in enumeration and computation than in effecting improvement in the state of the people among whom they lived." Particularly striking were comparisons between Manchester and non-manufacturing towns such as York and Hull, to show how far behind the public culture (e.g., schooling) industrial Manchester lagged.[57]

In this intricate dovetailing of inherited, ideological, moral, socio-economic, intellectual, and personal factors which actually produced a set of social-statistics institutions in England, no separation into intellec-tualist and externalist factors seems useful. There is no straight-line growth from the simplicities either of 17th-century "political arithmetic" or of 17th-century mathematics as displayed, for example, in A. De Moivre's *Doctrine of Chances* (London, 1718). No simple internal history of one subject, and no simple combination of two or three such internal his-tories, will suffice. A statistical approach forced its way past the Ricar-dians into social science because of the existence of a complex set of conditions, of which pre-existing special sciences were only a part. Can this complex be stated in a form useful as the basis for a complex historio-graphy? Not as yet.

To understand the introduction of a probabilistic approach into physics by Clausius and Clerk Maxwell is difficult enough, even given its prior use in astronomy. At least there the problem is, apparently, an intel-lectualist one internal to the physical sciences; as Mary Hesse has ob-served, there is no valid evidence of an influence *from* social statistics. A major complicating factor is that, as Lancelot Hogben says, the term "sta-tistics" has at least six meanings, two of which do not involve the calculus of probabilities and four of which do; and each has its unique relation to observable evidence. Collections of numerical data (as of population or exports), and devices based on them (life assurance tables, cost of living indices), are based on arithmetic. The Theory of Errors (as in astronomi-cal observations), Stochastic Reasoning (as in statistical mechanics), So-cial Statistics (as in Quetelet), and the Calculus of Judgment (as in Babbage on the number of witnesses needed to prove a miracle, or as in medical evidence for the validity of a vaccine) do now use the calculus of probabilities.[58] But the traditional idea was that a probabilistic approach is proper to the social sciences, as opposed to the true mathematical dem-onstration proper to the physical sciences; and this attitude may have made the task of the physicists psychologically more difficult than the task of the social reformers.

To understand the increased use of a probabilistic approach in social studies, the problem is that too few or too simple analytic categories are available for an explanation. I can only explain the phenomena in regular English. The best general statement I can make, so far, is that persons whose intellectual consciences were irritated by what they considered to be the deductive quasi-science of the Ricardians and whose social consciences were involved with "state of England" problems used the most prestigious kind of unexploited thought-weapon they could find, but sometimes de-graded it from being the guardian of exact theory (as it is in astronomy) into being an excuse for collecting horrible examples in order to influence Parliament. For this reason, parliamentary reports of this period have always been more useful to reformers, such as Marx and Engels, than to historians interested in a well-balanced descriptive analysis of the society.

These three examples I have given in increasing detail—the problem of Gauss; the great magnetic undertaking; the foundation of social-statistics institutions—show, I think, the need for detailed narratives before we are able to construct an outline of the development of science in the period. That narrative, at any adequate level, is largely missing. There exist in Britain large amounts of documents which are not merely unpublished but are also unread. Until many of them are read, narratives will simply leave out essential figures, acts, monographs. In my own work on John Herschel, I have found this to be strikingly true. Herschel simply does not appear in existing accounts of some matters in which he was an important—or, as in meteorology, the central—figure. Augustus De Morgan is a rather tangential figure in the accounts of 19th-century mathematics I have seen. The easiest way to find out the significance of "De Morgan's Rules" is to read a modern commercial textbook explaining "how to design an *efficient* digital computer."[59] Historians do not so much denigrate as ignore George Peacock's influence. But the parliamentary reforms of Cambridge in the 1850's were not those of Lord John Russell's idea of liberalism; they were largely those of Peacock's liberalism (insofar as it could triumph over William Whewell's conservatism).

The magnitude of exisitng records, while not usually quite as overwhelming as those in political history, causes comparable problems, given the manpower available. With more than 15,000 manuscript Herschel documents surviving, with more than 700 volumes of Airy papers at the Royal Observatory alone, where are the historians of science to read them all? With the presence of the penny post and/or an adequate messenger service, and the absence of the telephone, written communication was perhaps at an all-time high. (I don't count the majority of government forms and records of the present day as meaningful communications; too many of them, in my experience in the Civil Service, merely are an inadequate record of something which was actually done in some other way for some other reasons by some other persons—and often something other than what is recorded was actually done.) Airy, for example, wrote letters (that is, business notes) several times a day.

Perhaps the thorough study of, say, all the Airy papers would permit one to set up a historiographical model. This much, one might say, can be learned from the most complete documentation; but also, this other thing cannot. Thereafter we would have a test for other men and other matters. The best use of such a paradigm might be a negative one: by reference to the Airy model, we would see that even if we discovered more letters in a crate behind the back stairs, we still could not solve problem X. Therefore it would be time to do more analysis, more thinking.

Another such paradigmatic project, for intellectualist historians, would be to study *all* of the reactions to Darwin's *Origin of Species,* not merely in one country, but worldwide; and not merely among scientists, but among all groups. If the historian had first established what Darwin's position is (not an easy matter), then he would have a way of classifying all the groups with relation to one another. He could tell, for example, if Idealism was to England as Natur-Philosophie was to Germany or vitalism was to France; or if the debate between Idealism and Mill in England was equivalent to the debate between Natur-Philosophie and Marxism in Germany. Personally, I doubt that any of these equivalents are true, but I have seen all of them used. A worldwide comparative study would have enough levels of partial equivalence and partial difference to enable us to dispense with such simplicities. Once well done on one suitable subject such as Darwin, it would be available for use in considering each new subject.

Without some such justifications for large projects, it seems to me that in the 19th century we are likely to reach an impasse. No one cares enough about Airy to read all of the Airy papers for the sake of Airy alone. The best students will not consent to read manuscripts without stopping to think. But they must keep on reading if they are to get through all of the available manuscripts. And scholars will not consent to stop reading manuscripts, or to stop criticizing people who do stop.

The truth of the matter, as I see it, is that the study of private papers is one of the most suggestive ways of becoming aware of the problems in the field. But, unfortunately, the private papers do not solve many problems. Personal letters prove very little about science itself, even though they are invaluable about its politics and social setting. Letters can prove who engineered Ross's Antarctic expedition—Edward Sabine, Francis Beaufort, and John Herschel did. But the letters cannot by themselves prove what the permanent results of the expedition were, and whether Gauss's theories were really verified. It is very stimulating to read in John Herschel's private papers that he and Charles Babbage may have been the first to use an astatic magnetic needle.[60] But is it true? Was it important? These questions can be answered satisfactorily, if at all, only by working with published papers as well as private ones.

To this extent, then, the intellectualist emphasis is correct. The published papers—by which I mean circulated, not necessarily printed papers, for astronomers in the early 19th century still sometimes circulated detailed tables of observations by letter—and the existing instruments contain the essence of the history of science as a collection of special histories.

Letters, notebooks, early drafts of papers, etc., may indicate very interesting things indeed; but they are not themselves sufficient evidence from which to reconstruct the backbone of the subject.

But the published papers are mostly unread too. This is literally true in many cases; but it is figuratively true in many more. It is difficult to see what it is that a monograph says. Without some outside control, the historian will often see in it what he expects to see, that is, what an inadequate historical tradition has said is there. If he expects to see Nothing of importance, that is what he will see. Hence in practice even the most intellectualist historian needs the other papers to counteract his own too-simple expectations. Herschel's assertion regarding the astatic needle at least creates a problem for the historian, and by the time I finish with it, I will probably be wiser than when I began.

The same kind of intellectualist narrowness can occur in the study of instruments as occurs in the study of monographs, by giving their history from one successful device to another successful device. For example, a historian of meteorological instruments may speak of "the cup anemometer, invented by T. R. Robinson in 1846 after he had thought about some experiments by R. L. Edgeworth," without giving the title of Robinson's paper at the British Association meeting.[61] The title is, "On a Modification of Dr. Whewell's Anemometer." That Whewell's device was not practically successful should not obscure its role in generating a device that was.

The study of instruments is comparatively neglected for the modern period, and much of what has been done is of modest value. As this study develops, it should throw light on many interesting questions. Why were certain phenomena known but relatively neglected? Charles Babbage, for example, pointed out that Fraunhofer's lines were *hard to see* with existing instruments;[62] his friend Herschel had the great advantage of having been shown how to see them by Fraunhofer himself, and having been given some Fraunhofer glass. We assume that matters must have changed by 1880 when we find spectroscopy used in a popular novel as a symbol of certainty: "The Modern Blue-Stocking acknowledges few things that cannot be weighed in the balance or observed in the spectroscope."[63] In this case we have written testimony, but it is so incomplete as to suggest a variety of investigations.

It is not only modern simplifications which mislead historians. Scientists have been quite fertile in constructing their own myths, often for immediate polemical purposes. Charles Babbage's *Reflections on the Decline of Science in England* has tended to lead historians away from what may be the truth, that the British government patronised science perhaps more readily in the first half of the 19th century than in the half-centuries on either side (not counting the education of technicians for industry and the building of museums). Babbage's own government grant was large, by contemporary standards.

A more certain judgment can be made about the polemical assertion of Charles Lyell that theological ideas in geological thinking retarded the development of geology. Lyell presented this idea as an interpretation of the history of geology—it permeates the entire historical introduction to his *Principles of Geology* of 1830—but it was obviously designed for his own time as well, with especial reference to the ideas of his teacher, William Buckland. Some modern historians have unfortunately followed Lyell in applying the idea to geology contemporary with Lyell.

But that contemporary geology was not retarded. It was spectacularly progressive. So the myth that theological ideas retarded geology in this period is not only wrong; it is also an answer to the wrong question. When we see what it is that needs explaining—why were British geologists so good?—then we may still come upon ideas from theology, but in a very different role. For example: surely it cannot be said that Louis Agassiz's career retarded paleontology. Yet surely it is true that Agassiz's theological-philosophical ideas were important in guiding his research. And theological ideas were of importance in shaping Charles Darwin's thought, in part because he did not always recognize them as theological ideas, but only as obvious characteristics of the natural world which needed to be explained. Bernal has observed (but he regrets the situation): "It was not that Science had to fight an external enemy, the church, it was that the church—its dogmas, its whole way of conceiving the universe—was within the scientists themselves."[64] It is partly in this way, but partly also as a valid challenge, that theological ideas shaped Darwin's thought.

This reference to theology may serve to remind us that the historian of science must often delve into "ordinary" history; and he must do so in particular because he often needs to know about "ordinary" matters more accurately than he can determine from scholarly opinion in the field he is raiding. Thus in the original version of this chapter I criticised Pearce Williams for what I believed to be his errors concerning Coleridge's intellectual debts to Kant as opposed to Schelling.[65] Professor Williams has since shown that he knows more about the subject than I do; but what the affair also shows is that neither of us, as historians of science, could find the detailed analysis we needed in works by conventional literary-intellectual historians, and we both had to go to the sources ourselves.

Coleridge and Schelling found an ardent English disciple in Julius Hare, and it is useful to know something of Hare in order to see how far German Idealism penetrated into English intellectual life. Hare was a major agent in putting together the philosophy of the Broad Church movement, in introducing German historical scholarship into England, and in setting the tone of Trinity College, Cambridge, at the time (the 1820's) when most of the elements of advanced training were being brought together by the Trinity Fellows and their friends. Yet if the historian of science cannot rely on the established experts to explain the intellectual relationships and particular positions of Kant, Schelling, and Coleridge in enough detail for his purposes, still less can he turn to a standard church history to determine the importance of Julius Hare. The standard church history is often not even sure that the Broad Church movement *had* a philosophy, much less that Hare was one of its architects. (If philosophy must involve a metaphysics, then the Broad Church did not have a philosophy. But it had a morality and to a considerable extent an epistemology.)

Equally we cannot turn to a standard church history to determine the amount of money the nation, acting through church appointments, spent in subsidizing men of science with tutorships, fellowships, masterships, rectories, canonries, deaneries, and bishoprics. As Coleridge observed, "Among the numerous blessings of Christianity, the introduction of an established church makes an especial claim on the gratitude of scholars and philosophers." Or as Robert Peel wrote in 1844, cathedral preferment "is the chief source from which I am called upon to satisfy the claims founded upon great professional merit and learning, and upon services connected with the great public schools and universities." Thomas Macaulay in the 1850's calculated that the annual value of church livings in the gift of St. John's College, Cambridge, was twenty-four thousand pounds a year; for Trinity College, Cambridge, it was eighteen thousand. For St. Catherine's and Downing it was only a few hundred.[66]

There were at least three scientist deans: the geologist William Buckland, who was a reformer of some influence at Westminster, especially with regard to the School; the mathematician George Peacock, who was outstanding at Ely; and the geologist W. B. Conybeare, at Llandaff. I did not know of any scientist bishops, but Sir Edward Collingwood has drawn my attention to two in the Irish Anglican Church. John Brinkley, Astronomer Royal for Ireland and professor of astronomy at Dublin, became Bishop of Cloyne; and Charles Graves, professor of mathematics at Dublin, became Bishop of Limerick. The controversial Bishop Colenso of Natal in South Africa made his reputation with school texts on algebra (1841) and arithmetic (1843); his appointment counts as support of a scientific culture, though not of an individually creative scientist. The same is true of Henry Goodwin, Bishop of Carlisle, whose *Elementary Course of Mathematics*—that is, algebra, trigonometry, conic sections, mechanics, hydrostatics, optics, and Newton—had apparently become a common text for the Cambridge tripos around 1850 (first edition, 1846; sixth edition, 1866).

It is interesting to note that a Government's patronage of scholars was reasonably impartial in favoring ability, but openly partial in favoring men of its own party. The best Whig (Peacock) was made Dean of Ely when the Whigs were in power—and was an outstanding dean. The best Conservative (Whewell) was made Master of Trinity when the Conservatives were in power—and was an outstanding Master. Had the vacancies appeared in the reverse order, Peacock would probably have gotten Trinity, and Whewell might have gotten Ely. There was not, it seems to me, very much room for favoring mediocrity in a Government's manipulation of patronage concerning clearly scholarly positions in early Victorian England. But it would have seemed simply eccentric for a Conservative Prime Minister to put an ideological opponent into a position of public prestige and educational or political influence. Peacock's reforms at Ely, for example, were not limited to the cathedral; they involved the town of Ely as well.

If we turn from the government acting through the church to its action through other agents, it is equally true that we cannot find from single-agency administrative histories how much the government spent on science. To take one example: consider all of the government money in salaries in various departments that is implied by the following publication:

> Charles J. B. Riddell, Lieutenant, Royal Artillery, Assistant Superintendent of Ordnance Magnetical Observatories, *Magnetical Instructions for the Use of Portable Instruments Adapted for Magnetical Surveys and Portable Observatories and for the use of a set of small instruments adapted for a fixed magnetical observatory.* London, 1844. Her Majesty's Stationery Office. Printed by Order of the Lords Commissioners of the Admiralty. And a *Supplement to the*...London, 1846.

To take another example, how much of the Foreign Office budget should we charge to science when we find that one of Richard Burton's duties as consul at Fernando Po in 1861 was to make meteorological observations? How much when we decide that his appointment to Damascus was partly a reward for his exploring and anthropological studies?[67]

What base shall we use for alloting the expenses of the *Beagle's* voyage around the world among the categories of science, commerce, and maintenance of a navy in being? Darwin came aboard because Captain Fitzroy wanted a person acquainted with mineralogy and geology, who could look for metals and in general be qualified to "examine the land," so that he and the regular officers would not feel a need to do so. He himself thought the most interesting part of the voyage was carrying a chain of meridian distances around the world. Someone wanted a detailed coast survey of South America. Fitzroy eventually became chief of the Meteorological Office of the Board of Trade; did the *Beagle* voyage, and his meeting with John Herschel at the Cape of Good Hope, encourage his interests in that direction?[68] Once we have decided about the *Beagle,* we still have to appraise the voyage round the world of the *Sulphur,* 1836–42, under Edward Belcher, to survey the Pacific coasts of America and, as it turned out, to take part in a war in China.[69] Or, given twenty-six ships engaged in surveying between 1843 and 1846,[70] what percentage of their work constituted a contribution to scientific knowledge?

The historian of science needs to know *both* that Ross's Antarctic-bound ships were not primarily exploring vessels but were itinerant magnetic observatories, *and* that the British share of expenditures on the thirty or so co-ordinated observatories on land was divided among the Admiralty, the War Office, and the East India Company. For that matter, it is only the historian of science who is likely to know that the activities of the East India Company in sponsoring science are an obvious point of approach to the whole ideology of British rule. The Great Trigonometrical Survey of India shows the workings of British policy better than still another study of Macaulay's education minute; and its second great worker, who extended it north into unexplored regions, deserved to have Mount Everest named after him. As a scientific enterprise, it overshadowed all European geodetic operations.[71] If one is not attracted to terrestrial magnetism, here is an equally interesting and involved large-scale enterprise to study.

Indeed, the entire British Empire takes on a somewhat more permanent status than is now politically obvious, for one who is concerned with the concept of isostasy as based on Indian geodetic evidence, the origins of ethnology, observatories in Cape Colony and decisions about building the great Melbourne reflecting telescope, the *Flora Indica* of Joseph Hooke and Thomas Thomson, and the relation of pre-Cambrian geology to the rocky sheath of Canada.

If we consider political historiography, we find it has tended to minimize the differences between the parties, and to represent both sides as reasonably sincere men reacting to the same massive socio-economic situations. Tory men with Whig measures; honest Liberals who stumbled into an Empire. How refreshing, by contrast, it is to read the letters of the Ross conspirators (Frances Beaufort handling the Admiralty, John Herschel handling the Royal Society and the British Association, Edward Sabine handling everything and everybody) when the Conservative Robert Peel becomes Prime Minister in 1839.[72] How the suspense builds up as the conspirators react to the danger! How grateful they are when Peel is forced out by the Bedchamber Crisis! How determined they are not to relax a minute until Ross's flag is actually hoisted over his ship—then and only then will it be too late for Peel to return to office and stop them! And how their morale collapses, how coolness and bickering develop, once the unnatural tension is over and the expedition has put to sea.

In political history the Bedchamber Crisis is, as its name indicates, a joke, an embarrassment. Its only result was to permit Melbourne's fading Whig Government to drag out its unhappy life for another year or so. Its significance is in showing the residual powers, possibly unconstitutional, of a personally stubborn young Queen. Yet without the Bedchamber Crisis, "by far the greatest scientific undertaking which the world has ever seen" might never have had its climax, because a finance-minded Prime Minister might have decided to hold up a voyage whose justification he did not understand. By holding it up, he would have destroyed it; because Ross, the only suitable leader, would have had to accept another appointment.[73]

Since the publication of the earlier version of this chapter, William Aydelotte has produced an analysis of voting in the early Victorian period which supports the contention that party differences were more important than most political historians have emphasized in recent writing. He concludes, "The view that each party represented a hodge podge of opinions and objectives not clearly distinguishable from those of the other party—that the party conflict was a sham battle unrelated to questions of policy—seems altogether excluded.... The principal lines of cleavage were closely related to signficant ideological cleavages."[74] Interpreting the word "ideological" in a very broad sense, as including a basic approach or style as well as a thought-out intellectual position; and emphasizing that the opposite to the conservative Peel was not liberalism (political liberalism *was* Peelite) but Whiggism, in the cases I am concerned with; then I can say that my researches agree with Aydelotte's finding.

Thus if we ask, "Would Peel have held up the expedition?" we cannot tell by an analysis of Conservative party positions. Evidence from ordinary sources suggests that his feelings concerning science were strong but ambiguous, and were bound up with a conflict between Peel the self-made scholar and Peel the offspring of self-made businessmen. But the problem cannot be solved by reference to ordinary political history writings, whose authors do not always note that there *was* a conflict.[75]

What *is* certain is the contrast to the attitude of Melbourne and his Chancellor of the Exchequer, Spring Rice. Here is a Lieutenant of Engineers approaching the Chancellor to get money for magnetic apparatus: "I called upon Spring Rice last week and found him in the best humour possible for granting money." He asked to be sent a letter "which he would lay before the Lords of the Treasury and enforce with all his eloquence."[76] Perhaps scientists should not expect that kind of reception; perhaps after all it is the attitude of Peel and his disciples which is normal.

It was after thinking especially about the Airy papers and the great magnetic undertaking that I decided to call this chapter not "The History of Science in Depth" but "History in Depth." My remarks seem to have led some scholars toward despair, if R. P. Laidlaw's comment is any indication: "The warnings on the dangers of partial knowledge made by such as Michael Wolff and W. F. Cannon, which bibliographers love to quote, amount, whatever their truth, to a command to silence for several academic generations." However, I call for much *more* publication than at present, publication based on the enormous unused material I have drawn attention to. With so much to cover, one cannot wait for the niceties of perfect coverage in every case. I have called, to use Laidlaw's paraphrase, for "total war, proving techniques that can be used only partially elsewhere."[77] What one should avoid are more studies of the same old figures and events based on the same old sources and arranged in the same old way. To penetrate behind the stultifying phrases, "the *Westminster* said," or the *"Quarterly Reviewer's* remarks may be taken to represent," to the real idiosyncrasies of David Brewster and Thomas Galloway, or to the question of whether Charles Lyell's and John Herschel's positions in reviews were novelties to their audiences in the 1820's and the 1840's, is just what is good in order to see the complexities of thought in the period. To identify 100 unknown authors of reviews in only six months would seem to me an excellent return for scholarly investment; and I have been able to formulate a hypothesis as to the formation of Victorian scientific societies simply from an incomplete list of periodicals which Lionel Madden prepared.

It is not, to repeat, that massive new data will produce new and better history in some "Baconian" fashion. It is that new data will permit the prepared scholar to use his new guiding ideas in the production of better narratives. The history of science is the most interesting field of history at present. This is not because science at present is of interest, but because the history of science is now the most revolutionary field of history. The scholars connected with the name of *Isis* have completed the work of the Romantics in destroying, for rational historiography, the contempt for medieval civilization and the neglect of Islam which otherwise are endemic in modern Western society. The brilliant analysts who found their best vehicle of publication to be the *Journal of the History of Ideas* have forced the Scientific Revolution even into textbooks. We need not associate ourselves with the exaggeration of Herbert Butterfield, who stated that the Scientific Revolution "outshines everything since the rise of Christianity and reduces the Renaissance and Reformation to the rank of mere episodes, mere internal displacements within the system of medieval Christendom."[78] Nor need we accept as sufficient, any longer, the internal

253

intellectualist approach of Alexander Koyré. Yet we can still admit that the Scientific Revolution was in its more limited way as decisive a novelty as was the Reformation and perhaps even may prove to be as important as the emergence of the power-oriented sovereign state. Intellectually, the Scientific Revolution was itself an internal development from Christendom, rather like Pietism in not being limited to either Protestantism or Catholicism. Should we then pay careful attention to Bernard Semmel's *The Methodist Revolution* (New York, 1973) for our own subject?

It is not clear what journal or other avenue of publication will become the major vehicle for revolutionizing the history of the 19th century. For a while the *Proceedings of the American Philosophical Society* had, perhaps, a slight lead; and the *Notes and Records of the Royal Society* is in the competition. The *British Journal for the History of Science* went to the head of the class while Maurice Crosland was editor, and seems to be staying there. There are other possibilities. Book publication is becoming more feasible year by year, but this form traditionally increases greatly the ratio of information to new ideas; that is, it makes it much more laborious to present new theories or approaches. Put another way, it tempts the author to fill up space with "facts" traditionally accepted as valid and not the outcome of new sources and theories.

It is not part of my purpose here to go into the area of philosophy of science and its debates. For now I will associate myself with the essay of one of our master craftsmen, I. Bernard Cohen's "History and the Philosopher of Science," in Frederick Suppe's *Structure of Scientific Theories* (Urbana, 1974). On the practical side I will note that Robert Young has in various publications been able to approach our subject in the way I called for, and if not all of his results are definitive, then neither are those of anyone else. The best single publication I know of is Rejier Hooykaas, *Religion and the Rise of Modern Science* (Edinburgh, 1972); this, of course, is largely on the 17th century. One of the most important is the essay of Peter Mathias, "Who Unbound Prometheus? Science and Technical Change, 1600–1800," in the volume he edited, *Science and Society 1600–1900* (Cambridge, 1972). Two books show that the history of the geological sciences is coming of age: Gordon Davies, *The Earth in Decay* (London, 1969), a really challenging study; and Martin J. S. Rudwick, *The Meaning of Fossils* (New York, 1976), which makes one grateful that the author has written many articles of the same high quality as the chapters in this book. And now we can add Roy Porter, *The Making of Geology... 1660–1815* (Cambridge, 1977).

Through whatever avenues it is done, my predictions are that, when the history of the 19th century is re-written satisfactorily, it will be seen:

First, that in the history of science the massive watershed between

"Newtonian" and "modern" is not around 1900 but around 1850; and that the critical discoveries are not such things as X-rays but such things as the continuous spectrum. Already it is easy to explain relativity theory as the decision to choose Maxwell over Newton wherever a conflict between the two developed. Relativity was not itself a revolution, but the final success of a revolution: Yorktown, not Lexington or Saratoga. Likewise current quantum ideas can be seen as a decision to live with both Faraday's lines of force and Faraday's electrochemical discontinuities. Finally, given Babbage, Boole, De Morgan, and the Thomson brothers (James and William), the computer age follows merely with the addition of appropriate instrument-makers, and sufficient demand. (At least this is so on intellectualist grounds).

Second, that the history of science, which contains intellectual and personal and organizational and sociological materials in perhaps a purer distribution than most other specialties of history—thus there are as many stupid as brilliant scientists, but there seem to be few brilliant politicians—is the clearest and most direct vehicle for demonstrating that this period was the watershed for general history. With the help of Arthur Hughes' study of science in English encyclopaedias, we can even estimate the contours, in successive editions of the *Encyclopaedia Britannica.* In the seventh edition (1830–42) we are still in the old world. The eighth edition (1853–60) is the swing edition, registering the gains of the second quarter of the century; it is clumsy, sometimes contradictory, fascinating. "When we reach the ninth edition (1875–89), we are in a different intellectual climate.... To judge from the general tone of the articles in the ninth edition, one would assume that, for a sufficiently large proposition of educated people, the battles had all been won."[79] Chief science editors for this edition, we remember, were Thomas Huxley and James Clerk Maxwell.

To be sure, it has been apparent for some time that the era of the Revolution separates us from all previous history, and that many of our problems are new, whereas our value systems were generated (perhaps validly) in an earlier context. I like to think of the intervening period— from, say, 1840 to 1910 or 1920 as one existing in two worlds simultaneously: the world of the railroad, and the world of the horse. It is possible to suggest that the vexing problem: what *was* the Revolution? can be approached more freshly, not by studying still another cotton mill or arguing again over a set of population statistics, but by studying *all* of the problems which Charles Babbage encountered when he tried to have his calculating engine actually constructed.

Thus Babbage's scientific peers understood his calculating engines well enough. What they questioned was their feasibility and usefulness, given the state of technology, the available economic support, and Babbage's own personality. After fourteen years had been spent on a three-year project, Babbage's scientific sponsors (notably his good friend John Herschel) were acutely embarrassed about asking for still more money. George Airy acted as front man in damning the project to Melbourne's government, but it was not solely or even primarily his judgment that he was conveying. In this case, as I have also noted about the mid-Victorian rejection of the metric system, it is not satisfactory to condemn politicians and civil servants as backward or unsympathetic before you find out what kind of advice they had. Generally speaking, to agree with John Herschel was considered a mark of good sense in early Victorian England.[80]

What Airy as Astronomer Royal *could* speak of with authority was whether Greenwich Observatory would have any use for the calculating engine if it were perfected. His decision that it would not is an interesting one, which could well be the starting-point for a consideration of his character as well as the state of the art of astronomical calculation from Bessel on. Herschel, for example, did his own laborious calculations himself, presumably because of Bessel's influence.

I have found it helpful to memorize the *form* of the following statement:

"The reason why Babbage's scientific contemporaries did not recognize the usefulness of his calculating engines is that Babbage's calculating engines were not very useful."

This form gives other statements like the following:

"The reason why French phlogiston chemists did not recognize that air is essential to combustion is that air is not essential to combustion."

"The reason why British biologists in the early 19th century did not recognize that final cause is unscientific is that in biology final cause is not unscientific."

Many important statements in the history of science can be formed this way.

What strikes me as the most amusing fact I have come across recently is that the Government which most readily gave money for the building of this first modern computer was headed by—the Duke of Wellington![81] That fact alone should show that usual stereotypes of Toryism should be scrapped. It was Wellington, too, who in the spring of 1835 found time (in his role as Foreign Secretary) to set up arrangements with foreign governments for simultaneous observations of the tides in European states bordering on the Atlantic, and in the United States.[82] This international co-operation was to furnish data for William Whewell's studies of tidal phenomena.

Third, that the leadership in the re-writing and re-interpretation of the history of the 19th century will have to be assumed, not by other kinds of historians for whom the history of science is an esoteric specialty, and certainly not by historians of science for whom the history of science is an esoteric specialty, but by those who treat the history of science from the beginning—that is, in the selection of sources—not as an entity which has links with culture and society, but as an integral part of culture and society. The trouble with a study which calls itself, say, "The History of Geometrical Optics" is not in writing it; it is in trying to believe that what you are writing is history at all.

Much recent history of science seems determined to show, more than anything else, that scientific findings have never been objective, verifiable, reasonable, modest, or (some would say) true. I do not share this determination; scientific findings are, however, certainly limited and conditioned by the history of their time and society. Is, then, the approach indicated by Eric Cochrane in his title, "Science and Humanism in the Italian Renaissance," writing in the *American Historical Review* (*81*, 1976, 1039-1057), enough? Or is science still being treated as too separate a phenomenon? At any rate, before criticizing Cochrane's attempt, it is necessary to have a better procedure to recommend. The really satisfactory presentation of the subjective or sociological approach, to me, is Jerome Ravetz, *Scientific Knowledge and its Social Problems* (Oxford, 1971). That is, this is the position I enjoy arguing against.

To be sure, the history of science is no more the true center of British history in the 19th century than is the history of Parliament. But it is no less so. And historians of Parliament, or of politics, have amply had their chance, from Halévy on, to produce full and convincing history. And they have failed. So have "social" historians. And attempts to construct "cultural" history on the basis of literary culture alone are not very promising. The leadership in the re-writing of history will have to be assumed by historians for whom the totality of the past is an omnipresent pressure, limiting and defining the heroes at the same time that it gives a stage for their personal existence and gives a rationale for their all-too-human actions, just as it does for all humans.

Notes to Chapter Eight

The first version of this essay was a brief volunteered paper at the History of Science Society meeting in Philadelphia, 1963. An expanded version appeared in the journal *History of Science, 3* (1964), 20–38. This third version is revised and somewhat expanded.

1. John Herschel, "On the action of the rays of the solar spectrum on vegetable colors," *Philosophical Transactions, 132* (1842), 181–214; and "On certain remarkable Instances of deviation from Newton's Scale in the Tints developed by Crystals," *Trans. Camb. Phil. Soc., 1* (1819), 21–41.

2. J. L. E. Dreyer and H. H. Turner, *History of the Royal Astronomical Society* (London, 1923), p. 6.

3. Herschel to De Morgan, 16 March 1860 (John Herschel Papers, Royal Society).

4. Cf. George Airy to John Herschel, 1 February 1847 (John Herschel Papers, Royal Society), passing on Schumacher's request to Herschel. On Herschel's use of Bradley's observations in order to calculate orbits of long-period double stars, see Herschel to S. P. Rigaud, 15 August 1831.

5. Airy to Herschel, 24 April 1834; Herschel to Royal Society Committee of Physics, 8 July 1840 (John Herschel Papers, Royal Society).

6. Cf. Robert H. Scott, "The History of the Kew Observatory," *Proceedings of the Royal Society, 39* (1885), 37–86, esp. 45–64; and the file, "Establishment of Magnetic Observatories at Greenwich and in the Colonies to 1848" (Airy Papers, Royal Greenwich Observatory, Herstmonceux); Dreyer and Turner, *History of the Royal Astronomical Society,* pp. 109, 112–113, 155–156.

7. Report of the Astronomer Royal [Airy] to the Board of Visitors, 5 June 1841 (Airy Papers, Royal Greenwich Observatory, Herstmonceux).

8. Thomas Chalmers, *The Evidence and Authority of the Christian Revelation,* 2nd ed. (Edinburgh, 1815), pp. 174–175, 204; *On Natural Theology* (Glasgow, 1836), I, 229–230. On Chalmers generally, see William Hanna, *The Life and Writings of Thomas Chalmers* (New York, 1850–52).

9. Thirlwall to William Whewell, 31 October 1849 (Whewell Papers, Trinity College, Cambridge).

10. L. Pearce Williams, "Boscovich and the British Chemists," in *Roger Joseph Boscovich,* ed. L. L. Whyte (London, 1961), pp. 153–157, and his *Michael Faraday* (London, 1965). J. Brookes Spencer attacks Williams in "Boscovich's Theory and its Relation to Faraday's Researches: An Analytic Approach," *Archive for History of Exact Sciences, 4* (1967), 184–202; the methodology of this article is not completely convincing to me. See also Robert Siegfried, "Boscovich and Davy: Some Cautionary Remarks," *Isis, 58* (1967), 235. Following the examples of Robison and Leslie, I am willing to let a scientist tinker with Boscovich's atom (Davy added weight to it) and still be called a Boscovichian, but Siegfried won't allow this.

11. Robert Kargon, "William Rowan Hamilton, Michael Faraday, and the Revival of Boscovichian Atomism, *American Journal of Physics, 32* (1964), 792–795.

12. W. F. Cannon, "John Herschel," *Encyclopedia of Philosophy* (New York, 1967).

13. Robert L. Ellis, "Some Remarks on the Theory of Matter," *Trans. Camb. Phil. Soc., 8* (1842–49), 600–605.

14. Joshua C. Gregory, *A Short History of Atomism* (London, 1931), pp. 86–87, 91–93.

15. Richard Olson, "The Reception of Boscovich's Ideas in Scotland," *Isis, 60* (1969), 91–103. See also M. A. Sutton, "J. F. Daniell and the Boscovichian Atom," *Studies in History and Philosophy of Science, 1* (1971), 277–292, which raises the discussion, for Boscovichianism in London science, to a new level of precision. The "influence of Boscovich" debate seems to generate good articles.

16. John Robison, *A System of Mechanical Philosophy* (Edinburgh, 1822), I, 292, 297, 299.

17. R. P. Graves, *Life of Sir William Rowan Hamilton* (Dublin and London, 1882–89), II, 337. I don't mean to imply that Herschel was commenting directly on Leslie.

18. See Olson's comments in "Reception of Boscovich," pp. 91, 93–94, and notes 3, 4, 11, 12. It rather detracts from the purity of Olson's case that, as he says, at least one person who read Priestley on Boscovich was Dugald Stewart.

19. J. R. Partington, "Thomas Thomson, 1773-1852," *Annals of Science,* 6 (1948-50), 115-126. The 1st edition was 1802; 2nd edition, 1804; 3rd edition, 1807 (with Dalton's atomic theory now included); 4th edition, 1810; 5th edition, 1817; 6th edition, 1820. In 1830 it was split into four parts: Heat and Electricity (1830), Inorganic Bodies (1831), Mineralogy and Geology (1836), Organic Bodies (1838).

20. Arthur Hughes, "Science in English Encyclopaedias, Part I," *Annals of Science, 7* (1951), 348.

21. John Leslie, "Dissertation Fourth; Exhibiting a General View of the Progress of Mathematical and Physical Science, Chiefly during the 18th Century," in Dugald Stewart, James Mackintosh, John Playfair, and John Leslie, *Dissertations on the History of Metaphysical and Ethical, and of Mathematical and Physical Science* (Edinburgh, n. d.), p. 606.

22. Communication from Nathan Reingold, Joseph Henry Papers, Smithsonian Institution. The Papers have Henry's copy of Robison. Another copy marked "Secretary's Office" is in the library of the Smithsonian's Museum of History and Technology.

23. Lewis Campbell and William Garnett, *Life of James Clerk Maxwell,* 2nd ed. (London, 1884), pp. 109-110, 150, 156, 217, 273, 301; see also p. 89, on Kant.

24. John Herschel, *Preliminary Discourse on the Study of Natural Philosophy* (London, 1831), pp. 164, 198, 306; for two of his letters, see my article, "The Impact of Uniformitarianism," *Proc. Am. Philosophical Society, 105* (1961), 301-314.

25. Personal field research.

26. Campbell and Garnett, *Clerk Maxwell,* p. 178.

27. See Fig. 40, p. 132, in Simon Newcomb, *Popular Astronomy (2nd ed., New York, 1878);* Dreyer and Turner, *Astronomical Society,* p. 26.

28. James D. Forbes, "Report upon the recent progress and present state of meteorology," *Brit. Assoc. Report, 2* (1832), 196-258; and "Supplementary report on meteorology," *Brit. Assoc. Report, 10* (1840), 44-45. Various schemes for standardizing observations by amateurs were tried out during the century in various fields.

29. *Brit. Assoc. Report, 1* (1831), ix, 22.

30. *Cf.* H. C. King, *The History of the Telescope* (London, 1955), p. 192; Ernst Mach, *The Principles of Physical Optics,* tr. J. Anderson and A. Young (New York, n. d.), p. 301.

31. See William Thomson, "On geological time" and "Of geological dynamics," *Popular Lectures and Addresses* (London, 1889-94), II, 10-64 and 111-113; Thomas Huxley, "Address of the president," *Quarterly Journal of the Geological Society of London, 25* (1869), Part I, xxviii-liii.

32. "Topographical vision" is my term, to indicate Darwin's basic debt to Humboldt.

33. *Cf.* Baden Powell, "Report on the recent progress of discovery relative to radiant heat," *Brit. Assoc. Report, 10* (1840), giving the views of Melloni and Ampère (p. 10), Powell himself (p. 35), James David Forbes (p. 36). See also John W. Draper, *Scientific Memoirs* (New York, 1878), pp. 50, 79-80, 385, 402.

34. D. S. L. Cardwell, "Science and technology in the eighteenth century," *History of Science, 1* (1962), 30; A. R. Hall, "Merton Revisited," *History of Science, 2* (1963), 13 (Merton is here not the college, as one might have expected, but the American sociologist Robert K. Merton); W. F. Cannon, "History in Depth," *History of Science, 3* (1964), 27.

35. Henry Guerlac, "Historical Assumptions of the History of Science," in *Scientific Change,* ed. A. C. Crombie (London, 1963), p. 811 (his italics).

36. C. W. Dunnington, *Carl Friedrich Gauss* (New York, 1955), pp. 114–117 and chapter 10 generally; pp. 163–166, 174–190.

37. E. G. R. Taylor, *The Mathematical Practitioners of Tudor & Stuart England* (Cambridge, 1954), pp. 151–157. See also Eric G. Forbes, *The Birth of Scientific Navigation* (Greenwich, 1974)

38. *Ibid.*, p. 157 *vs.* pp. 155, 162. For one later set of instrument matters, see A. W. Skempton and Joyce Brown, "John and Edward Troughton," *Notes and Records of the Royal Society, 27* (1973), 233–262.

39. *Cf.* Royal Society Committee of Physics and Meteorology *Revised instructions for the use of the magnetic and meteorological observatories and for the magnetic surveys* (London, 1842), p. 8; and John Herschel to Gauss, 4 September 1842 (John Herschel Papers, Royal Society).

40. *Report of the Committee of Physics of the Royal Society* (London, 1840), p. 54; Napier Shaw, *Manual of Meteorology* (Cambridge, 1932), I, 8, 160–162. Herschel began his campaign in 1837, from South Africa.

41. W. E. Knowles Middleton, *Catalog of Meteorological Instruments in the Museum of History and Technology,* Smithsonian Studies in History and Technology #2 (Washington, 1969), p. 4, puts the transformation of meteorology "from a largely haphazard occupation of amateur observers, using simple instruments, into a more businesslike and systematic procedure pursued in permanent stations and using sophisticated instruments the results of which were automatically recorded or read at a distant point," as a phenomenon of the 1860's. Middleton seems to have overlooked the British scene earlier. See George B. Airy, *Description of the instruments and process used in the photographic self-registration of the magnetical and meteorological instruments at the Royal Observatory, Greenwich* (Greenwich, 1849; printed but not published); and George B. Airy, *Autobiography,* ed. Wilfred Airy (Cambridge, 1896), pp. 195, 203.

42. William Whewell, *History of the Inductive Sciences,* 3rd ed. (New York, 1890), II, 231; this section, first added in the 2nd edition of 1845 to Book II, Chapter II, is a useful short introduction to the enterprise.

43. Jacob Gruber, "Who was the *Beagle's* Naturalist?," *British Journal for the History of Science, 4* (1969), 266–282.

44. Oliver Warner, *A Portrait of Lord Nelson* (London, 1963), p. 28.

45. George W. Stocking, Jr., "What's in a name? The origins of the Royal Anthropological Institute (1837–71)," *Man, 6* (1971), 369–390.

46. Eric Roll, *History of Economic Thought* (London, 1961), pp. 206–211 (Malthus), 311–318 (Jones), 376 ff. (Jevons), 394 ff. (Marshall). For Whewell, see his "Mathematical exposition of some doctrines of political economy," and "Mathematical exposition of some of the leading doctrines in Mr. Ricardo's 'Principles of political economy and taxations,'" in *Trans, Camb. Phil. Soc., 3* (1830), 191–230, and *4* (1833), 155–198. For Babbage, see *On the Economy of Machinery and Manufacturers* (London, 1832), and his "Travelling Scribbling Book," pp. 17–87, 274 (Babbage Manuscripts, Science Museum, London).

47. Mill-Herschel Correspondence (John Herschel Papers, Royal Society).

48. W. Stanley Jevons, *The Theory of Political Economy* (London and New York, 1871), p. 141 (quotation), 141n (citation of Herschel); as contrasted to the simpler analogies in the preface, p. viii. Jevons' *Principles of Science* (London, 1874) was the philosophic work most directly descended from Herschel's *Preliminary Discourse on the Study of Natural Philosophy* of 1831.

49. J. W. Burrow, "Evolution and Anthropology in the 1860's: The Anthropological Society of London, 1863–71," *Victorian Studies, 7* (1963), 148.

50. On Lubbock (the father of Darwin's Lubbock), see as one example his letter to William Whewell, 25 June 1829 (Whewell Papers, Trinity College, Cambridge). Lubbock's papers at the Royal Society might well repay detailed study. On Galloway, see "Report of the Council—Obituary of Thomas Galloway," *Monthly Notices of the Royal Astronomical Society, 12* (1851–52), 87–89.

51. *Cf.* Quetelet to Herschel, 23 August 1834, and Herschel to Quetelet, 8 June 1837 (John Herschel Papers, Royal Society).

52. Charles Babbage, *Passages from the Life of a Philosopher* (London, 1864), pp. 474–476.

53. Thomas S. Ashton, *Economic and Social Investigations in Manchester, 1833–1933: A Centenary History of the Manchester Statistical Society* (London, 1934), p. 4.

54. John Lubbock to William Whewell, 25 June 1829 (Whewell Papers, Trinity College, Cambridge).

55. *Annals of the Royal Statistical Society* (London, 1934), pp. 4, 8–9; Isaac Todhunter, *William Whewell* (London, 1876), II, 161; *Lithographed signatures of the members of the British Association...with a report of the proceedings at the public meetings* (Cambridge, 1833), pp. 82, 90–92. I ave seen no valid evidence for the statement, in M. W. Flinn's introduction to Edwin Chadwick, *Report on the Sanitary Conditions of the Labouring Population of Great Britain*, 1842 (Edinburgh, reprint of 1965), that "this attempt to suppress the statistical study of social problems led to the breaking-away from the British Association of those interested in this aspect of statistical enquiry and to the creation, in 1834, of the Statistical Society of London." Flinn's citation is to a garbled account of "The Origin of the Royal Statistical Society" in B. Kirkman Gray, *Philanthropy and the State* (London, 1908), pp. 26–27. It is obvious that remarks by Sedgwick and by Whewell made at different times have been confused, and attributed to Whewell, who was *not* President of the Association.

56. *Annals of the Statistical Society*, pp. 10–13, 16.

57. Ashton, *Economic and Social Investigations*, pp. 2, 11, 31–33. For a much more detailed account, see now M. J. Cullen, *The Statistical Movement in Early Victorian Britain* (Hassocks, Sussex, 1975).

58. Mary Hesse, "Commentary", in *Scientific Change*, ed. A. C. Crombie (London, 1963), p. 476; Lancelot Hogben, *Science in Authority* (London, 1963), pp. 94–95.

59. Henry Jacobowitz, *Electronic Computers* (Garden City, New York, 1963), pp. 130–136.

60. Herschel to Airy, 22 December 1870 (John Herschel Papers, Royal Society).

61. Middleton, *Catalog of Meteorological Instruments*, p. 85.

62. Charles Babbage, *Reflections on the Decline of Science in England* (London, 1830), pp. 210–211.

63. Mrs. Edwards' *A Blue-Stocking*, quoted in Amy Cruse, *The Victorians and Their Reading* (Boston and New York, 1935), p. 361.

64. J. D. Bernal, in *Science and Ethics*, ed. C. H. Waddington (London, 1942), p. 114, as quoted in C. A. Coulson, *Science and Christian Belief* (Fontana Books, London, p. 25).

65. L. Pearce Williams, "The Physical Sciences in the First Half of the Nineteenth Century," *History of Science, 1* (1962), 7.

66. Samuel T. Coleridge, *Biographia Literaria*, ed. George Watson (London, 1960), ch. xi, p. 129; Charles Stuart Parker, ed., *Sir Robert Peel from his Private Papers* (London, 1899), III, 418; George O. Trevelyan, *Life and Letters of Lord Macaulay*, 2nd ed. (London, 1901), p. 649.

67. Fawn M. Brodie, *The Devil Drives: A Life of Sir Richard Burton* (New York, 1969), pp. 239, 292, 294.

68. Nora Barlow, "Robert Fitzroy and Charles Darwin," *Cornhill Magazine*, h. s. *72* (1932), 505, 497; Fitzroy-Herschel Correspondence (John Herschel Papers, Royal Society).

69. Edward Belcher, *Narrative of a Voyage Round the World, Performed in Her Majesty's Ship Sulphur, during the Years 1836–42, including details of the Naval Operations in China, from Dec. 1840, to Nov. 1841,* 2 vols. (London, 1843).

70. Christopher J. Bartlett, *Great Britain and Sea Power, 1815–1853* (Oxford, 1963), p. 153.

71. Charles Louis André and G. Roget, *L'Astronomie Pratique et Les Observatoires en Europe et an Amérique depuis le milieu du XVIIᵉ siecle.* Deuxième Partie, *Écosse, Irlande et Colonies Anglaises* (Paris, 1874), pp. 131–173; esp. 132–134, 146–147, 156–159. For the beginnings of scientific activity, see *"Madras.* East India Co.'s Observatory," Miscellaneous Manuscript Documents (Royal Astronomical Society Library).

72. The letters I have seen are in the John Herschel and Edward Sabine Papers at the Royal Society.

73. Sabine to Herschel, 3 December 1838, 1 April 1839, 4 September 1839 (John Herschel Papers, Royal Society Library).

74. William O. Aydelotte, "Parties and Issues in Early Victorian Politics," *Journal of British Studies, 5* (1966), 95–114, the quotation, 110. Problems of ideological terminology are well shown in Gordon B. Beadle, "George Orwell and the Victorian Radical Tradition," *Albion, 7* (1975), 287–299.

75. *Cf.* A. A. W. Ramsay, *Sir Robert Peel* (London, 1928), pp. 367–369; Norman Gash, *Mr. Secretary Peel* (Cambridge, Mass., 1961), pp. 2–4, 38–39, 47, 55–58, 281–282; Charles Stuart Parker, ed., *Sir Robert Peel from his Private Papers* (London, 1899), I, 16–18 (schooldays), 307–309 (pensions for George Airy and Mary Somerville); 363–365 (Davy's proposal for a natural history establishment); 387 (establishment of the Royal Medals); III, 417–418 (Buckland as Dean of Westminster); 422–424 (Whewell as Master of Trinity; Peel notes "his good temper and easy conciliatory manners"!); 433 ("two of the best portraits I have" are those of Cuvier and Owen); 441 (the £400 available for literary and scientific pensions in 1845 goes half to James David Forbes, half to young Alfred Tennyson); 444–447; 492.

76. W. Denison to George Airy, 5 May 1837, in "Establishment of Magnetic Observatories" (Airy Papers, Royal Greenwich Observatory, Herstmonceux).

77. R. P. Laidlaw, review of "Finding Out About the Victorian Period," conference at Victorian Studies Centre, University of Leicester, July 1971, in *Victorian Studies, 15* (1972), 378–379.

78. In the preface to his *Origins of Modern Science* (London, 1949).

79. Arthur Hughes, "Science in English Encyclopaedias, Part I," *Annals of Science, 7* (1951), 350.

80. Walter F. Cannon, "'Parliament and the Metric System'—A Further Comment," *Isis, 58* (1967), 235. These matters are rather garbled in a recent biography, Maboth Moseley, *Irascible Genius; a Life of Charles Babbage* (London, 1964).

81. *Cf.* "Extracts from Treasury Minute dated 28 April 1829" and "Mr. Stewart to Secretary, Royal Society, 24 December 1831." (Domestic Manuscripts, Vol. IV, Royal Society).

82. W. G. Farquharson, "International Hydrography 1835," *International Hydrographic Review, 30* (1953), 126.

Chapter Nine

The Truth-Complex and Its Demise

Read Nature; Nature is a friend to truth;
Nature is Christian; preaches to mankind;
And bids dead matter aid us in our creed.

—Young, *Night Thoughts*
Night IV, lines 703–705

It has been suggested that the history of science can help us and our children bridge the gap between the sciences and the humanities. It was the patron saint of the History of Science Society, the late George Sarton, who conceived of his mission as being that of apostle to the one-sided. None of his disciples can quite abandon the dream of what Sarton called "the new humanism," enshrining scientist, artist, and saint as equal heroes in a pantheon of creativity. I must admit, however, it is sometimes difficult to maintain this position in the flesh. Such an assignment may be bewildering to a young man who, in good Ph.D. fashion, is lucky if he is investigating so modern and broad a subject as, say, "The influence of 18th-century German theories of combustion on the ideas of Lavoisier's circle prior to 1775."

The historian of science, that is, is not really equipped to solve everybody's problems. What I shall do instead is what the historian usually does: suggest some ideas as to how we got where we are. This procedure will, as usual, suggest that "where we are" is a more complex place than is always admitted. The only general answer I know to the question, "Why study history?" is that, if trained professionals do not provide an accurate picture of the past, then the terrible simplifiers will move in and provide distorted accounts to be used to support their own schemes and propagandas.

The historian of science, however, whatever the scope of his Ph.D. thesis, is interested in science as a whole, with all of its internal relationships and all its external inter-relationships, with its people as well as its hypotheses, with its irrationalities as well as its proper monographs. The history of a science, a description which moves from discovery to discovery in one field in impersonal chronological fashion, is not what we mean by the history of *science.*

The English novelist and bureaucrat Sir Charles P. Snow, in his lecture of 1959 called "The Two Cultures and the Scientific Revolution," insisted on the differences between scientists and humanists. Sir Charles believed that these differences add up to two completely different cultures, which no longer can or want to understand each other. This mutual incomprehensibility is not based merely on technical difficulties in language. It would not be simply solved, for example, if all humanists learned calculus. It is rather that literature and the arts (except music) no longer seem relevant to scientists, not even in their personal concerns and certainly not in their professional ones. And the humanists, on their side, pretend that their traditional culture is all of culture, ignoring the scientific edifice, which is the "most beautiful and wonderful collective work of the mind of man." Coupled with this mutual lack of appreciation are the naive and stereotyped images that each group has of the other. In his novels, Sir Charles made it his part, in healing this breach, to humanize the image of the scientist, to show that he is not the insensitive brassy optimist that humanists think he is. Sir Charles therefore portrayed the scientist as one who shares to the full such typically human activities as: infatuation, jealousy, ambition, power politics, cheating, sibling rivalry, cruelty, divorce, and good straightforward treason. Sir Charles was indeed trying to show the scientist as human—human, that is, according to the theological doctrine of original sin.

That some such felt opposition between science and humanism exists is fairly obvious. Here is a statement so common that you have heard its equivalent before. An American businessman is speaking of education:

> Modern automatic equipment creates new problems. The scientist can devise the equipment, but only a man with broad understanding can put it effectively into use. So, from a purely practical standpoint, it seems to me that a course in the Humanities is the best one to serve as a foundation.

It will perhaps comfort the scientists among my readers, who thus allegedly do not have broad understanding, to remind them that the businessman is by no means implying that the professors of English or history have broad understanding. The miracle works only with the undergraduates.

C. P. Snow himself had a rather different view of the humanities. In his lecture there is an amazing passage. He reported that a scientist once asked him if the work of 20th-century writers, the Irish poet Yeats for example, did not lead to political wickedness, indeed to the gas chambers of Auschwitz. Sir Charles commented:

I thought at the time, and I still think, that the correct answer was not to defend the indefensible.... It was no use denying the facts, which are broadly true.... Literature changes more slowly than science...and so its misguided periods are longer. But it is ill-considered of scientists to judge writers on the evidence of the period 1914–50.[1]

When someone who himself writes novels calls the period 1914–50 a "misguided" period in literature, then clearly non-communication of the aims of artists has gone very far indeed. Like a number of ordinary folk who find "modern art" emotionally upsetting, a challenge to the stability of their universe, Sir Charles refused to accept the 20th century at any price. For this reason we may expect to find that his concept of "two cultures" was appropriate to some Victorian situation (that of Oxford around 1880, actually) but that our own situation is far more complex, as indeed those "misguided" poets and novelists of the period 1914–50 have been saying since 1914. I suggest that we live not with two cultures but in an n-dimensional clutter, where n is considerably greater than 2. I also suggest *when* this fragmentation began, when science and the other modes of truth turned each to go its own way. For English-speaking countries at least, the process began on November 24, 1859, with the publication of Charles Darwin's *On the Origin of Species by Means of Natural Selection.* Not merely did the book annoy some Christians; the reaction to it by scientists, poets, and other intellectuals shattered the meaning of truth in our culture. Whether natural science has ever regained its own unity is doubtful. Certainly Truth in general has not.

Historians of science always like to talk about Newton, and clearly if a system of truth was shattered in the time of Darwin, we would like to know what that system was. But please note that in what follows I am not discussing "Newtonianism" narrowly conceived, not a system of dynamics, not even truth in natural science—just plain truth. Furthermore, I am not talking about what Isaac Newton did, thought, or wrote; I am not talking about what may actually have happened in the 17th century, but rather what people in the 19th century thought had happened. I want to characterize Truth as it appeared to grown men when Darwin was a young man.

My case is based on events in Britain. Since the story opens with Robert Boyle and Isaac Newton, and ends with Charles Darwin and Lord Kelvin, this is appropriate. That things did not progress so smoothly in other countries only demonstrated, felt the English, the superiority of English civilization. The English Enlightenment in the 18th century proceeded with no such wrangles as disfigured the Enlightenment in France. Gerd Buchdahl has indicated how the French Enlightenment tended toward the breaking up of truth into mutually exclusive ideologies of necessitarianism and experimentalism and evolutionism.[2] It is little wonder that Anglican writers at the end of the 18th century were conscious not merely of a moral but even more of an intellectual superiority over the partial systems being produced on the continent. William Paley does not so much argue with these intellectual heretics as chide them for being unruly pupils who will not read the textbook.

Recent historians, such as Peter Gay in *The Enlightenment: The Rise of Modern Paganism,* have interpreted the Enlightenment by taking the tumultuous French scene rather than the serene English one as normal. No Voltaire hurling the volumes of Newton and Locke into the face of clerical authority needed to occur in England, because the Church of England was as fond of Newton and Locke as were the French *philosophes.* Indeed Newton became a high point of education at one citadel of the church, Cambridge, where even the classics withered except in a few colleges; and Locke shared the honors with Aristotle and the classics at the other citadel, Oxford.

The Truth-Complex, then, with which 18th-century divines could as easily rebuff Voltaire as deflect the scepticism of David Hume, which could absorb the revelations of German historical studies as well as the anti-Mosaic revelations of geology, which could motivate the art criticism of John Ruskin and lead romantic Wordsworth back into the paths of righteousness, which Charles Dickens could accept as naturally as he hated Evangelical preachers: this Truth-Complex, built on the corpuscular propaganda of Robert Boyle and the philosophic triumphs of Isaac Newton, is of some interest. As simplified and refined for over a century, it was the publicly available basis for debate. It contained four major sets of ideas, and it specifically excluded a fifth. The world can be understood by man in the early 19th century on the basis of the following four elements or elemental clusters:

1. Infinite space and time. The space was that of Euclid, though it did not need to be. The time was like space: it went to infinity in a continuous manner, and it went backward as easily as forward. If there should be evidence that time did not go back indefinitely, this would be interesting indeed, since time in itself has no mechanism for stopping.

2. Corpuscles, which are units of matter. If you call them atoms, be careful not to confuse them with the atoms of the Greeks, or with later electrical atoms. Corpuscles carry mass and impenetrability but not innate gravity or electricity: they are movable, but they do not possess motion of their own. Especially, they are "stupid"; indeed, they are "dead." They can take no notice of anything outside of themselves, be it force, other corpuscles, or laws. They can only be acted upon. They have no creative power. They cannot initiate organization or life.

Already we see why Greek atomic theory is incorrect. It is most improbable that a random meeting of dead corpuscles in the infinity of space would have happened in any given time so as to lead to our highly organized world. No one would believe in this "childish" doctrine, as Cicero had called it, unless he were trying to smuggle preconceived ideas (notably that of a spontaneous "creation") into our pure science. Lucretius was a fine poet and propagandist, but his notion of what constitutes a proper scientific explanation was juvenile, a Truth-Complex believer might feel.

3. The world contains regularities, which we can distinguish, and often express mathematically. Mathematics is a sign of truth; it is one of God's languages. Note that we call these regularities "laws," but we really shouldn't. We only observe what collections of corpuscles do. The so-called "laws" are not like human laws; they don't prescribe what the corpuscles *ought* to do. Hume refutes deists, who believe that God made the world and then left it to run on its own, since there is nothing in it to make it run on its own. Here is how Robert Boyle expressed the point:

> Nor will the Force of all that has been said for God's special Providence be eluded, by saying with some Deists, That after the first Formation of the Universe, all Things were brought to pass by the settled Laws of Nature.... For I look upon a Law as a Moral, not Physical Cause, as being indeed but a notional Thing, according to which an intelligent and free Agent is bound to regulate its Actions. But inanimate Bodies are utterly incapable of understanding what it is, or what it enjoins, or when they act conformably or unconformably to it: Therefore the Actions of inanimate Bodies, which cannot incite or moderate their own Actions, are produced by real Power, not by Laws.

John Ray put it even more forcefully: "And as for any external Laws or establish'd Rules of Motion, the stupid Matter is not capable of observing or taking any Notice of them...neither can those Laws execute themselves."[3]

We must not, that is, think that laws are causes, have power over corpuscles, or constitute a self-acting system. Hume, properly considered, is on our side. Newton has discovered the law of gravity, but he insists that he does not know the cause of gravity. Really, we ought to get rid of all examples of the pathetic fallacy. It is silly to speak of mass as "resisting" a change of motion. "Is a cup of tea to be accused of having an innate power of resisting the sweetening influence of sugar, because it persistently refuses to turn sweet unless the sugar is actually put into it?" asked James Clerk Maxwell.[4] The mass is accelerated just in proportion to the impressed force; the tea is sweetened just in proportion to the injected sugar.

The idea of law as a self-acting system presumably derives from Rome, perhaps from Stoics. At any rate, the Truth-Complex makes the idea nonsense. For it, the idea is a medieval superstition; a law implies a law-maker and a law-enforcer.

4. Our Truth Complex must contain God. In a universe where laws are only regularities, and corpuscles immersed in infinite space have no organizing power, God alone can explain organization. This is true even at the inanimate level: Sir Isaac himself has shown that the regularities of the solar system are highly unlikely on a chance basis. And if we go to the animate level, the naturalist is confronted by nothing so much as a vast collection of skillful adaptations—adaptations of mechanism to need, of physiology to environment, of one organism to another. These adaptations are enormously more complex, subtle, ingenious, and interrelated than any human being could conceive *a priori*. It is simply nonsense to believe that this organization has happened by the random clashing of dead corpuscles; it was planned by a mind infinitely greater than a human mind. Further, this organization is good. No organ of any creature is designed to give meaningless pain to that creature.

Mechanism, then, is not a sign of a dead universe but of a living power. A machine is a collection of dead corpuscles put together for a purpose. The universe as a whole is probably not a machine, but there are many purposeful machines in it which no human or sub-human intellect planned. Indeed, humans are themselves collections of these mechanisms, most of which we cannot duplicate even with ourselves as models available for us to copy.

So the only rational belief is that a good God has been the great organizer of the world, which left to itself would be a corpuscular chaos. Any other belief is not so much irreligious; it is recognizably unscientific. Indeed, in the 19th-century physicists find it desirable to use this insight to understand one of their more interesting laws. According to a usual explanation of the Second Law of Thermodynamics, the world left to itself *is now returning* to its natural corpuscular chaos, except in inexplicable foci of increasing organization such as living beings.

There were other contenders for elemental status. For example, several devout Newtonians insisted that force *is;* but the fact that they needed to insist shows that their position was not fully accepted. It is better to say that force belongs to God.

In these four elements, then—infinite space and time, corpuscular matter, empirical laws expressed mathematically, and God as a beneficient organizer—we have the basic elements of our Truth-Complex. There is here no conflict between science and religion. Nature, like the Bible, is an exposition by God:

> There is a book, who runs may read,
>   Which heavenly truth imparts,
> And all the lore its scholars need,
>   Pure eyes and Christian hearts.

The works of God above, below,
   Within us and around,
Are pages in that book, to show
   How God Himself is found.[5]

In 1786 the Quaker boy Thomas Young, leaving school at age 13, devoted himself "to the study of Hebrew and to the practice of turning and telescope making."[6] The tools for studying the Bible and the stars would open up the twin roads to God.

A period in which such a set of ideas dominates is a "religious" period, regardless of specific institutional reforms in society. Those historians who see church reforms in Britain in the first half of the 19th century as "secularization" are, in my opinion, merely imposing Tractarian opinions on history. Most such reforms were a form of "protestantizing" British Christianity, especially in increasing the separation of Church and State. To be sure, some Tractarians and some of their Anglo-Catholic descendants considered protestantism to be a form of secularism; but this position is not very useful for historical analysis.

In the Truth-Complex we have been discussing, there is no conflict between science and poetry. To read poetry giving accurate descriptions of the natural world is to commune with God, for everything in nature is, not by analogy but literally, God's message. There is no conflict between science and ethics; for both agree that God organized the natural world to be good.

We can now see how misleading was the late Alexander Koyré's characterization of the 17th-century revolution in science. He spoke of "the discarding by scientific thought of all considerations based upon value-concepts, such as perfection, harmony, meaning and aim, and finally the utter devalorization of being, the divorce of the world of value and the world of facts."[7] Perhaps this is what *ought* to have happened as a result of 17th-century dynamics. But British scientists in the 18th and 19th centuries by and large did not see any need for divorce. Even for inanimate nature it was clear that dynamics alone could not explain organization. And in natural history the concepts of harmony, meaning, and aim were fruitful working concepts for the scientist.

Our Truth-Complex, as I said, specifically excludes one element. One may read about nature, but one must not believe in her. The idea of Nature as a real entity is a pre-Christian and, what is more, a pre-scientific belief. One should really not even speak of the "laws of Nature;" they are God's laws for the world, or they are theories in science. One must not see in the world a self-organizing principle, or in dead corpuscles a plastic power or inherent virtue. To do this is to be a confused thinker. One must not postulate a quite unobservable formative power in organic matter, a

269

power of which real scientists can find no trace. All these are ways of smuggling back in the pagan idea of fecund Mother Nature. Nor should one really (though it may be harmless) use the old image from the 13th century, the image of the universe as a clock, with God as clockmaker. It may be misunderstood as endowing matter with at least a secondary level of self-action which our empirical Protestant brains cannot discover. It is proper to compare a particular organic structure to a watch or clock, in order to emphasize the concept of purposeful design. But the clock analogy for the whole universe is properly used only to emphasize that the inner mechanism God uses to run the world is *hidden* from us. If taken too seriously by some devotee of Cartesian or other incorrect philosophy, it is proper for a good Newtonian (like Samuel Clarke) to throw away the clock analogy altogether.[8] The great public clocks of the 14th–16th centuries, after all, as described by Lynn White in *Medieval Technology and Social Change*, not only were scientific instruments but also, with their whirling saints, cocks that flapped their wings, etc., were full of pageantry, fun, and surprises. God as scientist with a bit of buffoonery and magic thrown in: this may have suited the 14th-century Catholic imagination, but it could only play a carefully limited role in 18th–century Anglican natural theology.

There are people who commit these fallacies, of course. From the 17th through the 19th century this scientific heresy, this superstition of "Nature" keeps popping up, under different names and based on different overt philosophies; and each time it is put firmly in its place by the "proper" scientists. It is denounced as Platonism, as pantheism, as mysticism, as Romanticism, as idealism, as vitalism. In the 17th century the great naturalist John Ray believed in the idea of a "plastick Nature or Vital principle" in nature, but he could not make it stick in opposition to the Boyle-Newton corpuscular school. In the 18th century British divines often took the great pantheist Spinoza as their straw man; they didn't have to refute his philosophy, you see, they merely showed that it was not good science. The spread of Romantic tendencies was not a serious threat to the Truth-Complex, for Romantic attacks on Newton were mostly on Newton as presented by speculative French Enlightenment materialists; otherwise, Romanticism stressed the diversity of God's goodness to all his creatures. In the first decade of the 19th century the expert anatomist Lamarck proposed that his scheme of evolution worked by virtue of a property of organic matter to expand and complicate itself. Biologists observed rather tartly that they could find no evidence of any such inherent virtue in matter.

We can say that one of the effects of Darwin's "scientific" biology was to expose a whole group of biological notions—Lamarck's inherent tendency to expand, Richard Owen's archetypal forms, Louis Agassiz's Platonic ideas—for what they were: scientific heresies similar in kind to all of the other attempts to evade the "true" scientific system of Boyle and Newton. All of these escape mechanisms—and poets made attempts also—reduce to the desire to reinstate a "spirit of Nature" as an active agent in the world. This spirit, however derivative at first, might become a Goddess in competition with the Christian God. The constant reappearance of such heretics only helped to cement the alliance of natural scientists and theologians, both of whom knew how to distinguish scientific "truth" from vain philosophizing.

Nevertheless the Goddess Nature has proved to be Jehovah's most irrepressible competitor (one can respect, but hardly worship, Satan in our culture). The Truth-Complex kept her in check, but with its shattering she was free to roam abroad once more.

Since God is one, Truth is one, and God has given us two books of Revelation: the Bible, and the natural world. These two cannot be in conflict. If they seem to be, it is only because we have misinterpreted one or the other. In practice, it was the interpretation of the Bible that gave in. Here I think we should emphasize that behind Boyle and Newton stood an equally impressive culture hero: the figure of Galileo. "The persecution of Galileo in the name of religion must never be repeated," insisted speaker after speaker, preacher after preacher—with the fairly obvious implication, "Of course, if Galileo had lived in Protestant England, it would never have happened in the first place." Thus, generation after generation, the advanced school of Biblical interpretation found itself outmoded by the next generation, and found itself branded as "fundamentalist"—so its once-daring position seemed to the new generation. Now this verse, now this chapter, finally the whole of the Mosaic record was adjusted to satisfy science, not without pious objections, but still inexorably.

For example, when in the 1760's the founder of Methodism, John Wesley, proposed a meteorological explanation for Moses' flood, he was not being fundamentalist but progressive. Wesley's general position is illustrated by his translation of the New Testament. He took advantage of progress in textual criticism to make about 12,000 changes in the King James version so as to have a thoroughly modern translation. He also softened the Calvinist tone at some points: "atonement," for example, became "reconciliation." Having no scientific or historical reason to believe that the Biblical Flood had not actually happened, Wesley (like the astronomer Edmond Halley earlier) did not want to leave it as an incomprehensible physical miracle. He wanted to bring it within the pale of scientific understanding. God caused the Flood, but not by resorting to extra-physical agents. Wesley was, indeed, scolded for this intellectual venture by some of his more "fundamentalist" converts.[9]

Another important development in the Truth-Complex was the erection of a hierarchy of truth within natural science, so that the lower science could be adjudged as less worthy of attention than the higher in case of apparent conflict. I am not talking about any schemes set down by philosophers, but of the way scientists and intellectuals looked at the various disciplines. By the 19th century, among the major criteria of high rank were: the degree of use of mathematics in the science, both for deduction and measurement; the small number of assumptions necessary; and the degree of comprehensiveness of the science, so that many diverse phenomena were shown to be united under general laws, especially if the laws linked an otherwise unavailable part of the cosmos to our own level. On all of these counts, there was only one candidate for Queen of Truth. As we might expect, it was Newton's science. As one Victorian chemist put it, for the other natural sciences "the summit of their ambitions, and the ultimate aim of the efforts of their votaries, is to obtain their recognition as the worthy sisters of the noblest of these sciences—physical astronomy."[10]

Second place in England, however, varied in the period we are considering. A science that was barely in existence in Newton's time gradually worked itself into takeoff by the beginning of the 19th century and then shot up and up, past the biology of Linnaeus and the chemistry of Lavoisier. By about 1840, geology, with its associated paleontology, was probably in second place, and even in 1859 many educated persons would still have placed it there. Geology was certainly not as mathematical as one might wish, but in linking a vast amount of data concerning the living, the inorganic, and the extinct worlds, it was unrivalled. It is not inconsequential for the rest of our story that Charles Darwin started his scientific career, and first won professional status, as a geologist and not as a biologist.

Astronomy and geology, then, were Tennyson's overpowering "two shapes...taller than all the Muses."[11] By 1859, however, geology was deadlocked at its highest theoretical levels, and had been for too many years. An offshoot of 17th-century experimental philosophy, after developing sporadically in the 18th century, blossomed in France following the Revolution. In Ampère, Fresnel, Arago, Biot, Carnot, Fourier, and the like, physics finally found itself. It soon won eager converts in Britain and Germany, leaped into the vast generalizations of thermodynamics, and by 1859 was on the verge of attempting that great triple play (Faraday to Kelvin to Clerk Maxwell), the unification of optics, electromagnetism, and dynamics. Even with Maxwell's greatest decade yet to come, physics was working at a major transformation in the Truth-Complex. For the Victorian physicist, force was so real and tangible a thing that it was being transformed into an entity, a subtle fluid called "energy" with a reality status equal to that of matter. With its chief functions transferred to energy, force became toward the end of the century a pale ghost whose very existence was debated by the metaphysically-minded.[12] The alert scientific observer in 1859 might have awarded second place to physics, already, perhaps, displacing chemistry as the science of fundamental structural elements of reality.

This 19th-century science of "physics" was hardly an extension of Newton's work, as is sometimes claimed by modern physicists and philosophers. Most of it was new-minted since 1800; a principal tool was the analysis we associate with Lagrange. But it was a part of the Truth-Complex, and hence was assumed to be eventually compatible with Newton's part of the complex. In that sense physics was "Newtonian."

We may note that the conservation of matter had left no room for Jehovah to perform creation miracles. The bold new doctrine of the conservation of energy left no room for Dame Nature to perform vitalistic miracles; it did not much hinder Jehovah, since it was his energy that was being conserved, and it had long ago been established that he was a careful craftsman.

Chemistry was of widespread importance. For the amateur it was easy, it was fun. It had social prestige: the Royal Institution under Davy had sold the "great world" of London society on chemistry. The student of industry could see its value and lament that English universities were not turning out chemical technicians the way the German universities were doing. Most of the attacks in the second half of the century on Cambridge and Oxford as being scientifically backward, most of the demands for greater sponsorship of scientific education by the State, most of the praise of German modernity in this and that, relied on a simple train of thought which could appeal even to politicians. The best science is experimental science. Chemistry is the best experimental science. Therefore chemistry is the model science; and if Cambridge, say, is backward in chemistry it is backward in science.

Such a position, however, came into conflict with our Truth-Complex. After Faraday turned to electricity, chemistry was deficient in striking theories, in stable truths. And Victorian scientists fought hard against the notion that science is important in proportion to its industrial usefulness. Chemistry appealed to Prince Albert and Sir Robert Peel; but its prestige in the scientists' hierarchy of truth was not so great after mid-century.

Now let us consider sciences at the bottom of the hierarchy. These are the sciences of classification, such as botany. Classification, you see, is logical, not mathematical; Aristotelian, not Newtonian. The botanist is one who collects specimens to swell his herbarium, gives them barbarous names, and tries to arrange them in a system which some other botanist will then dispute. The unit of the science is the species, and the naturalist not only doesn't know what a species is; he can't even agree with other naturalists as to whether a specimen *is* a member of a given species or not. We remember Wordsworth's little tribute:

> Philosopher! a fingering slave,
> One that would peep and botanize
> Upon his mother's grave.[13]

It is true that some academics, such as William Whewell at Cambridge, thought it was good for everyone to study ideas of classification as part of a liberal education; but Whewell's suggestion only showed his liberal contemporaries that he was, indeed, a conservative schoolman.

Biology as a whole is not in much better shape than botany, mostly because biology is not a whole. Even the name is a new and perhaps unnecessary word. The naturalist is collector, explorer, classifier—is there a general science of natural history? Above the naturalist ranks the anatomist, to a considerable extent because of his importance in paleontology. And still unsettled, though very promising, is the status of the physiologist, especially the physiologist who has been seduced by thermodynamics and feels that the physiology of the brain is on the verge of becoming part of physics.

Who is the over-all theorist? It is not clear.

Now I am not going to try to convince you that in England in 1859 everyone really believed in the Truth-Complex. No phase of Western culture has been that conformist, and the early Victorian period was full of rebels of all kinds. It was, nevertheless, rather difficult to find acceptable words in which to express one's disbelief—except scientific words. Let me give you just one highly selected example.

From his Protestant Fundamentalist childhood through his Anglican early manhood to his Roman Catholic maturity, John Henry Cardinal Newman distrusted natural science and what it stood for. For the mature Newman, a scientist who tried to determine the evidence for Moses' flood was merely impertinent. Moses' flood being described in Scripture, the scientist or historian should simply accept it, and use its existence as a fact in his subsequent work. Yet in the very same book in which Newman stated this position, let us listen to him recommend, to a mid-19th-century Irish Catholic audience, the importance of theology: "Religious doctrine is knowledge, in as full a sense as Newton's doctrine is knowledge.... Theology has at least as good a right to claim a place [in a University] as Astronomy." This is true, says Newman, if one believes that religious facts are true "in the sense in which the general facts and the law of the fall of a stone to the earth is true."[14] Thus Newman, who does *not* believe in our Truth-Complex, cannot praise his own truth except by saying it is at least as true as Newton's truth. He even defines truth by the example that the expression $s = \frac{1}{2} gt^2$ is a true statement. There is no convincing way, in 1859, to say, "My truth is truer than... Truth."

Before 1859, then, God so loved the world that he made it good, and gave man a mind with which to investigate and display God's goodness in the form known as knowledge, scientific knowledge, even technical knowledge. In those far-off days it was not a polite fiction, it was God's own truth to see the railroad locomotive as an example of love. Some literary historians have expressed doubts that Victorian high culture ever admitted to an aesthetic based on the qualities of machines; but one need not base a cultural analysis on literary "high" sources alone. It is not necessary for a master builder to have written a book on the qualities of his "noble machine;" it is enough for him to have adorned it.[15] Standish Meacham has suggested that the Crystal Palace exhibition of 1851 spread "a general notion that God might have had a hand in the creation of machines." I agree, with the addition that the notion was orthodox much earlier.[16]

In our own time it might be sensible to talk of God's gift to the world through science and technology if Darwin had replaced the Truth-Complex with a complex of his own, or had shown the need for a modernization of the older complex, a modernization which was subsequently carried out by Einstein, or Fisher, or somebody. But Darwin merely shattered the Truth-Complex; and we are still playing with the pieces.

That this *is* what Darwin did was clear enough at the time. Let us listen to one of Darwin's old teachers, an excellent scientist in his own right, denounce his student:

> There is a moral or metaphysical part of nature as well as a physical. A man who denies this is deep in the mire of folly. 'Tis the crown and glory of organic science that it *does*... link material and moral; and yet *does not* allow us to mingle them in our first conception of laws.... You have ignored this link; and, if I do not mistake your meaning, you have done your best in one or two pregnant cases to break it. Were it possible (which, thank God, it is not) to break it, humanity, in my mind, would suffer a damage that might brutalize it....[17]

Now the author of this letter, Adam Sedgwick, has violated all "scientific etiquette"; Darwin bitterly resented the letter. Yet Sedgwick was correct except in one respect. Darwin may not have "done his best" to break the link between material and moral, but he certainly succeeded. The most obvious, and most immediate, and most important result of the *Origin of Species* was to effect a separation between truth in moral science and truth in natural science. It would be interesting to survey theologians, philosophers, etc., as to this point, but it is unnecessary. In combating all the little Social Darwinisms that sprang up after 1859, Thomas Huxley has given us eloquent statements on the subject. Between Darwinian nature, and man's hopes and values, the relation is that of flat contradiction. Man, according to Huxley, has for a little time the power to combat the cosmic processes which science reveals, and he should do it with all his heart, all his soul, all his mind, and for his neighbor as for himself.

And (although Huxley did not admit this) if the truths which science reveals are not moral, not lovable, it is not clear why their pursuit is moral or lovable. Indeed we know that the scientific quest for truth is not moral, unless it is undertaken morally. If you doubt this in spite of *Dr. Faustus,* consider the doctors who performed experiments on human victims in Hitler's concentration camps. So, it would seem, morality is an independent concept added to science from outside. But from where outside?

Now if you consider how much of what humanists study *is* concerned with moral science—that is, with human behavior—you see that Darwin drove out of the Truth-Complex not merely the religious thinkers, most of whom will now not rely on natural theology at any price, but also everyone else who wants to or has to make any commitment on human conduct. The social scientist, anthropologist or economist, for example, no longer can believe that "natural" behavior is a norm. Darwin has shown that, whatever it is, "nature" isn't any good. Indeed, the idea of "natural law" is so weak now, outside of Catholicism, that modern students have a hard time realizing that it ever really was believed by intelligent persons. The modern undergraduate simply recognizes that older peoples were naive.

Poets have been particularly affected. The poet has seldom discussed scientific ideas in detail for the very good reason, as William Wordsworth pointed out, that to do so would shrink his audience to scientists.[18] But the poet before 1859 knew what truth was, and where he stood with respect to it. He might attack received notions of truth, as Wordsworth sometimes did, but at least he knew what he was doing. The modern poet has no such possibility. To deny a scientific truth would be irrelevant, since science and human conduct have no relation. Nature poetry is meaningless, for nature is morally meaningless. Yet the poet is necessarily so connected with a system of truth that some of the best poets—Yeats and Eliot, for example—have found it more desirable to make up a private system of their own rather than to go without. Nature has been "moralized" by individual preference rather than by the culture.

Darwin's natural selection, then, was felt to be an unacceptable description of God's dealings with the world, or man's relation to the world, since natural selection is not intelligent, not good, and not even necessarily progressive. So what people did, after it was clear that Darwin in some sense was here to stay, was to try to hush up the scandal, repudiate the perceptive men who had cried out in alarm in 1859. But all they accomplished was to make matters worse.

Their tactic was very simple: to accept evolution in such terms of praise that no one would quite notice the aside in which one rejected natural selection. "What Darwin did," said many, "was to prove the theory of evolution and propose the hypothesis of natural selection to account for it, a very interesting hypothesis, if true (which I must admit seems to me rather unlikely)." In one of the really useful monographs published in the history of science, Alvar Ellegård has shown that this particular phraseology, "Darwin proved the *theory* of evolution and proposed the *hypothesis* of natural selection to account for the theory," is the phrase that Darwin's opponents assiduously promoted.[19]

Actually, the phrase is nonsense. If anything, Darwin proved that natural selection must operate in the world in an unsettling manner. Various previous "discoverers" of natural selection, which was an old idea since many classically-educated Englishmen of the 18th and 19th centuries had read it in Lucretius, had contended that natural selection would *preserve* the species by weeding out deviations from the central type. Others had suggested that natural selection would produce the variations within a species known as varieties or races. It is necessary to read authors on these points with a great deal of care, since each had his own idea of what he meant by species, what he meant by variety, and what he thought the difference was.

For example, according to Dr. W. C. Wells in the early 19th century, the dark skin of Negroes may have been produced by natural selection operating on ordinary men living in tropical sunlight. But Wells was talking about variety change, not species change. We can imagine his thought-processes as follows: the dark skin of Negroes is produced by natural selection. But natural selection produces only variation, not species change. Therefore Negroes are only a separate variety, not a separate species. Therefore the more extreme pro-slavery propagandists (Wells was from the American South) are wrong.[20]

Another position was that expressed by the ethnologist James Cowles Prichard. He admitted that the possibility of grouping similar species together, as in a genus, could lead to the opinion that one original structure had been modified into distinct species. But, he said, "we are unacquainted with any physical causes capable of producing such differences of structure.... This kind of speculation loses a great part of its probability when pursued to the extent to which those naturalists carry it, who maintain generally the transmutation of genera and species...." We are led to believe that the diversity of species dates from the creation of the genus, and is not caused by the subsequent influence of external agents, "by all that can be ascertained respecting the limits of species, and the extent of variation under the influence of causes at present existing and operating." Prichard was thus in 1826 well aware of an idea of species formation, but believed the scientific evidence was all against it.[21]

This position was developed by Charles Lyell in his *Principles of Geology* of 1830–33, and was thus one of those with which Charles Darwin had to contend. When he became personally convinced of the truth of descent with modification, Darwin was well aware that he must find some mechanism for species creation. Otherwise he would not have a theory of his own but would be just another supporter of a well-known and well-rejected speculation.[22]

It was Darwin, then, who showed that natural selection not only is a conservative process but also in some cases may be a process creative enough to produce new species. Then he showed how much order this idea would bring into the contemporary chaos or stalemates of biology, geology, and paleontology. Evolution appears as the result of Darwin's clearing-up; it does not furnish him with a plan or principle around which to organize his argument in his book. The basic new idea in Darwin's scheme is that natural selection will produce new species; that, not evolution, is after all the title of his book: *On the Origin of Species by Means of Natural Selection; or, the Preservation of Favoured Races in the Struggle for Life.* If that idea is not correct, then evolution was no more justifiable a generalization in 1859 than it had been for the previous century. We should therefore replace the sentence, "Darwin proved the theory of evolution and proposed the hypothesis of natural selection," with some such sentence as this: "Darwin corrected the theory of natural selection and with it made the inductive generalization, evolution, plausible for the first time."[23]

It was at this critical point—not at the idea of evolution, not at the existence of natural selection, but at the assertion that natural selection is creative enough to have produced new species—that Darwin's most dangerous scientific opponents struck. In doing so, they threw science herself into convulsion, and destroyed the hierarchy of truth. So it is not quite accurate to say that the moralists abandoned the Truth-Complex, for that complex had been shattered by the 1880's.

At first it seemed that the leading biologists of the period were Darwin's chief opponents, men such as Richard Owen and Louis Agassiz. But Darwin never feared them, and their attack withered. As I have mentioned, Darwin's "scientific" science (that is, scientific in the Boyle-Newton tradition) exposed the approaches of Owen and Agassiz to biology as scientific heresies, supportable only so long as the two men were thought to be the technical wizards of their own fields. And among geologists Darwin did not really have to worry, for Darwin himself was a distinguished geologist, among other things, and the *Origin of Species* is the great geological treatise of the period, among other things.

The dangerous attack on the *Origin* came from a more powerful source. The prestige of astronomy and physics was directed against Darwin. The astronomer John Herschel compared natural selection to the Laputan plan "for writing books by the concourse of accidental letters, and selection of such combinations as form syllables, words, sentences, &c."—although Herschel did not say in this passage whether Swift's satire undermined Darwin or Darwin's theory undermined Swift's satire.[24] The physicist Lord Kelvin showed that physical astronomy could not allow enough time for the age of the earth to permit natural selection to have taken place. Kelvin's collaborator in developing scientific inventions, the engineering professor Fleeming Jenkin, asserted that blending inheritance would quickly swamp any effects of natural selection working on Darwin's variations. James Clerk Maxwell, aided by contemporary discoveries in astronomy made with the recent invention of spectroscopy, denied that physical nature is basically evolutionary, and denied that molecular structure is complex enough to account for the infinite differences in heredity.

Kelvin had an easy time of it, for John Tyndall could not coach his friend Thomas Huxley into enough physics to argue convincingly with Kelvin on such recondite matters as the effects of the polar ice-cap on the earth's rotation. But Maxwell had his hands full with a bright young mathematician, William Clifford, who used the full resources of the empiricism present in the Truth-Complex to assert that Maxwell did not understand what a proof in physics proves. Jenkin's argument was one which Darwin had already considered and discounted privately; but he could not argue his case convincingly.[25]

I have given enough examples, I hope, to illustrate the general situation I want to draw attention to. Fights broke out all across the face of science. Not merely biologist against biologist, or geologist against geologist. These controversies solve themselves with time. But also geologist against astronomer, biologist against statistician, mathematician against physicist. How do you solve these fights? Notice that basically you had the naturalist—which is what Darwin as biologist was—against the truly prestigious sciences.

And Darwin wouldn't give in, as the lowly naturalist should. That he was worried, he has told us. That step by step, in edition after edition of *Origin,* he gave ground, we know. But, in conducting a confusing and worrisome delaying action, Darwin never really surrendered. He possessed in great measure a characteristic not always sufficiently praised in accounts of "how to conduct yourself scientifically": the characteristic of stubbornly maintaining his theories regardless of the valid arguments that could be brought against them. Darwin should have been put in his place, according to our Truth-Complex, if he disagreed with astronomy and physics: as naturalist, he was the bottom part of the hierarchy revolting against the top. But Darwin had the speculative self-confidence of the geologist in the great age of British geology, the 1830's and 1840's. He held on; and his revolt succeeded.

The result was that the truth-hierarchy eventually disintegrated. Biologists were still pro- or anti-Darwin, but on the basis of biological intuitions, not on the basis of extrapolations from physics. Physicists learned a momentary humility when the discovery of radioactivity destroyed one of the bases of Kelvin's arguments. But the damage had been done. No one could, any longer, trust the arguments of someone from another discipline. The Darwinian debates had let loose a series of arguments among scientists about the nature of science in a most unpleasantly immediate form, namely, what is proper scientific evidence, and who is to be the judge? It appeared that Darwin's idea of proof was quite opposed to that of Kelvin, as was Clifford's to that of Maxwell, while the view of the "real" philosopher John Stuart Mill seemed rather irrelevant to the whole discussion. An appeal to philosophy proved nothing, since the approved philosophy of the Truth-Complex was in dispute down to its very foundations.

I could go on and on with this, pursuing the breakdown into more and more configurations, but I would bore you; and that is what happened historically. After about twenty years of exciting themselves in trying to keep Truth alive, the late Victorians decided that Socialism and Imperialism, the income tax and pensions, Teddy Roosevelt and Rudyard Kipling, were more interesting subjects, so they left the heirs of Truth to give him a decent burial and settle the estate among themselves. Which they did: by adopting a policy of non-fraternization. It was agreed that truth in religion is no relation to truth in science, and vice versa. It was agreed that truth in science is no relation to truth in ethics, and vice versa. It was agreed that truth in ethics is no relation to truth in religion—which is a bit harder to swallow, but it does mean that preachers should endorse God but not labor unions. The preacher, like the physicist, is a specialist. Since he does not know about the physical realm (the area of the scientist), nor the social realm (the area of the sociologist), nor the economic realm (the area of the economist), nor the family realm (the area of the psychiatrist), his realm is clearly the realm of the non-existent.

In general, however, no one of these participants in the sharing-up of

Truth has really believed in it. The natural scientists, for example, believe that what they managed to seize for themselves is at least the best butter. But the poets were also ready with a doctrine of better butter, in part prepared for them by the Romantic poet Coleridge but held within moderate bounds (as were most Romantic extravagances) by the Truth-Complex as long as it reigned. The poets' theory classified natural scientists as mere anatomists of nature, whose very instruments condemned them to see not Truth as a whole but only artificial and isolated chunks of it.

The trouble with this procedure was that the theologians wanted Coleridge's theory also, but for themselves alone. So poetry, heretofore the most prestigious of the humanities, found itself all alone, with the task of asserting a whole universe from its own perceptions—a burden under which poetry has ben visibly staggering throughout the 20th century. Literary historians, however, unwilling to admit that a mere biologist could have led to such troubles, have blamed the isolation of poetry on the entire 19th century, or on the whole decline of European values since the Middle Ages—indeed on anything large enough to obscure the role played by the poets themselves in simply refusing to accept and deal with the Darwinian natural world.

In the meantime, the successes of the quantum mechanicians and the Einsteinians established physics as the most prestigious natural science; but it could tyrannize only over other sciences, not over theologians, poets, or psychiatrists. There was no possibility of a generally accepted new cosmology to replace the old one. The flight of the mathematicians into sheer fantasy deprived physics of its most useful ally for influencing non-scientists: what poet now admires set theory as even Wordsworth had to admire geometry? The attempt of physics to spin out a universe from its own assertions resulted in a naturalism even less appealing to many than that of Lucretius.

The Romantic attack on any given position is so convenient that it has recently been used even by astronomy against physics. Professor Fred Hoyle asserts that astronomers should receive more public money for their giant telescopes, which show the real cosmos, and physicists should receive less for their cyclo- and sychro-trons, which only anatomize the world into particles.[26]

Now let me specify what it is that shows that we are still far apart. The fragmentation I speak of shows itself in much more operational terms than in statements about the ancient history of Truth.

The current state of things is one in which we do not trust the other person's procedures. My method of obtaining truth I know; I know when to trust it, and how far to trust it. But your method I do not know. Its

281

results seem dogmatic. You easily assume things I have been laboring to find out. And every time you venture into my field, as you do sooner or later, you do not observe the rules of the game, and I cannot convince you that you have broken them. You are certain, even when I *know* that you are wrong.

I first ran into this mutual distrust in graduate school. I was the confidant of a physics student in his complaints as to the intellectual arrogance of chemists. "They pretend to understand molecules with 60 or 70 parts," he said, "and I have been trying for a year to understand an atom with just 3 parts, and I haven't succeeded yet." Intellectually, he felt, chemists were insincere, or just plain sloppy.

As historian, I ran into a somewhat different attitude. One day I remarked at dinner that I wished I knew what Kelvin really meant when he said that the word characterizing his career in science was "failure." To which the intelligent physics student on my right replied, "I'm sure there is a biography of Kelvin. Why don't you read it and find out?" The point is that he didn't mean to be insulting. There was nothing in his environment to suggest that historical facts are difficult and allusive, are indeed theoretical constructions which challenge the ingenuity of the person who invents them. It had never occurred to him that history is much more than a collection of events. But what bothered me most was this: what could this physics student have thought was the justification of history students in going through graduate training at all, if our procedure was simply to read books already written in order to find out?

For that matter, what does a theologian do? How does he find out? If he has no procedure as impressive as the use of oscilloscopes and mathematical symbols, why should we believe him? There is just as high a percentage of scientists today who are "religious" as in the past; but how many of these "religious" scientists believe that Karl Barth was as great a truth-finder as P.A.M. Dirac? The term just doesn't seem appropriate for a theologian, does it?

Now let us take an amusing example which popped up a few years ago within the humanities. For many years Edmund Wilson was literary critic and humanist-in-chief for the *New Yorker* magazine. Then came the discovery of the Dead Sea Scrolls, which had biblical scholars flocking from all corners to get into the act. In the midst of this bustle of scholarly activity, Wilson published a long article in which he asserted that none of these scholars could be completely trusted: they were all religious, you see. It was highly desirable that a non-religious biblical scholar (not merely a non-Christian one, for there were Jewish scholars at work) make sure that disreputable discoveries concerning the origins of Christianity were not being hushed up. In short, America's best-known literary critic apparently had no knowledge of, and could not imagine, the meticulous and carefully criticized procedures elaborated over the past hundred years which are used in biblical scholarship, the severity of the training required and the stiffness of the international competition, and how little a biblical scholar is interested in suppressing that truth which he has devoted his life (not merely his career) to finding out. Here in the heart of the humanities we have the same disbelief in the other person's procedures, and even in the other person's integrity, that my physicist friend expressed about chemists and that, *horribile dictu,* some young poets have expressed about physicists and chemists and literary critics and scholars and the establishment of knowledge in general. Yet, so long as we established ones disagree among ourselves, it is difficult to prove that astrologers should give up their own procedures and rely upon ours.

Finally, if this mutual distrust of one another's truth is our current situation, what can my own discipline, the history of science, do about it? Can it help us to communicate, one with another? Clearly if the situation is as I have described it, teaching calculus to all the humanists is not the solution. To be sure, I am in favor of teaching calculus to all the humanists, but for a somewhat different reason. Calculus is one of the two great achievements of the 17th century, the so-called "century of genius." With this creation, however logically imperfect it was, the West finally solved the great intellectual problem that the ancient world could never handle, the problem that vexed and tormented Greco-Western thought until the time of Leibniz and Newton, the problem, that is, of dealing with change, with impermanence. A person who has not understood this triumph is as ill-fitted to consider himself a full member of modern culture as is one who has not met Shakespeare or Racine.

Of course many science students have not studied calculus either. They have had it jammed into them as fast as possible so that they can start calculating as soon as possible. So I should say that I am all in favor of teaching calculus to the humanists *and* to the scientists.

If calculus is not the answer, neither is the answer to teach the history of science to undergraduates instead of Physics I or Biology I. To believe that you can teach science by teaching its history, that is, teach science by not teaching science, is an idea so odd that only our present plight could make it seem plausible. Such an idea does appeal, I will admit, to those spokesmen for educational science who feel that the elements of mathematics, and therefore real science, are so far beyond the powers of the freshman that we might just as well tell him anecdotes instead.

What one might do with Physics I or Biology I is to make them into courses in real science and not in TV cookery. On TV, as you know, the smiling hostess carefully follows the directions in the cookbook and, by the end of the allotted time, has produced a luscious cake every time. That was what experiments were like in my freshman courses. The manual told you what to do, and why; you did it; and you got the answer (as a matter of fact, I seldom did get the answer, but that only proved I was not a good experimenter). To this I can contrast my vicarious experiences with real experiments: the equipment never does work right, and the experiment never does come out quite as you expected, and if it did it would be fairly dull because then you wouldn't have discovered anything. The real spirit of experimentation, I take it, is that which I ran into in a friend at a large laboratory a few years ago. He was rubbing his hands with glee. "We've just done a whole set of experiments on...." he said, "and they all came out wrong! In some cases as much as 25 percent wrong! *That'll* send those *theoretischers* back to their blackboards!"

True science still recognizes, I believe, the Fourth Law of Thermodynamics, the law of the perversity of inanimate objects. Stated in terms of probability theory, this law says that you will probably be wrong more often than the laws of probability admit. This is one of the links between the sciences and the humanities, as a similar law was known in ancient Greek drama. We may call it Agathon's Law of Dramatic Action. Aristotle remarks in his *Poetics* that "such an event is probable in Agathon's sense of the word: 'it is probable,' he says, 'that many things should happen contrary to probability.'"[27] Another application of the law is well-known among modern playwrights as "Murphy's Law for Opening Nights." Murphy's Law is: "If something can go wrong, it will."[28]

Most freshmen know this law empirically, but they are still surprised to learn the saying the physicist J. J. Thomson was fond of, i.e., that the law of the constancy of Nature was never learned in a physics laboratory. To which I might add that the belief of some persons that an observation is repeatable is one of those illusions which no normal philosopher can admit.

This discussion brings us (by a rather devious route) to one thing that the history of science can do in helping our fragmented minds. In studying the work of scientists it can, indeed it must, explain their procedures, show why following just this procedure led this scientist to believe in just this result. And it will do this concretely, in terms of what the man did do and believe, as well as in terms of what he later said when he wanted to present a proper case for publication. And since the historian is interested in science, not in *a* science, he will compare the procedures of the physicist with those of the naturalist, or the mathematician, or the anthropologist. He will still find diversity, of course, if not always along conventional subject lines; historians of science have, I hope, discarded the myth that there is *a* scientific method. If there were, science would be an IQ test, and of little interest to the creative mind.

In doing all of this, the historian of science will have to explain the environment into which the particular incident fits. These procedures were indeed not sufficient to justify this belief all by themselves; they did so only as part of an existing complex of ideas, evidence, and men, a complex which was not usually limited to one natural science or even to natural science. Napoleon, for example, was a "cause" of the *Origin of Species*. It was the huge mob of naval officers left over from the Napoleonic Wars which helped to solidify the British Admiralty's love of scientific voyages; and H.M.S. *Beagle* was Darwin's best college.

Complexity, then, is necessary in our field, and it is worth looking at the output of a leading scholar such as Martin Rudwick to see the sheer *diversity* of approaches he has found necessary in the history of geology alone.

Such studies will not bring all of the fragments of truth into one focus, but they might be a start. As I said at the beginning of this chapter, it was George Sarton's dream to found a center which would bring together the achievements of scientist, artist, and saint for our appreciation. Historians of science are a bit more modest about their own capabilities than to try to do that in any rigorous fashion, but at least we don't feel that we are straying from our scholarly duty if we read some of the old theologians, or take seriously some of the Romantic poets. If we really went all out in Sarton's way, it would mean that we had discovered the new Truth-Complex, and that its name was...History. Personally I find this a delightful idea, but I fear that its promulgation would only increase the tendency of other disciplines to rush away, each from all the rest but especially from me.

Notes to Chapter Nine

The first version of this talk was given to the history department of the State University of New York, Long Island Center, in March 1962. The second version was the Sigma Xi lecture at the University of Pennsylvania in March 1966. The third version was published as "P. S. If I find out what Truth is, I'll drop you a line," *The Smithsonian Journal of History*, II (1967), 1–24. This fourth version is revised once again.

1. C. P. Snow, *The Two Cultures and the Scientific Revolution* (Cambridge, 1959), pp. 7–8.

2. Gerd Buchdahl, *The Image of Newton and Locke in the Age of Reason* (London, 1961).

3. Robert Boyle, *The Christian Virtuoso*, as quoted with approval by John Ray, *The Wisdom of God Manifested in the Works of the Creation*, 7th ed. (1717), p. 50; the quotation from Ray himself just precedes.

4. James Clerk Maxwell, *Scientific Papers* (ed. W. D. Niven, 1890), II, 779.

5. Hymn 168 in *Hymns Ancient and Modern*, often printed together with *The Book of Common Prayer*. The position was a standard Anglican one at least from 1600: see Richard Hooker, *The Laws of Ecclesiastical Polity*, Book I, *passim*, and the perceptive comments of H. L. Weatherby, "The Encircling Gloom: Newman's Departure from the Caroline Tradition," *Victorian Studies*, XII (1968), 57–71.

6. Alexander Wood and Frank Oldham, *Thomas Young* (Cambridge, 1954), p. 8.

7. Alexander Koyré, *From the Closed World to the Infinite Universe* (Baltimore, 1957, 1968), p. 2.

8. On Descartes' and Boyle's interpretation of the image, see Laurens Laudan, "The Clock Metaphor and Probabilism." *Annals of Science,* XXII (1966), 73–104; Samuel Clarke's reply to Leibniz is reported in Richard S. Westfall, *Science and Religion in Seventeenth Century England* (New Haven, 1958), p. 201. These may be used to modify the assertions in Francis C. Haber, "The Darwinian Revolution in the Concept of Time," *Studium Generale 24* (1971), 289–307, which nevertheless has an excellent discussion of the metaphor in connection with real late-medieval clocks. What Haber's evidence shows to me is that Newtonianism *defeated* "the mechanical philosophy" and therefore relegated its striking analogy to a secondary position.

9. *John Wesley's New Testament* (ed. George C. Cell, 1938), pp. vii–xiv; John Wesley, *Survey of the Wisdom of God in the Creation,* ed. B. Mayo, 3rd American ed., (New York, 1823), I, 438–439; II, 39; Edmond Halley, "Some Consideration about the Cause of the Universal Deluge," and "Some further Thoughts upon the Same Subject," read Dec. 12 and Dec. 19, 1694, in *Phil. Trans. Abridged,* VII (1724–34), 33–35 and 35–36.

10. Charles Daubeny, "Address to the meeting," *Brit. Assoc. Adv. Sci. Report,* VI (1836), p. xxiii.

11. Alfred Tennyson, "Parnassus," from *Demeter and Other Poems (Poetical Works,* London, 1953, p. 810).

12. Cf. Max Jammer, *Concepts of Force* (New York, 1962), p. 243. I go rather further than Jammer does.

13. William Wordsworth, "A Poet's Epitaph" (1800).

14. John H. Newman, *The Idea of a University* (London, 1899), pp. 42, 27.

15. *Cf.* John H. White, Jr., "Locomotives on Stone," *Smithsonian Journal of History,* I (1966), 49–60, which reproduces sources which probably give the shades and tones of colors actually used. Most surviving locomotives have been repainted several times; most printed reproductions are colored by the modern author using literary descriptions and his own sense of matching shades and tones to guide him.

16. Standish Meacham, "The Church in the Victorian City," *Victorian Studies,* XI (1968), p. 375.

17. Sedgwick to Darwin, November 24, 1859, in Charles Darwin, *Life and Letters* (ed. F. Darwin, New York, 1887), II, 44; Darwin's complaint to Richard Owen about the letter, Dec. 13, 1859 is given in Gavin de Beer, "Further unpublished letters of Charles Darwin," *Annals of Science 14* (1958), 104.

18. In the preface to the second edition of *Lyrical Ballads,* reprinted in William Wordsworth, *Poetical Works* (ed. T. Hutchinson, revised E. de Selincourt, London, 1950), p. 738.

19. Alvar Ellegård, *Darwin and the General Reader* (Göteberg, 1958).

20. William C. Wells, *Two Essays: One Upon Single Vision with Two Eyes; the Other on Dew …and an Account of a Female of the White Race of Mankind, Part of Whose Skin Resembles That of a Negro; With Some Observations on the Causes of the Difference in Colour and Form Between the White and Negro Races of Men* (London, 1818), pp. 425–439; See, however, Richard Shryock, "The Strange Case of Wells' Theory of Natural Selection," *Studies and Essays in the History of Science and Learning Offered in Homage to George Sarton* (1944), pp. 197–207, for a different evaluation of Wells' importance. In general see the satisfying article by Kentwood D. Wells, "William Charles Wells and the Races of Man," *Isis, 64* (1973), 215–225.

21. James C. Prichard, *Researches into the Physical History of Mankind,* 3rd ed. (London, 1836), I, 107–108; 2nd ed. (1826), II, 569, 583.

22. *Cf.* "Darwin's Notebooks on Transmutation of Species, Part III," ed. G. de Beer, *Bulletin of the British Museum (Natural History),* Historical Series, Vol. 2, No. 4 (London, 1960), p. 138: entry of September 7, 1838.

    I am especially indebted to several discussions with Dr. Sandra Herbert in developing this paragraph, as well as to an opportunity to read her unpublished Ph.D. thesis, "The Logic of Darwin's Discovery," (Brandeis University, June 1968).

23. I am indebted to communications with Professors Conway Zirkle and Richard Shryock which stimulated me to clarify my meaning in this section as compared to the version in my *SJH* article.

24. John Herschel, *Familiar Lectures on Scientific Subjects* (London and New York, 1866), p. 63 note.

25. In revising this account, I have been able to consult Peter J. Vorzimmer, *Charles Darwin: The Years of Controversy* (Philadelphia, 1970), Chapter 5: "Blending Inheritance." Joe D. Burchfield, "Darwin and the Dilemma of Geological Time," *Isis, 65* (1974), 301–321, shows how much Kelvin and the problem of time bothered Darwin. The whole controversy is described in Burchfield's *Lord Kelvin and the Age of the Earth* (New York, 1975).

26. Fred Hoyle, *Of Men and Galaxies* (Seattle, 1964), *passim.*

27. *Poetics,* XVIII, 6 (Butcher).

28. Jean Kerr, *The Snake Has All the Lines* (New York, 1962), p.75.

Library of Congress Cataloging in Publication Data

Cannon, Susan Faye.
Science in culture.

Includes index.
1. Science—England—History. 2. Physics—England—History. I. Title.
Q127.G4C36   509'.42   77-26004
ISBN 0-88202-172-9